T0343290

An Introduction to Atmospheric Gravity Waves

2ND EDITION

An Introduction to Atmospheric Gravity Waves

2nd Edition

Carmen J. Nappo

CJN Research Meteorology, Knoxville Tennessee 37919, USA

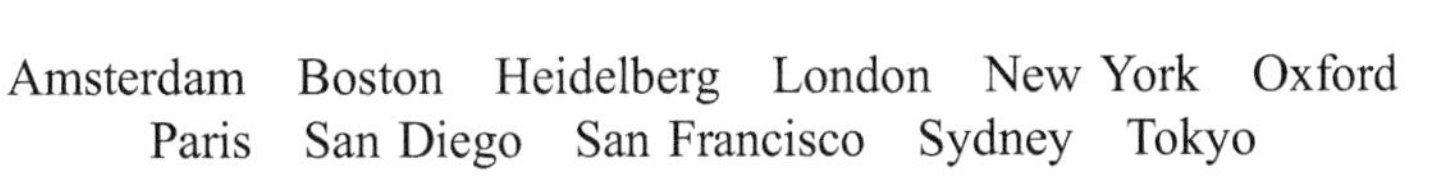

Elsevier

225 Wyman Street, Waltham, MA 02451, USA
The Boulevard, Langford Lane, Kidlington, Oxford OX5 1GB, UK
Radarweg 29, PO Box 211, 1000 AE Amsterdam, The Netherlands

Library of Congress Cataloging-in-Publication Data
Nappo, C. J.
An introduction to atmospheric gravity waves / Carmen J. Nappo. – 2nd ed.
p. cm.
Includes bibliographical references and index.
ISBN 978-0-12-385223-6
1. Atmospheric waves. 2. Gravity waves. I. Title.
QC880.4.W3N37 2012
551.51'5–dc23
2012018369

British Library Cataloguing in Publication Data
A catalogue record for this book is available from the British Library

Printed and bound in USA

13 14 15 10 9 8 7 6 5 4 3 2 1

ISBN: 978-0-12-385223-6

ACKNOWLEDGMENTS

No one got to where they are on their own. This is especially true for me. Since the first edition of this book, I've had the fortunate opportunity to teach introductory classes on atmospheric gravity waves at the University of Connecticut (thank you David Miller), the Meteorological Institute of Uppsala University (MIUU; thank you Cecilia Johansson and Anna Owenius Rutgerson), the Meteorological Institute of Stockholm University (MISU; thank you Gunilla Svensson and Michael Tjernström), Wageningen University (thank you Burt Holtslag and Ger-Jan Steeneveld), The University of Georgia (thank you Monique Leclerc), and NOAA's Atmospheric Turbulence and Diffusion Division, Oak Ridge, TN (thank you Bruce Hicks and Ray Hosker). I express my deep gratitude to those who answered my many questions and requests including, Ulrich Achatz, Joan Alexander, Ben Balsley, Bob Banta, Sukanta Basu, Anton Beljaars, Sean Burns, George Chimonas, Hey-Jun Chun, Dale Durran, Dave Fritts, Branko Grosogono, Ulf Högström, Qingfang Jiang, Martin Leutbecher, Larry Mahrt, Yannick Meillier, Thorsten Mauritsen, Eva Podgravist, Ron Smith, Ann-Sofi Smedman, Gert-Jan Steeneveld, Jielun Sun, Petra Thorsson, Michael Tjernström, Ivana Stiperski, Bas van de Wiel, and Jim Wilczak. My apologizes to those I may have missed.

A special thanks to Gunilla Svensson for letting me finish the first draft of the manuscript at MISU.

This book is dedicated to Prof. Em. Ann-Sofi Smedman, MIUU
and
Prof. Gunilla Svensson, MISU.

Thank you for the many years of help and encouragement.

CONTENTS

3

MOUNTAIN WAVES

4

DUCTED GRAVITY WAVES

5

GRAVITY WAVE INSTABILITY AND TURBULENCE

6

WAVE STRESS

7

GRAVITY WAVES IN THE MIDDLE AND UPPER ATMOSPHERE

8

Wave Stress Parameterization

9

Observations and Measurements of Gravity Waves

10

GRAVITY WAVE ANALYSES

APPENDIX A

THE HYDROSTATIC ATMOSPHERE

PREFACE

Since the first publication of *An Introduction to Atmospheric Gravity Waves* in 2002, more than 1840 articles containing the phrase 'gravity waves' have been published in the scientific journals of the American Meteorological Society. In the decade before that, approximately 1683 such articles were published, and from 1982 to 1992 approximately 1290 such articles were published. In 1992, membership in the American Meteorological Society was 10,814; in 2002 it was 11,163, and in 2011 it was 13,788. It would be tempting to correlate increasing interest in gravity waves with increasing membership in the American Meteorological Society. Regardless of this conjecture, we can assume that the increasing number of publications can be linked with the growing awareness of the roles gravity waves play on all scales of atmospheric motions. This increased interest in gravity waves can be attributed to increased technology in observation and measurement technology, the recognized importance of gravity waves in mesoscale numerical models, atmospheric chemistry, wind energy and air quality. Advances in computer technology are allowing detailed simulations of mountain waves, lee waves, wave-turbulence interactions using large eddy simulations (LES) and direct numerical simulations (DNS). Thus, it is likely that anyone interested in the atmospheric and oceanic sciences will encounter gravity waves at some time. In another context, gravity waves have tended do be a rug under which modeling errors are sometimes swept. However, gravity waves as a subject is seldom given as a formal university course but rather as a chapter in a dynamics book. In many cases, a researcher or student's knowledge of gravity waves is limited to their specific interests. Thus, for example, a climatologist might be interested in large scale inertia-gravity waves in the middle and upper atmospheres while a micro-meteorologist might be interested in short waves and their interactions with turbulence in the stable planetary boundary layer. These contrasting differences in scale point to the vast spectrum of atmospheric quasi-periodic disturbances which are ubiquitous and should be considered a branch of meteorology in itself.

The first edition of this book was motivated by an absence of a formal introduction to linear gravity wave theory. While there are many fine books on the subject, using these to learn the fundamentals of wave mechanics and the linear theory is arduous. I have intended to make this book a true introduction to atmospheric gravity waves. As in the first edition, I have tried to make the theoretical discussions as simple as possible, and to concentrate on physical insight rather than mathematical rigor. However, we recognize that waves are fundamentally a mathematical concept and construct. Some chapters are more mathematically concentrated than others. To ease the problem inherent in mathematical developments, I have tried to included many intermediary steps, and I have tried to avoid expressions such as "it can be shown..." or "it follows that...", etc. In some cases, the mathematical developments may not be essential, and one could say that the means do not justify the ends. In some cases, the derivation of certain results will be asked as exercises at the ends of chapters. My limited teaching experiences have shown that understanding is enhanced through solving problems. Thus, I have included problem sets for some chapters. The text is intended for graduate students and professionals with backgrounds in physics, meteorology, or fluid mechanics. An understanding of the basics of Fourier transforms is essential. I have included in the Appendices several atmospheric topics which may not be familiar to the reader. The ultimate intention is to provide the background necessary for continued independent study.

This edition has reorganized the material presented in the first edition so as to follow a more direct path from the basics of wave theory to analyses of wave characteristics. The subjects covered include linear wave theory, mountain waves, ducted waves, gravity wave instability, wave stress, wave stress parameterization, observation and measurement techniques, and data analysis. Elsevier has provided a web site that contains numerical models in FORTRAN and MATLAB, sample data sets, animations, videos, and corrections to errors at http://booksite.elsevier.com/9780123852236.

Chapter 1 presents a brief description of the types of gravity waves observed in the atmosphere and their related sources. We next introduce the basics of wave mechanics. Topics include wavenumbers, wavelengths, wave vectors, angular frequency, wave period, wave phase, wave fronts, wave dispersion, standing and traveling waves, phase velocity, wave packets, group velocity, buoyancy, static stability, slant-wise static stability, and the Boussinesq approximation. Chapter 2 introduces the basic mathematics and physics of linear gravity waves. The equation for linear gravity waves, the Taylor-Goldstein equation, is developed from basic principles, and solutions are discussed for idealized cases. Expressions are developed for wave dispersion and group velocities. Propagating and evanescent waves are defined and discussed. Wave energetics and wave action are discussed. In Chapter 3, the important topic of mountain waves is introduced.

'Mountain wave' is a general term for a topographically generated gravity wave. We shall examine the wave fields above two-dimensional infinite surface corrugations, isolated two-dimensional ridges or mountain ranges, and three-dimensional isolated mountains, and general heterogenous orography. Chapter 4 deals with the ubiquitous nature of wave ducting. We shall see that a gravity wave can become trapped between the ground surface and some upper level where wave reflections occurs. Under these conditions, the waves can travel long distances with relatively little attenuation. Examples of ducted waves include lee waves, solitary waves, undular bores, and jet streaks. Chapter 5 deals with the linear stability of gravity waves and its relation to turbulence, and is perhaps the most mathematically difficult chapter in the book. Indeed, if it were not for the ability of gravity waves to generate turbulence and thus modify the background flow, there would not much interest in their study. Waves and turbulence are often observed to exist simultaneously. In this chapter, we examine mechanisms leading to wave instability, wave breaking, and the resulting generation of turbulence. Topics include flow stability, Kelvin-Helmholtz waves, inflection point and stratified flow instability, stable and unstable modes, and wave-turbulence coupling, and critical levels where the Taylor-Goldstein equation has a singularity. In Chapter 6, we introduce the subject of gravity wave stress and stress divergence. Since the wave stress represents the vertical transport of mean-flow horizontal momentum and total wave energy, it is an important subject in momentum and energy balance calculations. When vertically-propagating gravity waves grow and break, resulting stress convergence acts as a drag on the atmospheric flow. The strongest source of wave stress are mountains. The wave stress balances the form drag due to pressure differences on upwind and lee side of mountains. Chapter 7 is a quick survey of gravity waves in the middle and upper atmosphere. In these regions, wave lengths range from tens to hundreds of kilometers, and wave periods range from hours to days. In the middle and upper atmosphere, the dynamics wave spectra rather than a single wave is important, and research on this topic continues. Chapter 8 discusses how the wave stresses can be parameterized in numerical models. Such parameterizations account for the momentum fluxes due to topography not realized by the model spatial resolution. Because midrange weather forecasts use relatively large grid cells, much topography is missed by the model; however, these terrains still impact with the real atmospheric flow. Thus forecast errors can result. Parameterizing the effects of sub-grid scale orography has greatly improved weather forecasts and global climate simulations. An important part of these parameterizations deal with wave breaking using the concept of wave saturation Chapter 9 reviews the many techniques for observing on all amtospheric scales. We describe observation methods including ground based instruments, tall towers, tethered and free-floating balloons, acoustic and electromagnetic radars, aircraft, satellite global positioning systems (GPS)-radio occultation sounding, and ionospheric airglow. However, these observations would be of little value if wave characteristics could not be calculated from the data. Chapter 10 describes numerical techniques for analyzing data for wave content, *i.e.,* phase speed, propagation direction, wavelengths, etc,. These techniques include correlations between pressure and wind speed perturbations, analysis of phase lag and cross correlations between wave observation stations, wavelet analysis, and beamsteering.

Many fine textbooks and monographs on gravity waves exist including, for example, Eckart (1960), Tolstoy (1963), Hines (1974), Gossard and Hooke (1975), Phillips (1977), Lighthill (1978), Gill (1982), Smith (1985) and Baines (1995), Sutherland (2010) and references contained therein.

Carmen J. Nappo
MISU
Stockholm, Sweden
1 December 2011.

1

FUNDAMENTALS

1.1 INTRODUCTION

We begin our study of gravity waves with a simple question. What is a wave? We all have an idea, perhaps intuitive, perhaps experiential, of a *wave*. Every one who has seen a body of water has seen a wave. But what in fact is a wave? We talk about wave motion, about wave speed, about wavelength as if the wave were tangible. As if we could pick it up, examine it, and place it somewhere else. Of course we cannot do this because a wave does not exist as a physical entity. A wave is made manifest by the periodic motions (oscillations) of fluid particles each moving with a phase slightly different than its neighboring particles. A similar effect is seen in the "moving" lights of a theater marquee (see EURL\Ch1\L1). Obviously, the light bulbs in the video are not moving; however, a sense of motion is produced by the proper phasing of the light switches. Generally, when we talk about waves we do not consider the motions of fluid particles. But it is these motions of the particles that is associated with wave energy. One of the things defining a wave is its ability to transport energy.

A stably stratified fluid is one in which the fluid density increases with depth. A characteristic of a stably stratified fluid is its ability to support wave motions. Except for a relatively thin layer in contact with the earth's surface, the *planetary*

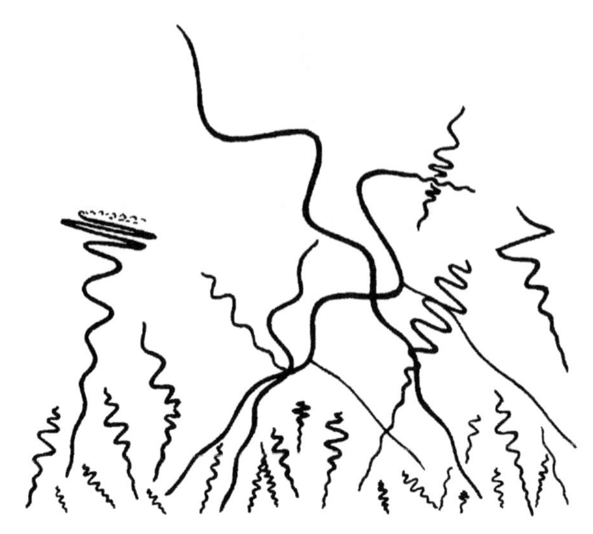

FIGURE 1.1 A surrealistic representation of atmospheric gravity waves. Taken from Hines (1974).

boundary layer (PBL), the atmosphere is almost always stably stratified, and it is reasonable to assume that gravity waves are to some degree always present. If it were possible to see these waves and to greatly speed up their motions, then we would see a wide variety of wave shapes moving in many directions. Hines (1974) presents a "surrealistic" representation of these waves, which is reproduced in Fig. 1.1. Many of the waves would be moving diagonally upward or downward across our field of view, but some would be moving horizontally. Some waves would extend through our whole field of view, and some would appear stationary as if frozen in space. We would see waves moving upward much as writhing snakes with their "wiggles" rapidly increasing in frequency and magnitude, and then suddenly be reflected downward. Some of these wiggly waves would not be reflected, but would instead seem to break apart into countless smaller waves which gradually fade from view. We would also see waves that follow curved paths, or are partially reflected downward and partially transmitted. Indeed, it would be a view of unending variety and action, but also a view of immeasurable complexity and puzzlement.

While we cannot see atmospheric gravity waves we can see the effects the waves have on the atmosphere.[1] For example, Fig. 1.2 is a photograph of a meteor trail taken by Anthony Nugnes in Silver Spring, Maryland, USA on January 18, 2010 at about 17:40 LST. The wiggles are due to perturbations of local winds by gravity waves. Figure 1.3 taken from Chimonas (1997) shows the wind profile

[1] Gravity waves exist in the oceans as well, and most of what we learn in this book applies equally well to deep waters. Indeed, it would perhaps be better to use the generic *fluids* instead of air or water.

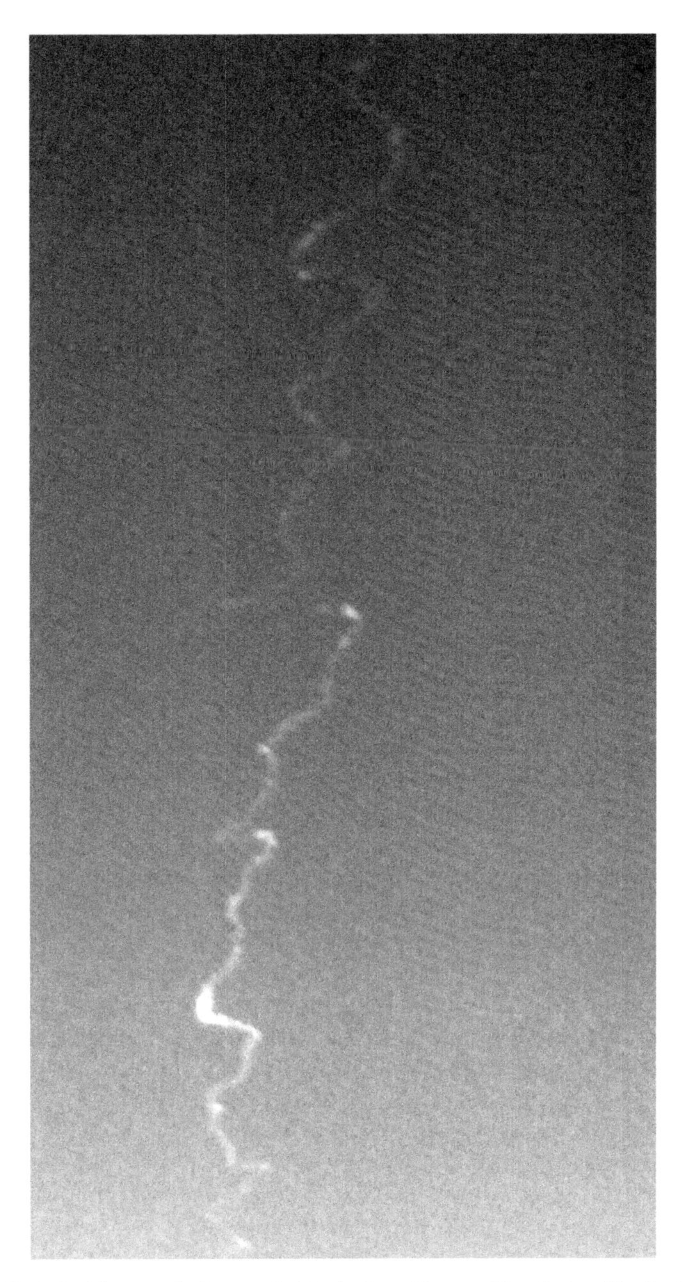

FIGURE 1.2 Meteor trail photographed by Anthony Nugnes, Silver Spring, Maryland, USA on January 18, 2010.

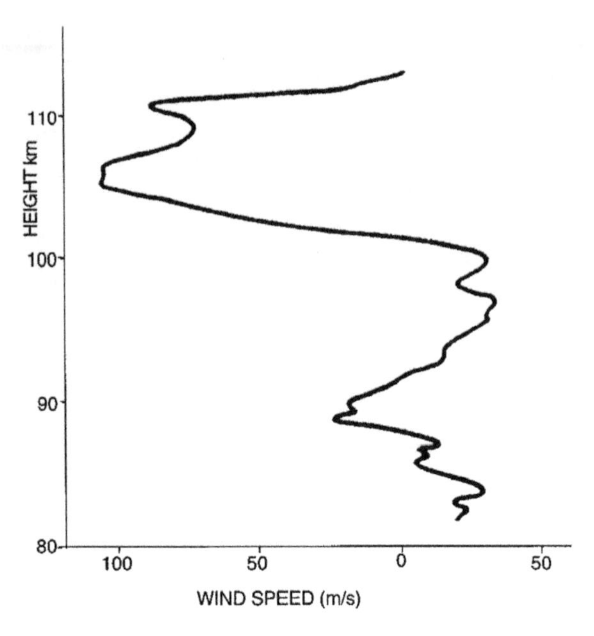

FIGURE 1.3 Meteor trail taken from Chimonas (1997).

in the atmosphere's E-region as revealed by a meteor trail reported on by Liller and Whipple (1954) and later analyzed by Hines (1960). In these high-altitude regions, the length scales of the vertical variations of the horizontal wind speed range from 1 km to 10–20 km (Chimonas, 1997). Figure 1.4, taken from Zamora (1983) is a "picture" of gravity waves in the planetary boundary layer obtained by an upward looking sodar. Sodar (see Chapter 8 for a description) is similar to radar except that sound waves are used instead of radio waves. The upward-moving sound pulses are partially reflected downward by layers of atmospheric turbulence. These reflected pulses are detected by the sodar and are represented by the dark bands in Fig. 1.4. These horizontal bands of turbulence are perturbed by passing gravity waves, thus revealing the wave's presence. In some cases, the waves break down and generate more turbulence. We see a wide range of wave frequencies, and in many cases we see high-frequency waves superimposed on low-frequency waves. Some waves last only a few minutes, while others persist for hours. Some waves appear to ascend or descend with time, and some seem to intermittently appear and disappear. Some waves have large amplitudes, while others are barely noticeable. The complicated images in Fig. 1.4 represent some of the fundamental characteristics and physics of atmospheric gravity waves. But however interesting the physics of these waves may be, unless the waves have a measurable effect on the atmosphere there is little reason for their study.

Although the characteristics of waves in stratified fluids have been known for many years, it remained a somewhat esoteric subject until Hines (1960) used

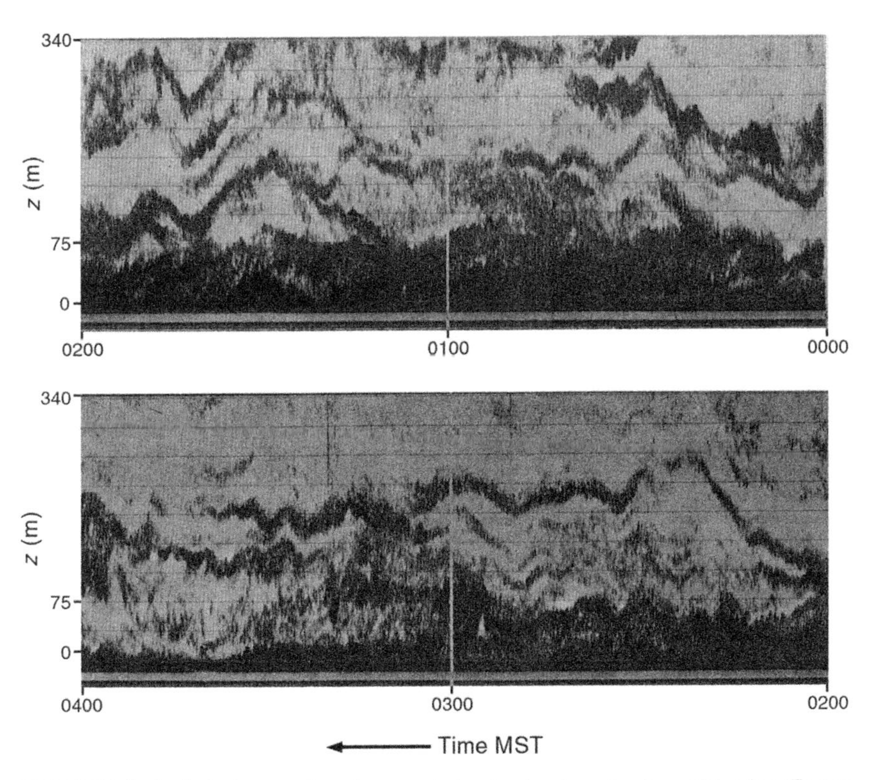

FIGURE 1.4 Sodar images of gravity waves in the planetary boundary layer. Taken from Zamora (1983).

linear gravity wave theory to explain the origins of irregular winds and turbulence observed in the middle and upper atmosphere as depicted in Figs. 1.2 and 1.3. Hines (1989) gives an historical perspective of this work, and the reader is encouraged to peruse this article. However, we must not overlook the earlier works on gravity waves done by, for example, Queney (1948), Scorer (1949), Martyn (1950), Gossard and Munk (1954), Palm (1955), and Sawyer (1959). The introduction of gravity wave theory into the field of meteorology initiated an avalanche of interest in the applications of the theory to atmospheric motions. Today it is recognized that gravity waves are essential parts of the dynamics of the atmosphere on all meteorological scales. On the largest atmospheric scales, the studies by, for example, Gossard (1962), Hodges (1967), Jones and Houghton (1971), Hines (1974), Lindzen (1981), and Holton (1982) examined the effects of gravity waves on the upper atmosphere and the general circulation see, for example, Fritts (1984) and Fritts and Alexander (2003), for a review of these and other studies. On the mesoscale, studies by Uccellini (1975), Stobie, Einaudi, and Uccellini (1983), Uccellini and Kock (1987), and Chimonas and Nappo (1987) examined the interactions between gravity waves and thunderstorms. Studies

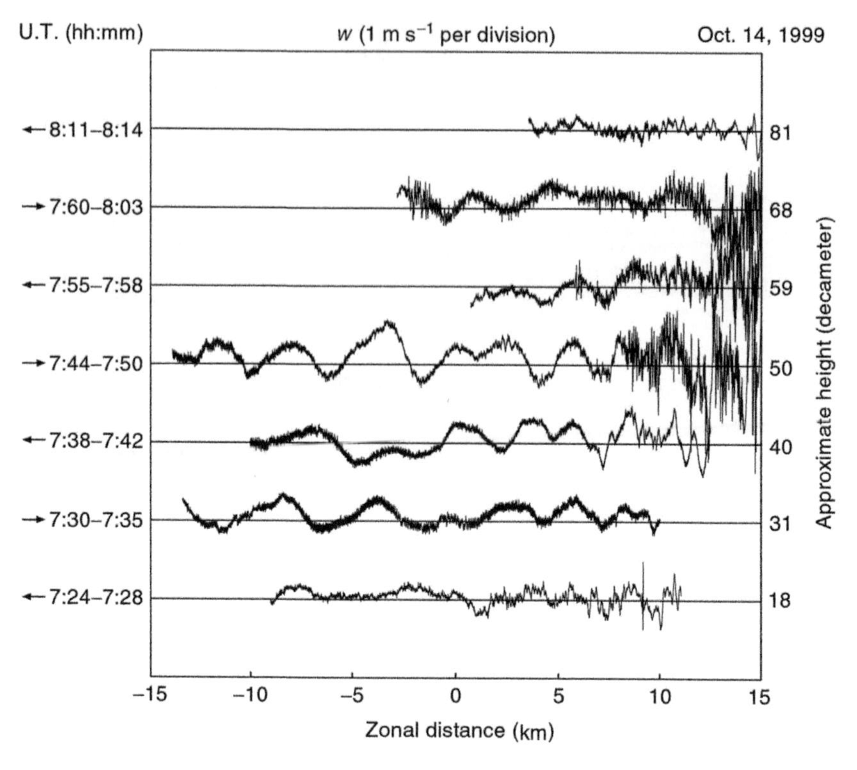

FIGURE 1.5 Vertical velocity recorded by the University of Wyoming King-Air B200 research aircraft on October 14, 1999 during the CASES-99 field campaign. The flights were in the east–west direction at various heights. The arrows on the left indicate the flight direction, and the time marks indicate the beginning and ending of data recording.

by Sawyer (1959), Bretherton (1969), Lilly (1971), Lilly and Kennedy (1973), Clark and Peltier (1977), Smith (1985), and Davies and Phillips (1985) examined the generation of gravity waves by mountains and the severe downslope winds and upper clear air turbulence (CAT) these waves can produce. Scorer (1949), Long (1953), Bretherton (1969), and Smith (1976) studied the generation of waves in the lee of mountains, *i.e.*, *lee waves.* Bretherton (1966), Shutts (1995), Grisogono (1995), Nappo and Svensson (2008), and Steeneveld, Nappo, and Holtslag (2009) examined the effects of wave drag in mesoscale numerical models. On the microscale, studies by Chimonas (1972), Einaudi and Finnigan (1981), and Fua *et al.* (1982) examined the interactions between gravity waves and turbulence in the stable planetary boundary layer; and Hines (1988), Chimonas and Nappo (1989), and Nappo and Chimonas (1992) examined the interactions of gravity waves generated by small-scale terrain features with the boundary-layer flow to produce turbulence in the upper regions of the stable PBL. The study of gravity waves and their effects on turbulence in the nighttime planetary boundary layer

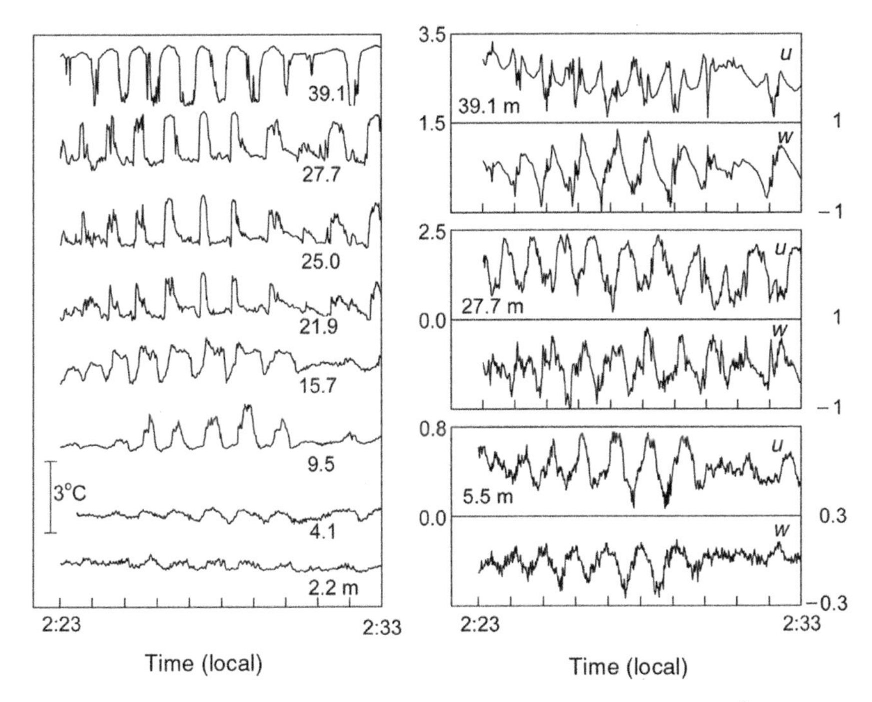

FIGURE 1.6 Ten-minute time series of temperatures (°C) and wind speeds (ms^{-1}) observed on July 13, 1994 in a boreal aspen forest. Canopy height was 21 m. Measurement heights are listed: u is the horizontal wind speed; w is the vertical wind speed. Taken from Lee *et al.* (1997).

(PBL) was a primary goal of field campaigns conducted in the planes of south-central Kansas, USA (CASES-99, Poulos *et al.*, 2001) and in the Salt-Lake City basin in Utah, USA (VTMX, Doran, Fast, and Horel, 2002). Figure 1.5 shows plots of vertical velocity observed by aircraft flights on October 14, 1999 during the CASES-99 campaign. The wave-like structures and turbulence seen between about 300 and 700 m AGL are typical. Tjernström *et al.* (2009) observed turbulence in the residual layer of the nighttime PBL where turbulence is not to be expected. They showed that this turbulence might be the result of breaking gravity waves. Lee *et al.* (1997), Lee and Barr (1998), and Hiscox, Nappo, and Miller (2006) examined gravity waves within and above a boreal forest canopy. An example of these *canopy waves* is shown in Fig. 1.6 taken from Lee *et al.* (1997). These waves are a common feature of the nighttime flow above forests.

Almost all of the fundamental theoretical studies of gravity waves have been done using the linear theory; a process that partitions the variables into constant or slowly varying background values and small first-order perturbations of these values. An example of small amplitudes disturbances is the surface of a pond as shown in Fig. 1.7. Linearization removes interactions of waves with

FIGURE 1.7 Small linear waves on a still pond.

waves and the resulting nonlinear transfers of energy. The *linear theory* allows analytical solutions of the wave equations. In the middle and upper atmosphere the background flows are often slowly varying and the linear theory is generally applicable. But in the troposphere and especially in the PBL the atmospheric flows are at best quasi-stationary, and the linear theory is uncertain. In the middle and upper atmosphere, gravity waves tend move in *wave packets* each packet composed of a spectrum of wavelengths. Nonlinear interactions among these packets can produce new packets with much shorter waves than contained in the original packets. These interactions cause wave energy exchange, but their mechanisms are yet to be understood (Fritts and Alexander, 2003). But in the troposphere, where many different wave frequencies can exit, nonlinear wave interactions are also important. Yet in spite of these limitations, the linear theory is still a robust tool for understanding and making first-order analyses of observations. However, it must be noted that many gravity wave characteristics exist which are beyond a linear analysis. Often, it is said that the linear theory "breaks down," but this does not imply a catastrophic failure; instead, the linear theory predicts nonphysical behavior such as unrealistically strong winds, large vertical temperature gradients, and unrealistically stable flows.

An important property of waves is their ability to transport energy away from disturbances (mountains, hills, thunderstorms, velocity jets, synoptic scale adjustments, *etc.*). Gravity waves act to distribute this energy throughout the atmosphere. The vertical distribution of energy and momentum is more rapidly done by waves than by the slow moving planetary and Rossby waves in the troposphere. Wave transport and subsequent deposition of energy is an important component of atmospheric dynamics. It is now recognized that much of the

FIGURE 1.8 "The Great Wave" by Hokusai (1760–1849). An example of a nonlinear wave.

turbulence in the nighttime PBL and CAT in the upper troposphere are due to breaking gravity waves. The roles of gravity waves in meteorology are continually being studied and expanded. Almost every issue of the *Atmospheric Research, The Journal of Atmospheric Science, The Journal of Atmospheric and Solar-Terrestrial Physics, The Journal of Fluid Mechanics, The Quarterly Journal of the Royal Meteorological Society, Monthly Weather Review, The Journal of Geophysical Research, Tellus, Boundary-Layer Meteorology, Atmospheric Environment, etc.*, and many Asian journals as well, contain articles on gravity waves. Considering the wide spectrum of the time and space scales of gravity waves and the complex interactions of these waves with themselves and the mean flow,[2] we expect the interest in gravity waves to increase in the future.

1.2 SOME WAVE MECHANICS

In this section, we define the space and time scales needed to describe and analyze waves. The fluid particles undergoing wave motions oscillate on surfaces that are perpendicular to the direction of travel of the wave. (However, for acoustic waves the fluid particles oscillate on surfaces that are parallel to the wave motion.)

[2] The author often draws examples from the lower troposphere and the PBL. This is because those regions of the atmosphere are more accessible to continuous measurements than the upper atmosphere. Also, the linear theory of gravity waves is applicable to all stably stratified flows.

FIGURE 1.9 Surfer riding a single wave. The surfer is at a phase point of the wave, and is moving at the phase speed of the wave.

We have all seen waves on calm water such as shown in Fig. 1.7, and these waves approximate the conditions for *linearity* which shall be discussed in this book; however, we will not discuss waves we see during storms such as rendered in Fig. 1.8. Such a wave is clearly nonlinear.

We must distinguish right off between *wave* and *waves*. If we consider a single wave, for example, a sine wave, then we are comfortable with the singular term, wave. We know what we are talking about. However, we often hear and use the plural in phrases such as "gravity waves," "Kelvin–Helmholtz waves," "unstable waves," *etc.* Sometimes the meanings of these phrases are clear, but often they can be confusing. For example, does a sine wave from 0 to 2π radians constitute a single wave? If a sine wave goes through several cycles, do we call these waves or "a" wave. Consider for a moment fish. In the English language, *fish* is the proper plural when applied to several members of the same species, for example, a school of fish; however, *fishes* is the proper plural when applied to several species of fish. Accordingly, we shall use the term *wave* to refer to a periodic disturbance described by a single frequency without regard to the number of cycles of the disturbance. We will also use the term to refer to a particular cycle of any periodic disturbance. Thus, for example, a surfer rides a wave as illustrated in Fig. 1.9. We shall use *waves* to refer to ensembles of disturbances with differing characteristics such as frequency, amplitude, length, or period. Sometimes we will use the term *wave train* or *plane waves*[3] to describe a series of repeated motions as shown in Fig. 1.10. Continuing our discussion, we see that the waves shown in Fig. 1.10 are long quasi-parallel lines with peaks, which are called *crests* and valleys, which are called

[3] See Hines (1955) and also page 233 of Hines (1974) for a precise definition of plane waves.

FIGURE 1.10 Two-dimensional waves on a calm sea.

troughs. These types of waves are *two-dimensional* because they are functions of two spatial dimensions and are assumed to extend indefinitely in the third. Waves at the beach are good examples of two-dimensional waves. A two-dimensional wave is sometimes called a plane wave. A *three-dimensional wave* is a function of all three dimensions. Examples of three-dimensional waves are the *ring wave* on the surface of a calm pond as shown in Fig. 1.11 and *lenticular clouds* as shown in Fig. 1.12, which are often seen above mountains. In the atmosphere and oceans, three-dimensional waves are spherical not unlike the layers of an onion. In this book, we shall study mostly two-dimensional waves; however, in many cases what we learn from two-dimensional waves can be extended to three dimensions. The waves shown in Figs. 1.7 and 1.10 appear on the surface of water and are called *surface waves*; however, the motions of the fluid particles which create the wave extend to some degree throughout the fluid. Thus, we can speak of a *wave field* which permeates the fluid much as an electromagnetic field permeates space. In the absence of boundaries where wave reflections can occur, a wave field for some variable $q(x, y, z)$ will exist everywhere as, for example, an electric or magnetic field; however in regions far from the wave, wave amplitudes can be vanishingly small.

1.2.1 WAVES, HARMONICS, AND MODES

Wave structures are often described in terms of *modes* and *harmonics*. These terms sometimes appear to be used arbitrarily. Harmonic pertains to or denotes a series of oscillations in which each oscillation has a frequency that is an integral

FIGURE 1.11 A ring wave on a still pond.

multiple of some basic or fundamental frequency. This terminology also is used in Fourier series. A mode is a standing wave between two fixed points. A vibrating sting is a good example. The fundamental mode or *gravest* mode is a standing wave with a length one-half of a full wave. The fundamental mode has one antinode and two nodes (the end points). The second mode has the length of a full wave cycle and two antinodes or three nodes. A confusion between modes and harmonics can be resolved if we consider modes to be all waves which end with nodes at boundaries. Such will be the case when waves are reflected in a way to cause resonance. Animations of these modes harmonics other wave dynamics can be seen in EURL\Ch1\WAVES. Note that the degree of the harmonic counts the number loops or antinodes between the ends of the string. Thus, for example, the second harmonic has two antinodes. The important point is that modes and harmonics will always refer to standing and trapped waves.

1.2.2 FRAMES OF REFERENCE

Because waves move in a medium that might also be moving, confusion can arise if we are not clear as to the frames of reference of observations and theoretical analyses. In a *Lagrangian* reference frame, fluid motions are observed in a coordinate system moving with the flow, generally taken to be the mean flow. In this frame, some waves will be moving faster than the observer and others will be moving slower. Most atmospheric observations are made in a coordinate system fixed to the earth. A reference frame that is stationary relative to the flow is called

FIGURE 1.12 A lenticular cloud produced by a three-dimensional wave over a mountain.

an *Eulerian* reference frame. Observations in Lagrangian and Eulerian reference frames are related through appropriate mathematical transformations along with the constraint that the physics in both frames must be identical. However, it often happens that the descriptions of certain motions are more conceptual and more easily described in one frame rather than another. In this book, we shall use mostly the Eulerian reference and exceptions to this rule will be noted.

1.2.3 WAVE SCALES

We shall use a Cartesian coordinate system (x, y, z) with x and y in the horizontal plane, and z in the vertical direction. The coordinates have unit vectors $(\hat{x}, \hat{y}, \hat{z})$. Unless otherwise noted, the horizontal directions of wave motion will be along the x-axis. The *wavelength* λ is the distance between successive crests[4] of a wave as illustrated in Fig. 1.13. In this book, we shall consider mostly waves with horizontal wavelengths less than 1000 km so that the effects of the earth's curvature may be neglected (Hines, 1968). Accordingly, we can consider horizontal planes as being flat. The mathematical description of a wave involves trigonometric functions, for example, $\sin(2\pi x/\lambda)$, *etc.*, and it is convenient to define the *wavenumber* as

$$\kappa = \frac{2\pi}{\lambda}. \tag{1.1}$$

[4] Actually, the distance between any two identical points on the wave is a wavelength.

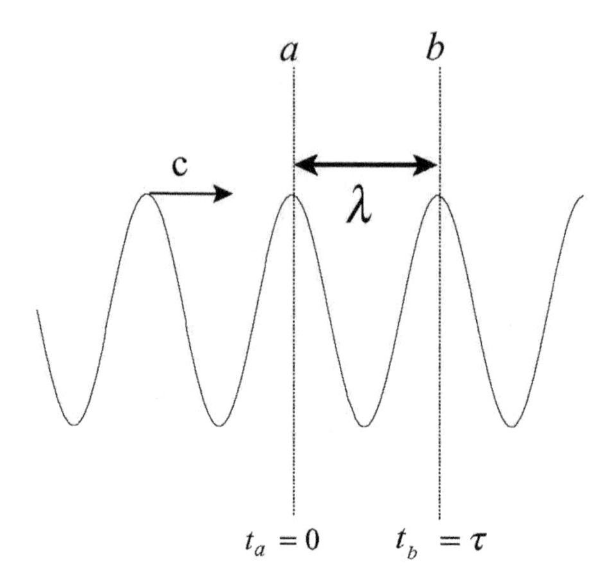

FIGURE 1.13 A wave with wavelength λ moves to the right with speed c. The wave crest moves from point a to b in time τ which is the period of oscillation of the wave as seen by a stationary observer.

We can think of the wavenumber as 2π times the number of wave oscillations per unit length, or wavelength per unit radian. The wavenumber is a fundamental property of a wave. We will label the wavenumbers in the x-, y-, and z-directions as k, l, and m, respectively. These wavenumbers are defined as

$$k = \frac{2\pi}{\lambda_x}, \quad l = \frac{2\pi}{\lambda_y}, \quad m = \frac{2\pi}{\lambda_z}, \tag{1.2}$$

where λ_x, λ_y, and λ_z are the wavelengths of the wave in the x-, y-, and z-directions, respectively. The *wave vector*, $\vec{\kappa}$, points in the direction of the traveling wave, and is given by

$$\vec{\kappa} = k\hat{x} + l\hat{y} + m\hat{z}. \tag{1.3}$$

The *wave period*, τ, is the time required for the fluid particles to make one oscillation. If the wave is moving in a still medium, then the wave period τ is the time required for successive wave crests to pass a stationary observer, as illustrated in Fig. 1.13. For waves with periods less than a few hours, the effects of the earth's rotation (the Coriolis force, see the Appendix) can be ignored. The *wave frequency*, ω, is 2π times the number of wave oscillations per unit time or radians per second *i.e.*,

$$\omega = \frac{2\pi}{\tau}. \tag{1.4}$$

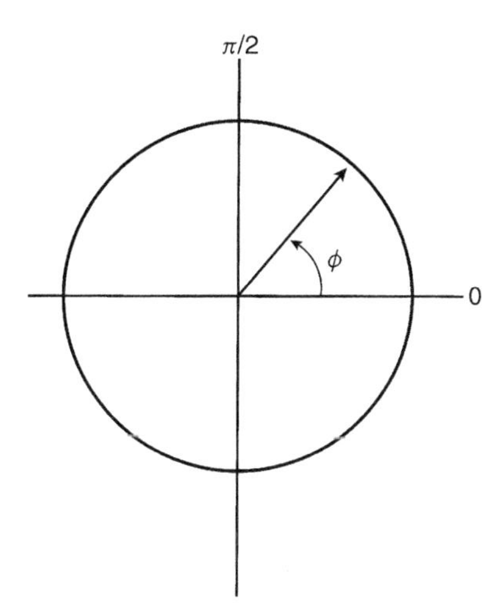

FIGURE 1.14 An illustration of phase angle in polar coordinates.

1.2.4 WAVE PHASE AND WAVE SPEED

Let $A\cos(kx - \omega t)$ describe a wave with amplitude A, wavenumber k, and frequency ω. As we shall see, the minus sign indicates a wave moving in the positive x-direction. A single oscillation of the wave either in space or time is a *cycle* of 2π radians. Each point in the cycle is a *phase point*. If the wave cycle is represented in *polar coordinates*, as illustrated in Fig. 1.14, then wave phase, ϕ, is represented by the positive angle between the radius vector and the horizontal axis that corresponds to $\phi = 0$. Thus, wave phase is often referred to as *phase angle*, *i.e.*, $\phi = kx - \omega t$. In the two-dimensional case,

$$\phi = \vec{\kappa} \bullet \vec{r} - \omega t = kx + mz - \omega t, \tag{1.5}$$

where $\vec{r}$ is the radius vector defined by

$$\vec{r} = x\hat{x} + z\hat{z}. \tag{1.6}$$

Now consider a wave in the (x, z) plane. Then,

$$q(x, z, t) = \Re A\, e^{i\phi} = A\cos(kx + mz - \omega t), \tag{1.7}$$

where $\Re$ indicates the real part of the complex number.[5] From (1.7) we see that at a fixed point in space, the function q oscillates with angular frequency ω, and at any instant of time q will have a wave structure of the form $\cos(kx + mz)$. In a two-dimensional plane, lines of constant ϕ connect the same points in the wave field. If the wave is a sine wave, then all points in the wave field where ϕ is a multiple of

[5] We shall use mostly exponential notation in describing waves. Since this results in complex numbers, we must select at the very end of an analysis the real parts of the numbers in order to get physically meaningful results.

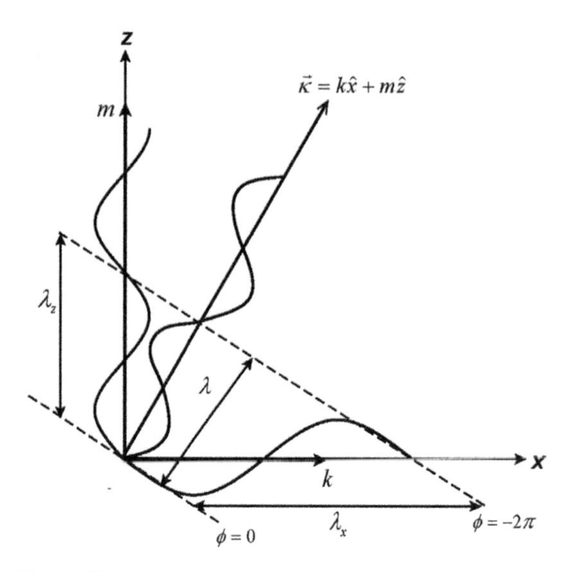

FIGURE 1.15 An illustration of wave fronts and wave vectors for a two-dimensional wave. The wave fronts are perpendicular to the wave vector. The negative values of phase angle ϕ indicate that these wave fronts passed a stationary observer earlier than the following fronts.

$\pi/2$ will be wave crests. The straight lines connecting these points represents the *wave fronts*. Figure 1.15 illustrates the wave field. The wavenumbers k and m are parallel to the $\hat{x}$- and $\hat{z}$-axes respectively and are the components of wave vector $\vec{\kappa}$. The dashed lines show the wave fronts corresponding to the nodes of the wave. A *node* is that part of the wave where $q(x, z, t) = 0$. Note that some authors restrict fronts to wave crests, but this is not necessary. The equation for a family of wave fronts is

$$\phi(x, z) = kx + mz = \phi_c, \tag{1.8}$$

where ϕ_c is a constant value of wave phase. If the $m \to 0$, then $\lambda_z \to \infty$, and the wave fronts become increasingly parallel to the vertical axis. Likewise, if $k \to 0$, then $\lambda_x \to \infty$, and the wave fronts become increasingly parallel to the horizontal axis. From (1.8) we can write

$$z_\phi = -\frac{k}{m}x + \phi_c, \tag{1.9}$$

where z_ϕ is the height of the wave phase point ϕ_c relative to some reference height. We can define a *phase vector* as

$$\vec{\phi} = x\hat{x} + z_\phi\hat{z}. \tag{1.10}$$

Then using (1.9) in (1.10) gives

$$\vec{\phi} = x\left(\hat{x} - \frac{k}{m}\hat{z}\right). \tag{1.11}$$

It follows that

$$\vec{\kappa} \bullet \vec{\phi} = (k\hat{x} + m\hat{z}) \bullet x\left(\hat{x} - \frac{k}{m}\hat{z}\right) = 0, \tag{1.12}$$

and we see that the wave vector is perpendicular to the wave fronts as illustrated in Fig. 1.15. The magnitude of the wave vector is

$$|\vec{\kappa}| = \sqrt{k^2 + m^2}, \tag{1.13}$$

where we have rotated the coordinate axis so that the x-axis is in the direction of horizontal wave motion. It follows from (1.13) and (1.2) that

$$\left(\frac{1}{\lambda}\right)^2 = \left(\frac{1}{\lambda_x}\right)^2 + \left(\frac{1}{\lambda_z}\right)^2. \tag{1.14}$$

It is clear from (1.14) and Fig. 1.15 that λ must be less than either λ_x or λ_z, and hence λ is not a vector.

To determine the wave speed, we pick a point on the wave, for example, a wave crest, and follow it along the direction of the translating wave as illustrated in Fig. 1.16 which shows the wave at times τ and $\tau + \Delta t$. The *phase speed*, c, of the wave is the speed at which a point of constant phase moves in the direction of the traveling wave. It is important to keep in mind that we are talking about the speed of a disturbance (the wave) moving through a fluid, not the speed of the fluid. The surfer seen in Fig. 1.9 is moving horizontally at the wave phase speed. If we differentiate (1.5), holding the phase point ϕ constant we get

$$\left.\frac{\mathrm{d}\phi}{\mathrm{d}t}\right|_{\phi} = \vec{\kappa} \bullet \frac{\mathrm{d}\vec{r}}{\mathrm{d}t} - \omega = 0, \tag{1.15}$$

and consequently

$$\frac{\mathrm{d}|\vec{r}|}{\mathrm{d}t} = c = \frac{\omega}{\kappa} = \frac{\omega}{\sqrt{k^2 + m^2}}. \tag{1.16}$$

The speed of the wave along the horizontal x-axis, c_x, is obtained by differentiating (1.5) with respect to t while holding z constant, *i.e.*,

$$\left.\frac{\mathrm{d}\phi}{\mathrm{d}t}\right|_{\phi,z} = k\frac{\mathrm{d}x}{\mathrm{d}t} - \omega = 0, \tag{1.17}$$

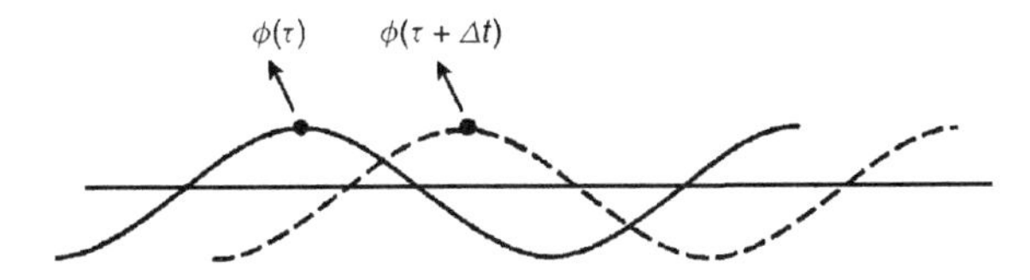

FIGURE 1.16 A wave moving in the x–z plane seen at times t and $t + \Delta t$.

and then

$$c_x = \frac{\omega}{k}. \tag{1.18}$$

Likewise, the phase speed in the z-direction is

$$c_z = \frac{\omega}{m}. \tag{1.19}$$

Note also that if we multiply (1.14) by ω^2 and use (1.16) we get

$$\left(\frac{1}{c}\right)^2 = \left(\frac{1}{c_x}\right)^2 + \left(\frac{1}{c_z}\right)^2 \tag{1.20}$$

indicating that phase speed is not a vector. It is interesting to note from (1.16) if ω is a slowing varying function of k, then the phase speed increases with decreasing wavenumber. Thus, we have perhaps the counter-intuitive result that long waves travel faster than short waves.

In text books and scientific literature, we see references to either phase speed or phase velocity. For example, Gossard and Hooke (1975), Booker and Bretherton (1967), and Baines (1995) use "phase velocity" while Hines (1960), Lighthill (1978), and Gill (1982) use "phase speed." The distinction between the terms "speed" and "velocity" is trivial if we agree that c is the speed of a wave in the direction of the wave vector. Another term that can be confusing is *propagation*. We shall see that propagation refers to the transport of energy. We shall use wave speed to refer to the movement of wave fronts.

1.2.5 GROUP VELOCITY

One of the most important concepts in gravity wave theory is *group velocity*. We shall see that the rate of energy propagation by waves is the group velocity. In the real world, all waves must have a source, a point of creation. There are many sources of waves. These can be instantaneous, for example, at the point of a large explosion or a lightening strike. Other sources can be continuous in time over extended space, for example, flow over a mountain range or a baroclinic instability of a Rossby wave. But in any case, the energy created by these disturbances must move away from the source. In the stably stratified atmosphere, gravity waves carry this energy away. In many books on gravity waves, the wave characteristics, *i.e.*, wavenumber, phase speed, frequency, *etc.* are treated as if they are independent of wave energy. We have noted that expressions such as (1.7) imply a wave field everywhere in space, but this is an idealization. It is only when the origin of the wave and the transport of the wave energy are considered that the problem becomes physical. Indeed, it is the energy transport that creates the wave. The wave does not create the energy transport. The speed and direction of energy transport is determined by the group velocity.

Consider now the example proposed by Lighthill (1978). Imagine a stone thrown into a pond of still water. After the initial splash, a ring wave expands away from

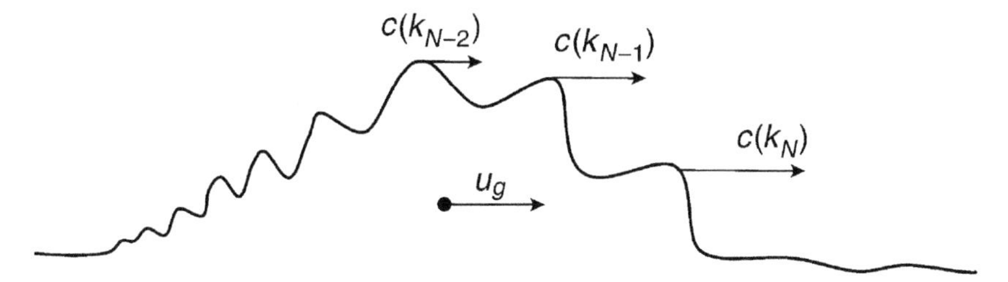

FIGURE 1.17 A ring wave on the surface of a calm pond moves with group velocity u_g, but the wavelets move with their individual phase speeds, c, which are functions of their wavenumbers, i.e., $k_N, k_{(N-1)}, k_{(N-2)}$.

the center as illustrated in Fig. 1.11. If we could follow the wave, and examine it closely we would see that the surface of the wave is continually disturbed by a series of smaller waves or wavelets, which move from the rear to the front of the ring wave as illustrated in Fig. 1.17. These wavelets were created when the stone struck the water surface. The energy of that initial disturbance is distributed over a *spectrum* of waves spread over a small range of wavenumbers with the largest wavelength being about twice the size of the stone. The ensemble of the wavelets forms a bundle of waves called a *wave packet*. Each wavelet is moving with a speed proportional to its wavelength. The wavelets at the front have larger wavelength than the waves at the rear, and consequently have greater speed than those at the rear. We would find it quite difficult (actually impossible) to follow a single wave from the rear to the front of the ring wave. This is because as the wave moves to the front its wavelength increases. Thus, the wave we initially identified at the rear of the wave packet no longer resembles the wave at the front of the packet. It is clear that the wavelets are moving faster than wave group, which we observe to have speed u_g. The wavelets appear neither ahead nor behind the wave packet, and we must conclude that the energy transport is being done by the wave packet rather than by the individual waves forming the packet. (These are sometime referred to as *pilot waves*.) The wavelets have no energy content. If they did, then we would be able to observe them as they moved ahead of the wave packet. Thus, the energy of the disturbance is contained in the wave packet which we can observe on the water's surface. But what is the speed of this packet?

There are several ways of developing the equation for group velocity, and these can be found in, for example, Hines (1960), Bretherton (1966), Gossard and Hooke (1975), and Lighthill (1978). We envision a two-dimensional wave packet moving horizontally well after its initiation. The packet is composed of a narrow spectrum of linear waves with wavelengths varying slowly from wave to wave so that the dominate wave, which defines the packet, has local phase ϕ. The rate of change of ϕ with time is $2\pi/\tau$, *i.e.*, the wave frequency ω. However, the phase ϕ is also a function of distance x. At a given time, the phase decreases in value, relative to its value at wave crest, with increasing x. The rate of this decrease is equal to $2\pi/\lambda$, *i.e.*, the wavenumber k. Thus, the phase $\phi(x,t)$ satisfies the equations

$$\partial\phi/\partial x = -k, \quad \partial\phi/\partial t = \omega. \tag{1.21}$$

Equation (1.21) can be combined to give

$$\partial k/\partial t + \partial\omega/\partial x = 0. \tag{1.22}$$

Lighthill (1978) refers to (1.22) as an equation of the continuity for phase. This is because k represents in some sense a phase density (phase per unit length), and ω represents in some sense a phase flow rate (phase passing a fixed point per unit time). If we now regard ω as a function of k, *i.e.*, $\omega = \omega(k)$, (1.22) becomes

$$\frac{\partial k}{\partial t} + \frac{d\omega}{dk}\frac{\partial k}{\partial x} = 0, \tag{1.23}$$

if we define $\partial\omega/\partial k$ as u_g, then (1.23) can be written as

$$\frac{\partial k}{\partial t} + u_g\frac{\partial k}{\partial x} = \frac{D_g k}{Dt} = 0, \tag{1.24}$$

and we see that the dominant wavenumber is constant when moving at the group velocity.[6]

We can develop the group velocity by considering a wave packet composed of only two waves traveling in the horizontal direction. Let the waves have equal amplitudes, A, but slightly different frequencies, *i.e.*, $\omega + \delta\omega$, $\omega - \delta\omega$, and slightly different wavenumbers, *i.e.*, $k + \delta k$, and $k - \delta k$. The superposition of these waves is given by

$$\zeta = A\cos[(k+\delta k)x - (\omega+\delta\omega)t] + A\cos[(k-\delta k)x - (\omega-\delta\omega)t]. \tag{1.25}$$

With some trigonometry, (1.25) can be written as

$$\zeta = 2\alpha\cos(\delta kx - \delta\omega t)\cos(kx - \omega t). \tag{1.26}$$

Figure 1.18 shows a graph of (1.26) at times $t = 0$, 30, and 60 s for the case $k = 2\pi/60\ \text{m}^{-1}$, $\delta k = 2\pi/400\ \text{m}^{-1}$, $\omega = 2\pi/10\ \text{s}^{-1}$, and $\delta\omega = 2\pi/400\ \text{s}^{-1}$. The two waves combine to form an amplitude-modulated carrier wave $\cos(kx - \omega t)$. The sloping lines in Fig. 1.18 connect the crests of the carrier wave and the node points of the wave packet at various times. The slopes of these lines are proportional to the speeds of the crests and the nodes, and it is clear that the carrier waves are moving faster than the packet. The amplitude modulation is given by $2A\cos(\delta kx - \delta\omega t)$. The phase speed of the carrier wave is $\omega/k = 6\ \text{m s}^{-1}$. Note that wave energy cannot pass through the node points of the carrier wave since the amplitude is zero. The wave energy is trapped in the wave packet, and is constrained to move with the wave packet at the group velocity. In the same way that we developed an expression for the phase speed, we calculate the group velocity in the x-direction as:

$$u_g = \frac{\delta\omega}{\delta k}. \tag{1.27}$$

[6] Note that $\frac{D_g k}{Dt}$ is the total derivative, using group velocity rather than fluid velocity.

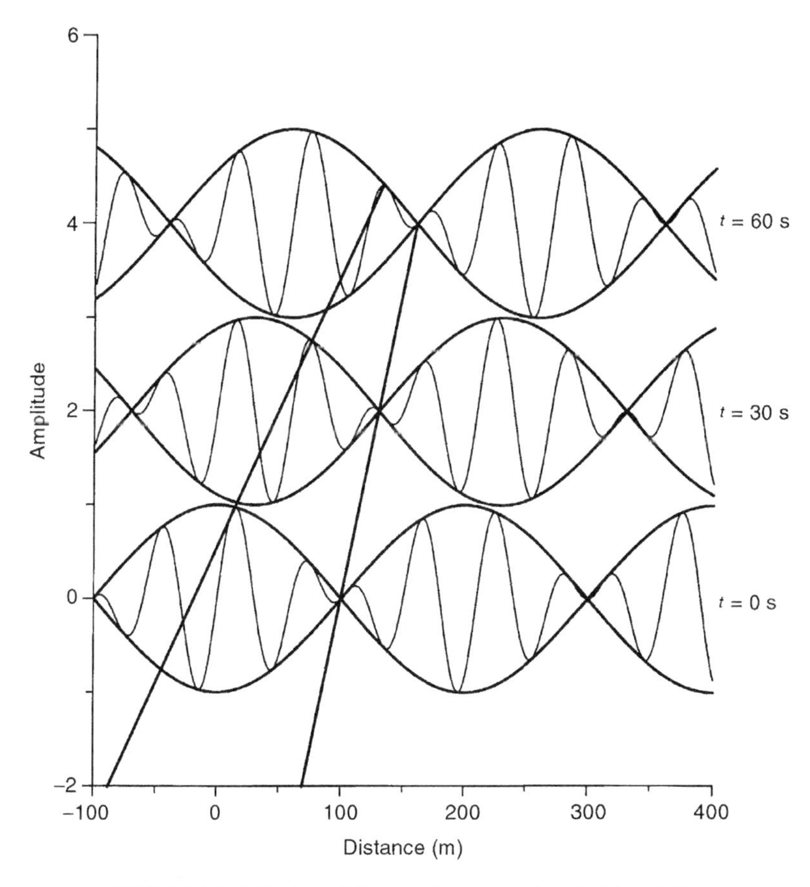

FIGURE 1.18 A modulated carrier wave at times 0, 10, and 60 s.

In the limit of small values of $\delta\omega$ and δk, (1.27) becomes:

$$u_g = \frac{\partial \omega}{\partial k}. \tag{1.28}$$

Likewise the group velocity in the vertical direction is

$$w_g = \frac{\partial \omega}{\partial m}. \tag{1.29}$$

The group velocity for the wave in Fig. 1.18 is $u_g = 1\ \mathrm{m\,s^{-1}}$. We shall see that the group velocity is always less than the phase velocity. A link to an interactive program for illustrating phase speed and group velocity is found at EURL\Ch1\L2.

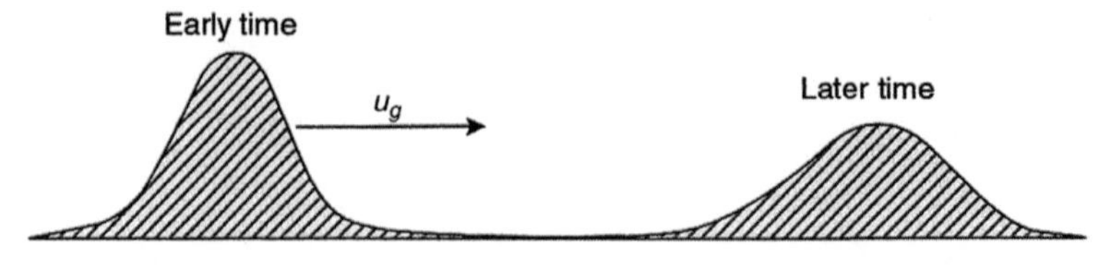

FIGURE 1.19 A cross-section of a ring wave seen at early and late times.

1.2.6 WAVE DISPERSION

Phase speed is a function of wave frequency and wavenumber, *i.e.*, $c = \omega/k = \lambda/\tau$. For a given wave period, τ, long waves will travel faster than short waves, and this leads to wave *dispersion*. Recalling the example of a surface ring wave created by a stone thrown into a calm pond (see Fig. 1.11), we observe that initially, the disturbance is sharply peaked as illustrated in Fig. 1.19. As the ring expands, the width of the disturbance increases. This is because the waves that comprise the disturbance are moving at different phase speeds. Because the long waves move faster than the short waves, the disturbance broadens. Note that the total energy of the wave packet remains constant, but disperses horizontally. The relation between phase speed and wavenumber is called the *dispersion relation*. However, it must be said that as the ring expands the energy content of the wave is spread over an increasing area and the amplitude of the packet decreases accordingly.

1.3 THE BUOYANT FORCE

The term gravity wave implies that gravity is the restoring force acting on a fluid parcel that has been displaced from its equilibrium position. Obviously, in the absence of gravity these waves would not exist. However, it is not gravity but rather the fluid *buoyancy* that is acting. Consider an atmosphere at rest and an air parcel of mass m_p in equilibrium with its environment at height z_e. Let the parcel be displaced upward a small distance δz from z_e. We assume that during this displacement the air in the parcel does not mix with its surroundings, and that the process is adiabatic, *i.e.*, there is no net transfer of heat across the surface of the air parcel. The buoyant force, $\vec{F}_\mathrm{b}$ acting on the fluid parcel is

$$\vec{F}_\mathrm{b} = -g(m_\mathrm{p} - m_\mathrm{a})\hat{z}, \tag{1.30}$$

where m_a is the mass of air displaced by the fluid parcel and g is the acceleration of gravity[7] (positive upward). From the second law of motion we have,

$$m_\mathrm{p}\frac{\mathrm{d}^2(\delta z)}{\mathrm{d}t^2} = -g(m_\mathrm{p} - m_\mathrm{a}). \tag{1.31}$$

[7] Note that formally $g = g_0/(1+z/a)^2$ where a is the mean radius of the earth, z is the height above the ground surface, and $g_0 = 9.81\ ms^{-}1$.

The mass of each air parcel is

$$m_{\mathrm{p}} = \rho_{\mathrm{p}} v_{\mathrm{p}}, \tag{1.32}$$

and

$$m_{\mathrm{a}} = \rho_{\mathrm{a}} v_{\mathrm{a}}, \tag{1.33}$$

where ρ is density and v is volume. We assume the volumes of the air parcel and the displaced air are equal, and that the pressure in the air parcel is always equal to the environmental pressure. The *equation of state* for dry air is given by the ideal gas law, *i.e.*,

$$P = \rho R T, \tag{1.34}$$

where P is the atmospheric pressure expressed in Pa,[8] T is the temperature measured in Kelvin (K), and R is the specific gas constant. For dry air R = 287 J kg^{-1} K^{-1}. Using (1.34), (1.31) becomes

$$\frac{\mathrm{d}^2(\delta z)}{\mathrm{d}t^2} = -g\frac{\rho_{\mathrm{p}} - \rho_{\mathrm{a}}}{\rho_{\mathrm{p}}} = -g\frac{T_{\mathrm{a}} - T_{\mathrm{p}}}{T_{\mathrm{a}}}, \tag{1.35}$$

where ρ_{a} is the environmental air density, and T_{a} and T_{p} are the environmental and parcel temperatures respectively. We now expand T_{a} and T_{p} to first order, *i.e.*,

$$T_{\mathrm{a}}(z_{\mathrm{e}} + \delta z) = T_0 + \left.\frac{\partial T_{\mathrm{a}}}{\partial z}\right|_{z_{\mathrm{e}}} \delta z + \cdots, \tag{1.36}$$

$$T_{\mathrm{p}}(z_{\mathrm{e}} + \delta z) = T_0 + \left.\frac{\partial T_{\mathrm{p}}}{\partial z}\right|_{z_{\mathrm{e}}} \delta z + \cdots, \tag{1.37}$$

where T_0 is the temperature at equilibrium height z_{e}. Using (1.36) and (1.37) in (1.35) and noting that the change of environmental temperature due to the vertical displacement is small, *i.e.*,

$$T_0 \gg \frac{\partial T_{\mathrm{a}}}{\partial z}\delta z. \tag{1.38}$$

Equation (1.35) becomes

$$\frac{\mathrm{d}^2(\delta z)}{\mathrm{d}t^2} = -\frac{g}{T_{\mathrm{a}}}\left(\frac{\partial T_{\mathrm{a}}}{\partial z} - \frac{\partial T_{\mathrm{p}}}{\partial z}\right)\delta z. \tag{1.39}$$

Because the motion of the air parcel is adiabatic, its change of temperature with height is

$$-\frac{\partial T_{\mathrm{p}}}{\partial z} = \frac{g}{c_p} = \Gamma, \tag{1.40}$$

where c_p is the *specific heat capacity* at constant pressure and Γ is the *adiabatic lapse rate* (see Appendix). For dry air, c_p = 1005 J kg^{-1} K^{-1} so that $\Gamma \approx$

[8] A Pascal, Pa, is a stress or pressure of 1 N m^{-2}; 10 μbar = 1 Pa.

10 K km^{-1}. We also set $\partial T_a/\partial z = \gamma_a$, *i.e.*, the *atmospheric temperature gradient.* Equation (1.39) can now be written as

$$\frac{d^2(\delta z)}{dt^2} = -\frac{g}{T_a}(\Gamma - \gamma_a)\delta z. \tag{1.41}$$

The *potential temperature* (see Appendix) is defined as

$$\theta = T_a \left(\frac{1000}{P}\right)^{\frac{R}{c_p}}. \tag{1.42}$$

The potential temperature is the temperature an air parcel would have if it were brought down adiabatically from a height where the pressure is P to the height where the pressure is 1000 hPa, *i.e.*, the ground surface. It is straightforward to show that if the logarithmic derivative[9] of (1.42) is taken, and if use is made of (1.34) and the *hydrostatic approximation* (see Appendix), *i.e.*,

$$\frac{\partial P}{\partial z} = -\rho g, \tag{1.43}$$

then

$$\frac{1}{\theta}\frac{\partial \theta}{\partial z} = \frac{1}{T_a}\left(\frac{\partial T_a}{\partial z} + \frac{g}{c_p}\right) = \frac{\Gamma - \gamma_a}{T_a}. \tag{1.44}$$

Using (1.44) in (1.41) gives the result

$$\frac{d^2(\delta z)}{dt^2} = -\frac{g}{\theta}\frac{\partial \theta}{\partial z}\delta z. \tag{1.45}$$

If $\partial\theta/\partial z > 0$, then (1.45) represents simple harmonic motion in the vertical direction. If the air parcel is displaced vertically and released, then its motion is described by

$$\delta z(t) = A\,e^{iNt} + B\,e^{-iNt}, \tag{1.46}$$

where A and B are constants and

$$N = \sqrt{\frac{g}{\theta}\frac{\partial \theta}{\partial z}}. \tag{1.47}$$

If N is real, then it is the frequency of oscillation of the air parcel, $2\pi/\tau_N$, and is called the Brunt–Väisälä frequency, and τ_N is the Brunt–Väisälä period of oscillation. If N is imaginary, *i.e.*, $\frac{\partial \theta}{\partial z} < 0$, then

$$\delta z = A\,e^{-N_I t} + B\,e^{N_I t} \tag{1.48}$$

[9] The logarithmic derivative of $f(x)$ is $\frac{d \ln f}{dx} = \frac{1}{f}\frac{df}{dx}$.

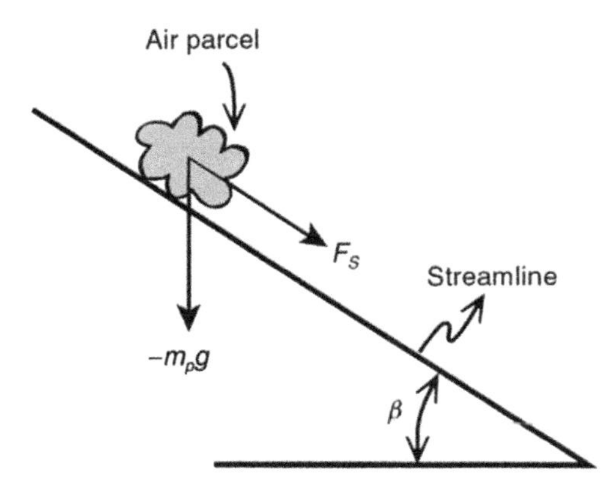

FIGURE 1.20 Air parcel displaced from equilibrium along an inclined plane.

where $N = iN_I$ and subscript "I" denotes an imaginary number. Solution B in (1.48) represents unbounded growth of the air parcel's vertical displacement, *i.e.*, an *instability*. This type of instability is due to the thermal properties of the atmosphere, and is referred to as *convective instability*. Thus, we see that wave motion is possible only when the atmosphere is *stably stratified*, *i.e.*, $\frac{\partial\theta}{\partial z} > 0$; however, this does not mean that all gravity waves have frequency N. We shall see that the Brunt–Väisälä frequency represents the maximum frequency for vertically propagating gravity waves. In an *isothermal* atmosphere where $\partial T/\partial z = 0$, $N \approx 0.02\ \mathrm{s}^{-1}$, and the buoyancy period is about 5 min. At $z \approx 100$ km, $N \approx 0.023\ \mathrm{s}^{-1}$. Holton (2004) gives $N \approx 0.012\ \mathrm{s}^{-1}$ for average tropospheric conditions, so that the buoyancy period is about 8 min.

The buoyant force described by (1.30) is restricted to vertical displacements of fluid parcels. Purely vertical displacements of air parcels due to waves occur only when the wave train is moving horizontally, for example, surface waves on water. However, gravity waves always propagate at an angle to the vertical, and so the fluid parcels will be displaced at an angle to the vertical. To examine this case, let δs be the displacement of a fluid parcel from its equilibrium position on a surface inclined an angle β to the horizontal as shown in Fig. 1.20. As we shall see, air parcels displaced by a gravity wave oscillate along streamlines that are perpendicular to the path of the wave, and so we can consider the inclined surface to be a streamline. The buoyant force acting on the displaced air parcel is

$$\vec{F}_\mathrm{s} = m_\mathrm{p}\frac{\mathrm{d}^2(\delta s)}{\mathrm{d}t^2} = -g\sin\beta(m_\mathrm{p} - m_\mathrm{a}), \tag{1.49}$$

where $-g\sin\beta$ is the component of gravity along s. From (1.49) it follows that

$$N' = \left[\frac{g}{\theta}\frac{\partial\theta}{\partial z}\sin^2\beta\right]^{\frac{1}{2}} = N\sin\beta. \tag{1.50}$$

If $\beta = 0°$, then the motion is horizontal but not oscillatory since then $N' = 0$. We see that the range of buoyancy frequencies extends from 0 to N depending on the angle of propagation relative to the horizontal plane, β.

1.4 THE BOUSSINESQ APPROXIMATION

In the next chapter, we derive the equations for gravity waves from the Euler equations of motion for an irrotational, frictionless atmosphere, *i.e.*,

$$\rho \frac{\mathrm{D}\vec{U}}{\mathrm{D}t} = -\nabla p + \rho \vec{g}, \tag{1.51}$$

$$\frac{\mathrm{D}\rho}{\mathrm{D}t} + \rho \nabla \bullet \vec{U} = 0, \tag{1.52}$$

$$\frac{\mathrm{D}P}{\mathrm{D}t} - c_\mathrm{s}^2 \frac{\mathrm{D}\rho}{\mathrm{D}t} = 0, \tag{1.53}$$

where $c_\mathrm{s}^2 = \frac{c_p}{c_v}\frac{p}{\rho}$ is the speed of sound, c_v is the specific heat capacity for dry air at constant volume, and

$$\frac{\mathrm{D}}{\mathrm{D}t} = \frac{\partial}{\partial t} + \vec{U} \bullet \nabla. \tag{1.54}$$

Equations (1.51)–(1.53) represent the conservation of momentum, the conservation of mass, and the conservation of thermal energy respectively. From the thermal energy equation (1.53), we see that the density and pressure are related, and in general $\rho = f(P, T)$. But in such an atmosphere, *acoustic* waves will exist as well as gravity waves, and this can lead to complications. It is through the *Boussinesq approximation* (see, for example, Spiegel and Veronis, 1960) that we can eliminate the acoustic waves.

Consider (1.51), and make the substitutions $\rho = \rho_0 + \rho_1$ and $P = P_0 + P_1$ where subscript 0 indicates a mean or background value, and subscript 1 indicates a small departure from the mean state, *i.e.*, a *perturbation*. Then (1.51) becomes

$$(\rho_0 + \rho_1)\frac{\mathrm{D}\vec{U}}{\mathrm{D}t} = -\nabla P_0 + \rho_0 \vec{g} - \nabla P_1 + \rho_1 \vec{g}. \tag{1.55}$$

We assume the background state is in hydrostatic equilibrium (1.43), so that (1.55) becomes

$$\left(1 + \frac{\rho_1}{\rho_0}\right)\frac{\mathrm{D}\vec{U}}{\mathrm{D}t} = -\frac{1}{\rho_0}\nabla P_1 + \frac{\rho_1}{\rho_0}\vec{g}. \tag{1.56}$$

The Boussinesq approximation assumes that $|\frac{\rho_1}{\rho_0}| \ll 1$, and accordingly fluctuations in density affect the buoyancy term much more then the inertial term. Thus, density fluctuations are considered only when they occur in combination with g.

The condition that $|\frac{\rho_1}{\rho_0}| \ll 1$ is satisfied when the vertical scale of variations of the mean variables, λ_z, is much less than the isothermal *scale height* of the atmosphere (see Appendix) *i.e.*,

$$\lambda_z \ll H_s = RT_0/g, \tag{1.57}$$

where H_s is the scale height of the isothermal atmosphere. To visualize this requirement, consider the vertical variation of density in an isothermal atmosphere, *i.e.*,

$$\rho = \rho_s \, \mathrm{e}^{-z/H_s}, \tag{1.58}$$

where ρ_s is the density at the ground surface. Then

$$\frac{\partial \rho}{\partial z} = -\frac{\rho_s}{H_s} \mathrm{e}^{-z/H_s} = -\frac{\rho}{H_s}. \tag{1.59}$$

If we replace the derivative in (1.59) by differentials, we can write

$$\frac{\delta \rho}{\rho} = -\frac{\delta z}{H_s}, \tag{1.60}$$

where $\delta\rho$ is a small change in density due to a small vertical displacement δz. Now, if we identify the small density change $\delta\rho$ with the density perturbation ρ_1, and if we identify the small vertical displacement δz with the scale of the wave motion λ_z, then we can write

$$\left| \frac{\rho_1}{\rho_0} \right| = \frac{\lambda_z}{H_s}. \tag{1.61}$$

We see from (1.61) that if $\lambda_z \ll H_s$, then $|\frac{\rho_1}{\rho_0}| \ll 1$.

As demonstrated by Spiegel and Veronis (1960), a result of the Boussinesq approximation is that the fluctuating changes in density due to local pressure variations are negligible. In this case, we can treat the fluid as being incompressible, and acoustic waves are eliminated. A consequence of the separation of the pressure and density changes is that as $c_s \to \infty$, *i.e.*, the fluid becomes incompressible, and (1.53) can be broken into two terms. Then the Euler equations take the form

$$\frac{\mathrm{D}\vec{U}}{\mathrm{D}t} = -\frac{1}{\rho_0}\nabla p + \frac{\rho_1}{\rho_0}\vec{g}, \tag{1.62}$$

$$\nabla \bullet \vec{U} = 0, \tag{1.63}$$

$$\frac{\mathrm{D}\rho}{\mathrm{D}t} = 0. \tag{1.64}$$

In the Appendix, it is shown that under the Boussinesq approximation,

$$\frac{1}{\theta_0}\frac{\partial \theta_0}{\partial z} = -\frac{1}{\rho_0}\frac{\partial \rho_0}{\partial z}. \tag{1.65}$$

Another result is,

$$\frac{\theta_1}{\theta_0} = -\frac{\rho_1}{\rho_0}. \tag{1.66}$$

Note that because of (1.65) the Brunt–Väisälä frequency (1.47) can be written as

$$N^2 = -\frac{g}{\rho_0}\frac{\partial \rho_0}{\partial z}. \tag{1.67}$$

PROBLEMS

1. Calculate the Brunt–Väisälä frequency at 5 km, 15 km, 50 km, and 100 km. Assume a constant temperature atmosphere with $T = 280$ K.
2. Aircraft warning lights atop two co-located tower flash at different rates. One light flashes every 2 s and the other flashes every 1.5 s. How many seconds does it take for both lights to flash at the same time? How many seconds does it take from the time of simultaneous flashing to the time of opposite flashing?
3. A constant density balloon in a constant lapse-rate atmosphere oscillates up and down with a period of 8 min. What is the Brunt–Väisälä frequency?
4. Using (1.53) show that $\frac{\theta'}{\theta_0} = -\frac{\rho'}{\rho_0}$.
5. A 10 cm wide rock is thrown into a deep pond of still water. What is the group velocity of the resulting surface wave?
6. The Brunt–Väisälä frequency is 0.02 s^{-1}, and a gravity wave is moving upward making an angle of 32° with the horizontal plane. What is the maximum frequency of the wave?
7. A thunderstorm traveling at 20 ms^{-1} generates an atmospheric bow wave. The bow wave makes an angle of 60° with the direction of the storm. What is the phase speed of the wave?
8. A jet liner is flying level at a speed of 350 km h^{-1} when it encounters a gravity wave. If the period of vertical oscillations of the plane is 2 s, what is the wavelength of the wave?
9. If a wave has $\lambda_x = 150$ km and $\lambda_z = 6$ km, what is the wavenumber of the wave and its angle to the horizontal?
10. On a balloon drifting toward the east at 5 ms^{-1} a wave is observed moving to the northeast at 10 ms^{-1}. What is the speed and direction of the wave as seen on the ground surface?
11. A two-dimensional, (x, z), wave is propagating to the east, yet a meridional wave perturbation v_1 is observed. Explain how this can be.
12. Assuming constant Brunt–Väisälä frequency, N, show that low-frequency waves move more vertically than horizontally and high-frequency waves move more horizontally than vertically.
13. What is the maximum frequency of a two-dimensional internal gravity wave when it's group velocity vector makes an angle of 10° with the horizontal plane?
14. A meteor explodes in the stratosphere, sketch the group velocity vectors and the phase lines of the gravity waves produced by the explosion.
15. A wave is moving in a direction 30° above the horizontal plane. If $N = 0.02$ s^{-1}, what is the frequency of the wave?

2

The Linear Theory

2.1 Introduction

Almost all of what we know about the nature of atmospheric gravity waves derives from the *linear theory*. From a computational consideration, the simplified equations describing linear systems are solved much more rapidly than nonlinear ones.[1] But perhaps more important, is that linear systems are more comprehensible and understandable than nonlinear systems. For example, in the system of three simultaneous, nonlinear differential equations describing pure katabatic flow over a simple slope (see, for example, Nappo and Rao, 1987) it is difficult to visualize the interactions between momentum, turbulence stress, and heat flux.

The recipe for linearization is simple. Some variable q is expanded into a *background state* q_0 and a first-order *perturbation* q_1. The background state is taken to be steady or slowly varying and horizontally uniform, but varying in the vertical direction. Except for the very long horizontal waves, the effects of the earth's rotation make a small contribution to the wave dynamics and is often ignored as are air viscosity and thermal conductivity, except for very short horizontal and vertical wave lengths. The perturbation q_1 is assumed much smaller than q_0 and

[1] However, with ever-increasing computer speeds, this advantage is becoming less significant. For example, Fritts *et al.* (2009) used a direct numerical simulation (DNS) model to examine the details of the interactions of a gravity wave with turbulence.

does not affect the background state. An example is a comparison of the height of surface waves with the depth of the sea. But if the perturbations are small, then the products of the perturbations, *i.e.*, $q_1 \times w_1$ are smaller and are neglected. The perturbations can be caused by several mechanisms including turbulence, density currents, thermal plumes, *etc.*; however, here we shall consider only perturbations due to gravity waves. Neglecting products of the wave perturbations negates the interactions of waves with waves. The waves can add or subtract to form packets of waves or standing waves, but they cannot interact to form new waves. Each wave of a packet behaves independently. The accuracy and practical limitations of the linear theory were examined by Dörnbrack and Nappo (1997) by comparing the results of a linear wave model with a nonlinear, time-dependent, hydrodynamic numerical model (Dörnbrack, 1998). They concluded that essentially similar results are obtained from each model for important wave parameters and first-order dynamic effects.

The simplicity of the linear theory is not without cost. In the real atmosphere, waves interact with other waves as well as with turbulence which itself might have been produced by interactions between waves and the mean flow. Observations of gravity waves made by, for example, Gossard and Munk (1954), Caughey and Readings (1975), Bedard, Canavero, and Einaudi (1986), Hooke and Jones (1986), Koch and Golus (1988), Einaudi, Bedard, and Finnigan (1989), Hauf *et al.* (1996), Lee and Barr (1998), Smith *et al.* (2002), Sun *et al.* (2004), and Nappo, Miller, and Hiscox (2008) show complex wave structures, with time-varying amplitudes usually with several frequencies present. A nearly *monochromatic* wave with nearly constant amplitude lasting more than several cycles is seldom observed. Applying linear analyzes to observations of wave phenomena is often frustrating as discussed by Hunt, Kaimal, and Gaynor (1985) and Finnigan (1988). For example, wave amplitudes change with time; waves in the nighttime PBL are often nonlinear because of the presence of the ground surface; and it is generally difficult to distinguish between gravity waves and turbulence (see, for example, Stewart, 1969; Finnigan, 1988; Sun *et al.*, 2004). Yet in spite of these difficulties, the linear theory is still useful, and provides a first-order estimate of most wave phenomena. Also, the simplicity of the linear theory gives an understandable picture of the wave processes and observations. Relatively few people can think and work in a nonlinear world.

Linear theory describes the possible disturbances of an existing state of uniform motion. A dynamic theory describes the consequences of forces acting to change the motions of a system. Thus, a dynamic theory is nonlinear. What this means is that a linear analysis must begin with a known uniform state of motion and energy. Linear theory cannot predict the motions of waves in a fluid, it can only predict possible wave states in the fluid. However, the possibility of waves is not a sufficient condition for waves. This is important. We shall see there are conditions where the linear theory describes waves, but observations show no evidence of waves as described by Smith *et al.* (2002) in connection with mountain lee waves. In this case, the linear theory either fails or some acting forces have not been considered.

In other cases, observations show waves where the linear theory describes no waves. This leads to an 'updating' of the linear theory. An example of this is the observations of wave disturbances in the ionosphere and their explanation by Hines (1974).

2.2 THE TAYLOR GOLDSTEIN EQUATION

The *Taylor–Goldstein* equation is the *wave equation* for linear gravity waves. We consider the Euler equations for irrotational, frictionless, and non-heat conducting, *i.e.*, adiabatic flow. The velocities in the x, y, z directions are u, v, w respectively. Here and throughout most of this book we shall use a two-dimensional (x, z) reference plane. Extension to three dimensions is straightforward. Under the Boussinesq approximation the governing equations are:

$$\frac{\partial u}{\partial t} + u\frac{\partial u}{\partial x} + w\frac{\partial u}{\partial z} = -\frac{1}{\rho}\frac{\partial p}{\partial x}, \tag{2.1}$$

$$\frac{\partial w}{\partial t} + u\frac{\partial w}{\partial x} + w\frac{\partial w}{\partial z} = -\frac{1}{\rho}\frac{\partial p}{\partial z} - g, \tag{2.2}$$

$$\frac{\partial u}{\partial x} + \frac{\partial w}{\partial z} = 0, \tag{2.3}$$

$$\frac{\partial \rho}{\partial t} + u\frac{\partial \rho}{\partial x} + w\frac{\partial \rho}{\partial z} = 0. \tag{2.4}$$

Equations (2.1) and (2.2) express the momentum in the x and z-directions respectively. Equation (2.3) is the continuity equation, and (2.4) represents the conservation of mass. We linearize (2.1)– (2.4) according to

$$q(x, z, t) = q_0(z) + q_1(x, z, t). \tag{2.5}$$

We also assume that the background flow is in hydrostatic balance. Then (2.1)– (2.4) become

$$\frac{\partial u_1}{\partial t} + u_0\frac{\partial u_1}{\partial x} + w_1\frac{\mathrm{d}u_0}{\mathrm{d}z} = -\frac{1}{\rho_0}\frac{\partial p_1}{\partial x}, \tag{2.6}$$

$$\frac{\partial w_1}{\partial t} + u_0\frac{\partial w_1}{\partial x} = -\frac{1}{\rho_0}\frac{\partial p_1}{\partial z} - \frac{\rho_1}{\rho_0}g, \tag{2.7}$$

$$\frac{\partial u_1}{\partial x} + \frac{\partial w_1}{\partial z} = 0, \tag{2.8}$$

$$\frac{\partial \rho_1}{\partial t} + u_0\frac{\partial \rho_1}{\partial x} + w_1\frac{\mathrm{d}\rho_0}{\mathrm{d}z} = 0, \tag{2.9}$$

where $\rho_0(z)$ is the background atmospheric density. We next assume wave-like solutions of the form

$$u_1(x, z, t) = \tilde{u}(z)e^{i(kx-\omega t)}, \tag{2.10}$$

$$\rho_1(x, z, t) = \tilde{\rho}(z)e^{i(kx-\omega t)}, \tag{2.11}$$

$$p_1(x, z, t) = \tilde{p}(z)e^{i(kx-\omega t)}, \tag{2.12}$$

$$w_1(x, z, t) = \tilde{w}(z)e^{i(kx-\omega t)}. \tag{2.13}$$

Equations (2.6)–(2.9) now become

$$-i\omega\tilde{u} + iu_0k\tilde{u} + \tilde{w}\frac{\mathrm{d}u_0}{\mathrm{d}z} = -\frac{i}{\rho_0}k\tilde{p}, \tag{2.14}$$

$$-i\omega\tilde{w} + iu_0k\tilde{w} = -\frac{1}{\rho_0}\frac{\mathrm{d}\tilde{p}}{\mathrm{d}z} - \frac{\tilde{\rho}}{\rho_0}g, \tag{2.15}$$

$$ik\tilde{u} + \frac{\mathrm{d}\tilde{w}}{\mathrm{d}z} = 0, \tag{2.16}$$

$$-i\omega\tilde{\rho} + iu_0k\tilde{\rho} + \tilde{w}\frac{\mathrm{d}\rho_0}{\mathrm{d}z} = 0. \tag{2.17}$$

Note that because $\tilde{p}$, $\tilde{w}$, $\tilde{u}$, and ρ_0 are functions only of z, we can write their derivatives as total rather than partial. We now define the *intrinsic frequency*, Ω, as the frequency of a wave relative to the flow, *i.e.*, the wave frequency measured by an observer drifting with the mean flow, then

$$\Omega = \omega - u_0k. \tag{2.18}$$

The frequency ω is the *apparent frequency* measured by an observer in a fixed coordinate system, for example, by a *microbarometer* or a *sodar* at the ground surface. Chimonas and Hines (1986) refer to Ω as the *Doppler-shifted* intrinsic wave frequency. The wind speed, u_0, in (2.18) is the component of the background wind in the direction of the wave vector. If we consider both horizontal directions then

$$\Omega = \omega - u_0k - v_0l = \omega - \vec{v}_{\mathrm{H}} \bullet \vec{\kappa}, \tag{2.19}$$

where $\vec{v}_{\mathrm{H}}$ is the horizontal background wind vector. If we write (2.18) as

$$\omega = \Omega + u_0k \tag{2.20}$$

then as illustrated in Fig. 2.1, ω is greater than Ω if the wave is moving with the wind and smaller than Ω if the wave is moving against the wind. This is the same kind of frequency shift found in acoustic waves, for example, the change in frequency of a passing train. From (2.20) the apparent horizontal phase speed is

$$c = \frac{\Omega}{k} + u_0 = c_{\mathrm{I}} + u_0, \tag{2.21}$$

where $c_{\mathrm{I}} = \Omega/k$ is the *intrinsic* phase speed. In Chapter 4, we shall see that Ω plays an essential role in the trapping or ducting of gravity waves. Using (1.67)

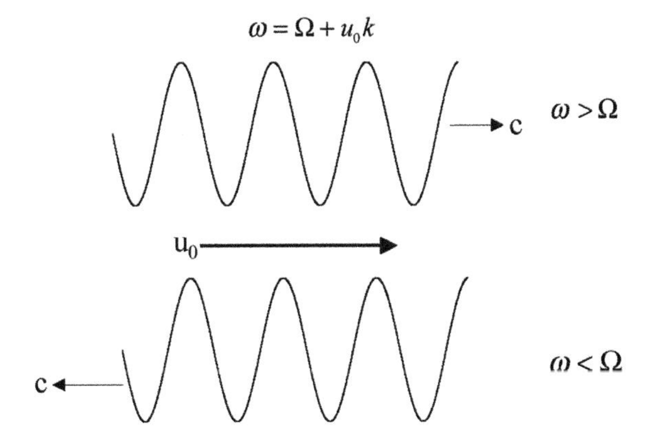

FIGURE 2.1 Illustration of Doppler shifted frequency.

and (2.18) equations (2.14)–(2.17) become

$$i\Omega\tilde{u} - \tilde{w}\frac{du_0}{dz} = \frac{i}{\rho_0}k\tilde{p}, \tag{2.22}$$

$$i\Omega\tilde{w} = \frac{1}{\rho_0}\frac{d\tilde{p}}{dz} + \frac{\tilde{\rho}}{\rho_0}g, \tag{2.23}$$

$$ik\tilde{u} + \frac{d\tilde{w}}{dz} = 0, \tag{2.24}$$

$$i\Omega\tilde{\rho} - \tilde{w}\frac{\rho_0}{g}N^2 = 0. \tag{2.25}$$

Hines (1960) refers to (2.22)–(2.25) as the *polarization equations* because they give the relative phases and amplitudes of various wave quantities. Note that $\tilde{w}$ appears in every equation; therefore, we will consider the phases of the other variables relative to it. Assume a solution of the form

$$\tilde{w}(z) = Ae^{imz} = A(\cos mz + i\sin mz), \tag{2.26}$$

where m is a vertical wavenumber. Note that the real part of $\tilde{w}$ is $A\cos mz$ and the imaginary part is $A\sin mz$. Expanding (2.24) into real and imaginary parts gives

$$ik(\tilde{u}_R + i\tilde{u}_I) = -Am(\cos mz + i\sin mz), \tag{2.27}$$

where $\tilde{u}_R$ and $\tilde{u}_I$ are the real and imaginary parts of $\tilde{u}$ respectively. Then

$$ik\tilde{u}_R - k\tilde{u}_I = -Am\cos mz - iAm\sin mz. \tag{2.28}$$

Taking the real part of (2.28) gives

$$\tilde{u}_R = -A\frac{m}{k}\cos mz. \tag{2.29}$$

Depending on the sign of m, $\tilde{u}_R$ is either in phase with or 180° out of phase with $\tilde{w}$. Proceeding as above, we see that $\tilde{p}_R$ is in phase with $\tilde{u}_R$ and either in phase with or 180° out of phase with $\tilde{w}_R$. Expanding (2.25) gives

$$i\Omega\tilde{\rho}_R - \Omega\tilde{\rho}_I = \frac{A\rho_0 N^2}{g}\cos mz + i\frac{A\rho_0 N^2}{g}\sin mz. \tag{2.30}$$

Dividing (2.30) by i and taking the real part gives

$$\tilde{\rho}_R = \frac{A\rho_0 N^2}{g\Omega}\sin mz. \tag{2.31}$$

Because $(A\rho_0 N^2)/(g\Omega) > 0$, $\tilde{\rho}_R$ is always in quadrature with $\tilde{w}_R$, *i.e.*, they are always 90° out of phase. Because of this phase relation, the vertical perturbation mass flux is

$$\text{vertical mass flux} = \frac{1}{\lambda}\int_{-\lambda/2}^{\lambda/2}\rho_1 w_1 \mathrm{d}z = 0, \tag{2.32}$$

where λ is the horizontal wavelength of the wave. Note also that because $\rho_1/\rho_0 = -\theta_1/\theta_0$ there can be no net transport heat by a linear gravity wave.

Now solving (2.22)–(2.25) for $\tilde{w}$ gives

$$\frac{\mathrm{d}^2\tilde{w}}{\mathrm{d}z^2} + \frac{1}{\rho_0}\frac{\mathrm{d}\rho_0}{\mathrm{d}z}\frac{\mathrm{d}\tilde{w}}{\mathrm{d}z} + \left[\frac{k^2N^2}{\Omega^2} + \frac{k}{\Omega}\frac{\mathrm{d}^2u_0}{\mathrm{d}z^2} + \frac{k}{\Omega}\frac{1}{\rho_0}\frac{\mathrm{d}\rho_0}{\mathrm{d}z}\frac{\mathrm{d}u_0}{\mathrm{d}z} - k^2\right]\tilde{w} = 0. \tag{2.33}$$

The second term in (2.33) represents the effects of changing atmospheric density on wave amplitude. We can remove this height dependence of the base-state density by assuming an exponentially decreasing atmospheric density of the form $\rho_0 = \rho_s \mathrm{e}^{-z/H_s}$ where ρ_s is the atmospheric density at the ground surface and $H_s = RT/g$ is a scale height usually identified with the scale height of an isothermal atmosphere where R is the universal gas constant for dry air, and T is the mean atmospheric temperature. Then, (2.33) becomes

$$\frac{\mathrm{d}^2\tilde{w}}{\mathrm{d}z^2} - \frac{1}{H_s}\frac{\mathrm{d}\tilde{w}}{\mathrm{d}z} + \left[\frac{k^2N^2}{\Omega^2} + \frac{k}{\Omega}\frac{\mathrm{d}^2u_0}{\mathrm{d}z^2} - \frac{k}{\Omega}\frac{1}{H_s}\frac{\mathrm{d}u_0}{\mathrm{d}z} - k^2\right]\tilde{w} = 0. \tag{2.34}$$

We can simplify this equation by using the integration factor $\mathrm{e}^{\int 1/2H_s \mathrm{d}z}$ to define a new variable, $\hat{w}$, by

$$\tilde{w} = \mathrm{e}^{z/2H_s}\hat{w}. \tag{2.35}$$

Substitution of (2.35) into (2.34) leads to the Taylor–Goldstein equation (Taylor, 1931; Goldstein, 1931)

$$\frac{\mathrm{d}^2\hat{w}}{\mathrm{d}z^2} + \left[\frac{k^2N^2}{\Omega^2} + \frac{k}{\Omega}\frac{\mathrm{d}^2u_0}{\mathrm{d}z^2} - k^2 - \frac{k}{\Omega}\frac{1}{H_s}\frac{\mathrm{d}u_0}{\mathrm{d}z} - \frac{1}{4H_s^2}\right]\hat{w} = 0. \tag{2.36}$$

Note that if (2.35) is used in (2.22) one gets

$$i\Omega\tilde{u} - \frac{\mathrm{d}u_0}{\mathrm{d}z}\hat{w}\mathrm{e}^{z/2H_s} = \frac{ik}{\rho_0}\tilde{p} \tag{2.37}$$

and if (2.35) is used in (2.25) one gets

$$i\Omega\tilde{\rho} - \hat{w}\mathrm{e}^{z/2H_s}\frac{\rho_0}{g}N^2 = 0. \tag{2.38}$$

To have a consistent notation, we must define new variables for $\tilde{u}$ and $\tilde{p}$, and $\tilde{\rho}$, *i.e.*,

$$\tilde{u} = \mathrm{e}^{z/2H_s}\hat{u}, \tag{2.39}$$
$$\tilde{p} = \mathrm{e}^{z/2H_s}\hat{p}, \tag{2.40}$$
$$\tilde{\rho} = \mathrm{e}^{z/2H_s}\hat{\rho}. \tag{2.41}$$

The Taylor–Goldstein equation then takes the form

$$\hat{w}'' + \left[\frac{N^2}{(c-u_0)^2} + \frac{u_0''}{(c-u_0)} - \frac{1}{H_s}\frac{u_0'}{(c-u_0)} - \frac{1}{4H_s^2} - k^2\right]\hat{w} = 0, \tag{2.42}$$

where we have replaced the derivatives by primes. If we replace the bracketed factor by m^2, then (2.42) becomes

$$\hat{w}'' + m^2\hat{w} = 0. \tag{2.43}$$

If m is constant, then

$$\hat{w} = A\mathrm{e}^{imz} + B\mathrm{e}^{-imz}, \tag{2.44}$$

where A and B are constants. The basis of linear gravity-wave theory is (2.44). If we know A, B, and m we know essentially all we can know about the gravity wave. The amplitudes A and B are determined by the boundary conditions of the problem. We must keep in mind that although (2.44) is simple it is not naive. Indeed, it is quite sophisticated with many subtleties, assumptions, and nuances.

If m is real, (2.44) tell us that the amplitude of the vertical component of the wave-perturbation velocity varies sinusoidally with height with vertical wave number $m = 2\pi/\lambda_z$ where λ_z is the vertical wave length. We shall see that when m is real, gravity waves transport energy vertically. Energy-transporting waves are usually referred to as *internal* or *propagating*. If, however, m is complex, *i.e.*, $m = m_R + im_I$ where m_R is the real part and m_I is the imaginary part of m, then the wave is not sinusoidally varying with height but exponentially varying with height. Gravity waves with complex m do not vertically transport energy; these waves usually are referred to as *external* or *evanescent*. Gossard and Hooke (1975) in a section titled "Names and nonsense" express some well-established opinions about the use and misuse of nomenclature that has grown with the development of gravity wave theory. Accordingly, we shall refrain from using the terms external and internal in describing gravity waves, and instead use propagating and evanescent which are more descriptive of the wave characteristics. We shall use the terms propagate and propagating to refer only to the transport of energy. Sometimes these terms are applied to the movements of wave fronts. In these cases, to say a wave is 'propagating vertically' is not clear as to what is being propagated, energy or wave phase? Thus, we shall use propagate or propagating to refer only to wave energy. Then an upward moving wave has upward moving wave fronts, while an upward propagating wave has energy moving upward.

The first term in the brackets in (2.42) is called the *buoyancy* term. This is the most important term in the wave equation because it strongly influences whether m is real or complex. Also, we see that there is a second-order singularity associated with this term if $c = u_0$. The second term is called the *curvature* term. If the background wind profile has a jet, this term becomes important. The third term is the *shear* term; because H_s is about 10 km in the troposphere and about 7 km in the stratosphere, this term is usually small in the lower atmosphere. In upper atmosphere studies, the fourth term is sometimes called the *compressibility* because it is a function of density. The last term in the brackets is called the *non-hydrostatic* term. This is the second most important term in m. If the background wind and stratification are constants and $1/(4H_s^2) \ll k^2$, then if the buoyancy term is greater than k, m is real, but if the buoyancy term is less than k, m is complex. However, for waves with very long horizontal wavelengths $\mathrm{d}\hat{p}/\mathrm{d}z \approx \hat{\rho}g$, *i.e.*, the vertical velocity perturbations are in hydrostatic balance. In this case, k will not appear in (2.42), for this reason the term it is called the non-hydrostatic term. The height z_c of the level where a singularity exists, *i.e.*, $c - u_0(z_c) = 0$ is called a *critical level.* Obviously for solutions to (2.42) critical levels will have to be considered if they are present. The treatment of critical levels is given later in Chapter 5.

Analytic series solutions of (2.42) are possible, and in some cases solutions in terms of special functions (for example, Bessel functions) are known; however, solutions in terms of plane-waves require the bracketed term in (2.42) be constant. These plane wave solutions in physical space will have the form

$$w_1(x, z, t) = \hat{w}(z)\mathrm{e}^{z/2H_s}\mathrm{e}^{i(kx-\omega t)}. \tag{2.45}$$

However, using (1.58) an alternate form is

$$w_1(x, z, t) = \hat{w}(z)\left(\frac{\rho_s}{\rho_0(z)}\right)^{1/2}\mathrm{e}^{i(kx-\omega t)}. \tag{2.46}$$

From (2.46) we see that if the wave fronts move upward, then the background air density seen by the wave, ρ_0, decreases. But from (2.46), this means that the wave amplitude, w_1, increases with height. It will be shown in Chapter 6 that associated with a vertically moving gravity wave is a vertical flux of mean-flow horizontal momentum or *wave stress* given by

$$\tau(z) = -\rho_0\overline{u_1 w_1}, \tag{2.47}$$

where the overline indicates a horizontal average which is usually over one horizontal wavelength, *i.e.*,

$$\rho_0 u_1 w_1 = \frac{1}{\lambda}\int_0^{\lambda} \rho_0 u_1 w_1 \mathrm{d}x. \tag{2.48}$$

It is important to note that in integrals such as (2.48), u_1 and w_1 are real values.

It will be shown in Chapter 6, that in the absence of wave dissipation, this momentum flux is constant. But if ρ_0 decreases with height, then the momentum being transported upward must also decreases. The only way the momentum can be constant is for $\overline{u_1 w_1}$ to increase in such a way as to balance the decrease in density. The factor $(\rho_s/\rho_0)^{1/2}$ accounts for this increase. An analogous effect causes the 'crack' of a bullwhip (see How to Make a Bullwhip and Cracking a Bullwhip at EURL). The thickness of a bullwhip decreases from about 5 cm at the handle to about 0.1 cm at the end. The decrease in linear density causes the amplitude of wave initiated near the handle to grow as the wave moves down the whip. By the time the impulse reaches the end of the whip the perturbation velocity of the whip end exceeds the speed of sound causing the loud crack sound.

2.3 A SIMPLE SOLUTION

We begin our examination of the solutions of the Taylor–Goldstein equation (2.42) with the simple case of constant background stratification, N, and wind speed u_0. Because the background flow is constant, its vertical scale of variation is zero and thus, $\lambda_z \ll H_s$.

2.3.1 NO BACKGROUND WIND SPEED

In the case of no background wind speed, (2.42) becomes

$$\hat{w} + \left[\frac{k^2 N^2}{\omega^2} - k^2\right] \hat{w} = 0. \tag{2.49}$$

The general solution in Fourier space is

$$\hat{w}(z) = A\mathrm{e}^{imz} + B\mathrm{e}^{-imz}, \tag{2.50}$$

where m is the vertical wavenumber given by

$$m^2 = k^2 \left[\frac{N^2}{\omega^2} - 1\right]. \tag{2.51}$$

Note that A and B can be complex numbers. Solving (2.51) for ω gives the *dispersion relation*

$$\omega = \pm \frac{kN}{(k^2 + m^2)^{1/2}}. \tag{2.52}$$

The dispersion relation is perhaps the most important element in the linear wave theory because it relates the angular frequency of the wave to the wave structure and the local environment. The dispersion relation fixes all the variables of the wave field, *i.e.*, k, m, and ω. Consider the negative branch of (2.52), then

$$\omega = ck = -\frac{kN}{(k^2 + m^2)^{1/2}}, \tag{2.53}$$

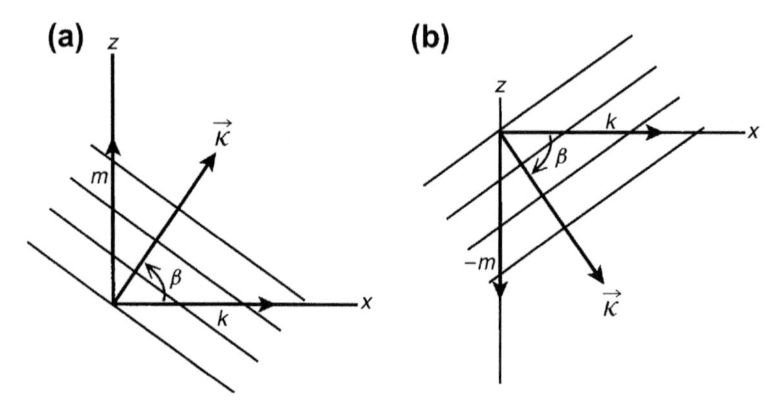

FIGURE 2.2 Wave vectors and wave fronts for an upward propagating wave (a) and a downward propagating wave (b).

which tells us that the horizontal phase velocity, c is negative, and thus the wave must be traveling in the negative x-direction. But if the wave is traveling in the negative x-direction, then $k < 0$ so that $\omega > 0$. Thus, if we keep to this convention, then $\omega > 0$ for all conditions.

The general solution of (2.49) in physical space is

$$w_1(x, z, t) = A e^{z/2H_s} e^{i\phi_+} + B e^{z/2H_s} e^{i\phi_-}, \tag{2.54}$$

where the wave phase function for the A solution is

$$\phi_+ = kx + mz - \omega t, \tag{2.55}$$

and the phase function for the B solution is

$$\phi_- = kx - mz - \omega t. \tag{2.56}$$

As shown in Section 1.2.3, the lines of constant phase, *i.e.*, the wave fronts for ϕ_+ solution, move upward with speed $c_z = \omega/m$, and slant in the $-x$ direction. The wave fronts for the ϕ_- solution move downward with speed $c_z = -\omega/m$, and slant in the $+x$ direction. The wave vectors and wave fronts for these cases are illustrated in Fig. 2.2. From the dispersion relation (2.52) we see that

$$\omega = \frac{kN}{(k^2 + m^2)^{1/2}} = N \cos\beta, \tag{2.57}$$

where β is the angle the wave vector makes with the horizontal as illustrated in Fig. 2.2. Several important aspects of wave motion are implied by this simple result. For example, if N changes with height, then waves of constant frequency move vertically along curved paths.[2] The intrinsic wave frequency (in this case $\Omega = \omega$ since $u_0 = 0$) cannot be greater than N. When $\omega \to N$, $\beta \to 0$, and the wave

[2] Strictly speaking the general solution (2.50) holds only when N and u_0 are constant, but we can use these results to approximate the behavior of slowing varying N.

fronts tend to move only horizontally because from (2.61) $m \to 0$. Then the fluid particles tend to oscillate vertically, and there is little propagation of energy either upward or downward. When $\omega/N \to 0$ corresponding to strong stratification or low frequency waves, $\beta \to \pi/2$, and $k \to 0$. Now the wave propagates energy almost totally vertically, and the fluid particles oscillate almost horizontally. We see that the spectrum of gravity waves that can propagate vertically is constrained to frequencies between these two limits, *i.e.*, $0 < |\beta| < N$.

It is of interest to consider for a moment the upper and lower limits of propagating gravity wave frequencies. What physical process limits ω to values less than N? As $|\beta| \to 0$, the wave fronts become increasingly vertical, and the fluid particles oscillate in increasingly vertically oriented planes. In Chapter 1, we saw that the frequency of oscillation of a vertically displaced parcel in a stably-stratified fluid is N. This is the resonant frequency of the fluid, and even though one could imagine some process that would force a vertical oscillation at a frequency greater than N, this vibration would not be supported by the fluid buoyancy, and the amplitude of the oscillation would decay rapidly with distance from the forcing point. The same effect is realized when one attempts to increase the oscillation frequency of a pendulum by shaking its pivot point. Thus, the maximum frequency for propagating gravity waves is N; however, a wide spectrum of wave frequencies exists below this value. It is sometimes erroneously assumed that all gravity waves have frequencies equal to N. Clearly, this is not correct. When $|\beta| \to \pi/2$, the wave fronts become increasingly horizontal. If the wave fronts were indeed perpendicular to the gravity force, the fluid particles would experience no buoyancy force because there would be no vertical displacement in their motions.

Let us now consider the group velocities for this case. Using (1.29) and (1.31) we have

$$u_g = \frac{Nm^2}{(k^2 + m^2)^{3/2}}, \tag{2.58}$$

and

$$w_g = -\frac{Nmk}{(k^2 + m^2)^{3/2}}. \tag{2.59}$$

An interesting and paradoxical result is that the vertical phase speed, c_z, and the vertical group velocity, w_g are always of opposite sign. To see this more clearly, use $c_z = \omega/m$, and (2.57) in (2.59) to get

$$w_g = -c_z \sin^2 \beta. \tag{2.60}$$

Because $\sin^2 \beta$ is always positive, c_z and w_g must be of opposite sign. Recall that the group velocity represents the speed of wave energy propagation, and so from (2.60) we see that if the wave fronts are moving upward, then wave energy must be propagating downward. Hines (1989a) describes a film made with Dave Fultz of the University of Chicago in 1967.

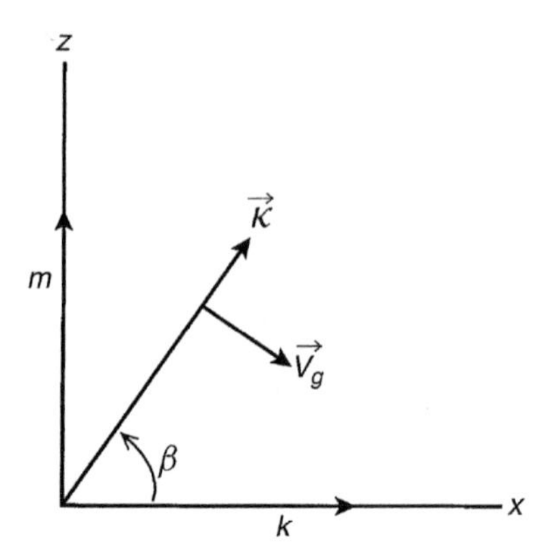

FIGURE 2.3 Relationship between phase velocity and group velocity for an internally propagating gravity wave. The flux of wave energy is in the direction of the group velocity vector.

> "This film exhibited a laboratory tank simulation in which water having a height-varying salinity modeled the atmosphere's height-varying density. A rocking paddle at the top of the tank acted as the source of waves and so produced a downward energy flow. It was seen to be producing ripples of phase that progressed downward – in complete accord with normal experience but in complete contradiction of my accompanying patter on gravity waves! Or so it seemed, until a burnt-out match appeared on the screen, collected smoke from thin air, burst spontaneously into flame, and then was struck back into its pristine, virginal state. This entropy experiment was then repeated, but with time now progressing and the phase ripples ascending toward the source, as advertised."

Later in this Chapter we shall discuss the propagation of energy by waves, and it is in this context that the c_z, w_g sign paradox is resolved.

Consider now a downward moving wave, but with upward propagating wave energy. Then

$$\vec{k} = k\hat{x} - m\hat{z}, \tag{2.61}$$

and

$$\vec{v}_g = \frac{Nm}{(k^2 + m^2)^{3/2}} \left(m\hat{x} + k\hat{z}\right), \tag{2.62}$$

where $\vec{v}_g$ is the vector group velocity. Then

$$\vec{v}_g \bullet \vec{k} = \frac{Nm}{(k^2 + m^2)^{3/2}} (km - mk) = 0, \tag{2.63}$$

and we see that the flux of wave energy is perpendicular to the wave vector and

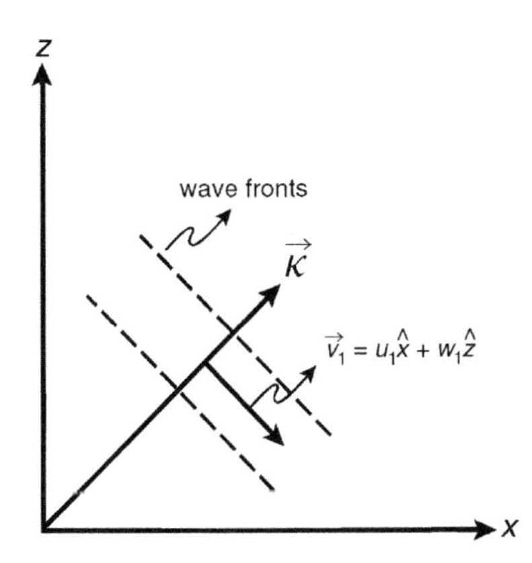

FIGURE 2.4 The motions of fluid particles are parallel to the wave fronts and perpendicular to the wave vector.

parallel to the wave fronts as illustrated in Fig. 2.3. From the continuity equation (2.16), using the B-solution of (2.52) we have

$$\hat{u} = \frac{m}{k} w_1, \tag{2.64}$$

or equivalently

$$\frac{\mathrm{d}x}{\mathrm{d}z} = \frac{m}{k}. \tag{2.65}$$

Thus, fluid particles oscillate in planes parallel to the wave fronts and perpendicular to the wave vector as illustrated in Fig. 2.4.

It is of interest here to consider the *monochromatic* gravity wave. Such a wave consists of single horizontal and vertical wavenumbers k and m respectively, and a single frequency, ω. Phase speeds, ω/k and ω/m are well defined; however, it would appear that the group velocities $\partial\omega/\partial k$ and $\partial\omega/\partial m$ are not well defined since ω, k, and m are constants. Because, for example, ω and k are constants, one could conclude that $\partial\omega/\partial k = 0$. However, from Section 1.2.4, we can think of a wave as a packet of wavelets with a narrow spectrum of frequencies. In such a group, ω will be a slowly varying function of k. If this packet consists of two waves with $k_1 - k_2 = \delta k$, then $\partial\omega/\partial k \approx \omega/\delta k$. We see that as $\delta k \to 0$ $u_g \to \infty$. This leads to the paradox that the group velocity of a monochromatic wave is infinitely large. Mathematically speaking, a monochromatic wave will extend indefinitely in space and persist indefinitely in time. Such a wave has neither beginning nor ending points. A wave of infinite extension and duration will have transported energy throughout its length, and this requires an infinite group velocity. But all "real" waves are composed of wavelets which act to form the starting and ending points of the wave. To be initiated, a gravity wave requires a disturbance to the stratification, and it is unlikely that only a single wave will be excited by the disturbance. Indeed, a spectrum of waves will always be excited. The fact that waves have a beginning tells us that a spectrum of waves is required *a priori* so that at some point and at some time these waves can interact to cancel each other thus making an ending. To be accurate, we should say "nearly-monochromatic" rather than monochromatic.

From (2.51) we see that the vertical wavenumbers for the general solution to (2.49) are

$$m = \pm k \left[\frac{N^2}{\omega^2} - 1 \right]^{1/2}, \tag{2.66}$$

where the positive branch indicates an upward moving wave, and the negative branch indicates a downward moving wave. If $\omega < N$, then m is real, and the wave is propagating. If $\omega > N$, then m is imaginary, and the wave is evanescent. Consider the case when $\omega > N$. Then, we can write

$$m = \pm ik \left[1 - \frac{N^2}{\omega^2} \right]^{1/2} = \pm iq. \tag{2.67}$$

The general solution is then

$$w_1 = A\mathrm{e}^{-q(z-z_\mathrm{L})}\mathrm{e}^{i(kx-\omega t)} \quad \text{for } z > z_\mathrm{L}, \tag{2.68}$$

$$= B\mathrm{e}^{+q(z-z_\mathrm{L})}\mathrm{e}^{i(kx-\omega t)} \quad \text{for } z < z_\mathrm{L}, \tag{2.69}$$

where z_L is the height where the evanescent wave is created. Since we require the solutions to be bounded, we use solution A or B depending on the sign of $z - z_\mathrm{L}$. The waves then resemble those illustrated in Fig. 2.5. Wave amplitude decreases exponentially with distance from the level of wave generation. Evanescent waves propagate only horizontally. It is possible for a wave to switch from propagating to evanescent if the stratification is changing with height. At the point where the wave switches, wave reflection occurs. Wave reflection will be treated in Chapter 4.

2.3.2 CONSTANT BACKGROUND WIND SPEED

We next consider the effect of a constant background wind on wave propagation. We are reminded that the wind speed, u_0, is the component of the background wind velocity in the direction of wave propagation. The Taylor–Goldstein equation is now

$$\frac{\mathrm{d}^2\hat{w}}{\mathrm{d}z^2} + \left[\frac{k^2N^2}{\Omega^2} - k^2 \right] \hat{w} = 0. \tag{2.70}$$

The phase angle (1.8) is now given by,

$$\phi = kx + mz - \Omega t. \tag{2.71}$$

The vertical wavenumber is

$$m = \pm \left[\frac{k^2N^2}{(\omega - u_0k)^2} - k^2 \right]^{1/2}, \tag{2.72}$$

FIGURE 2.5 Evanescent wave generated at height z_L.

where we have used $\Omega = \omega - u_0 k$. Again we take the positive branch in (2.72) to signify an upward moving wave and the negative branch to signify a downward moving wave. For $m > 0$ (and correspondingly $w_g < 0$)

$$\omega = u_0 k + \left[\frac{N^2 k^2}{m^2 + k^2} \right]^{1/2}, \tag{2.73}$$

and

$$c = u_0 + \frac{N}{k} \cos \beta, \tag{2.74}$$

$$c_z = u_0 \frac{k}{m} + \frac{N}{k} \cos \beta, \tag{2.75}$$

$$u_g = u_0 + \frac{m^2}{N^2} (c - u_0)^3, \tag{2.76}$$

$$w_g = -\frac{k|m|}{N} (c - u_0)^3. \tag{2.77}$$

Note that c_z and w_g are of opposite sign as expected. For $m < 0$ (and correspondingly $w_g > 0$)

$$\omega = u_0 k - \left[\frac{N^2 k^2}{m^2 + k^2} \right]^{1/2}, \tag{2.78}$$

and

$$c = u_0 + \frac{N}{k}\cos\beta, \tag{2.79}$$

$$c_z = -u_0\frac{k}{|m|} - \frac{N}{|m|}\cos\beta, \tag{2.80}$$

$$u_g = u_0 + \frac{m^2}{N^2}(c-u_0)^3, \tag{2.81}$$

$$w_g = \frac{k|m|}{N}(c-u_0)^3. \tag{2.82}$$

Again we see that c_z and w_g are of opposite sign. The vertical and horizontal group velocities are functions of $(c-u_0)^3$ and therefore are dependent on the relative values of c and u_0. We will examine how these values effect w_g.

Case 1a: $m > 0$ and $c > u_0$

- $c = u_0 + \frac{N}{k}\cos\beta$,
- $u_g = u_0 + \frac{m^2}{N^2}|c-u_0|^3$,
- $c_z = u_0\frac{k}{|m|} + \frac{N}{|m|}\cos\beta$,
- $w_g = -\frac{k|m|}{N}|c-u_0|^3$,
- $k > 0$.

Case 1b: $m > 0$ and $c < u_0$

- $c = u_0 - \frac{N}{|k|}\cos\beta$,
- $u_g = u_0 - \frac{m^2}{N^2}|c-u_0|^3$,
- $c_z = -u_0\frac{|k|}{|m|} + \frac{N}{|m|}\cos\beta$,
- $w_g = \frac{|k||m|}{N}|c-u_0|^3$,
- $k < 0$.

Case 2a: $m < 0$ and $c > u_0$

- $c = u_0 + \frac{N}{k}\cos(-\beta)$,
- $u_g = u_0 + \frac{m^2}{N^2}|c-u_0|^3$,
- $c_z = -u_0\frac{k}{|m|} - \frac{N}{|m|}\cos(-\beta)$,
- $w_g = \frac{k|m|}{N}|c-u_0|^3$,
- $k > 0$.

Case 2b: $m < 0$ and $c < u_0$

- $c = u_0 - \frac{N}{|k|}\cos(-\beta)$,
- $u_g = u_0 - \frac{m^2}{N^2}|c-u_0|^3$,
- $c_z = u_0\frac{|k|}{|m|} - \frac{N}{|m|}\cos(-\beta)$,
- $w_g = \frac{|k||m|}{N}|c-u_0|^3$,
- $k < 0$.

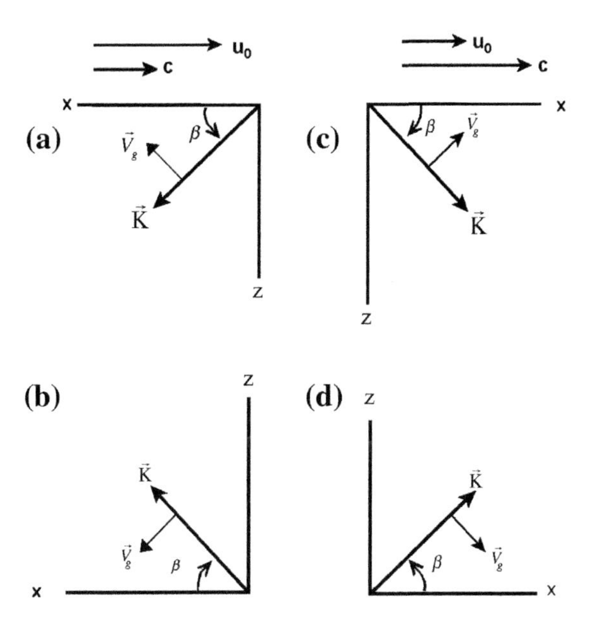

FIGURE 2.6 Wave vectors and group velocity vectors for (a) upward propagating energy and wave speed less than background wind speed; (b) upward propagating energy and wave speed greater than background wind speed; (c) same as (a) but with downward propagating energy; (d) same as (b) but with downward propagating energy.

These cases are illustrated in Fig. 2.6. We can visualize the wave motions sketched in Fig. 2.6 with an illustration. Imagine you are in a balloon drifting at a constant height h where the wind speed is $u_0(h)$ in an atmosphere with positive wind sheer, *i.e.*, horizontal wind speed increasing with height. You are observing an upward moving wave with horizontal phase speed c, but the phase speed you see is the intrinsic speed $c_{\mathrm{I}} = c - u_0(z)$. Below you let $c > u_0(z)$ and hence $c_{\mathrm{I}} > 0$, you observe the wave moving ahead of you (Fig. 2.6d). As the wave continues upward, it will appear to slow down, stop at $z = h$, and then above you move away with increasing speed (Fig. 2.6c).

The background wind also affects the polarization equations. For example, consider (2.22) which relates the horizontal wind perturbation, $\tilde{u}$, to the pressure perturbation, $\tilde{p}$. For constant background wind, the equation is

$$u_1 = \frac{P_1}{\rho_0(c - u_0)}, \tag{2.83}$$

where we have replaced the wave amplitudes by the perturbation values. If $c > u_0$ the pressure and wind speed perturbations are in phase, but if $c < u_0$ then the perturbations are 180° out of phase. However, in the case of a height varying background wind (2.83) no longer applies. Then from (2.22) we have

$$u_1 = \frac{P_1}{\rho_0(c-u_0)} - \frac{u'_0}{k}\Re(\mathrm{i}w_1). \tag{2.84}$$

Gossard and Hooke (1975) suggest that for reasonable values of u'_0/k the effect of the wind shear is small.

2.4 THE WKB OR SLOWLY VARYING METHOD

We have seen that the Taylor–Goldstein equation (2.42) has plane-wave solutions only when the atmospheric variables are constant. However, one may ask if an approximate solution is possible when u_0 and N are slowly varying? Consider the equation

$$\frac{\mathrm{d}^2\psi}{\mathrm{d}z^2} + Q^2(z)\psi = 0. \tag{2.85}$$

The WKB method (see, for example, Pitteway and Hines, 1965; Einaudi and Hines, 1970; Gill, 1982; Laprise, 1993; Baines, 1995; Bender and Orszag, 1999) provides an approximate solution to (2.85) when $Q(z)$ is slowly changing. The WKB approximation (developed by Wendel, Kramers, and Brouillian) to the solution to (2.85) is

$$\psi(z) = \psi_0 Q^{-1/2}\mathrm{e}^{\pm i\int_0^z Q\mathrm{d}z}, \tag{2.86}$$

where ψ_0 is a constant. Equation (2.86) satisfies the equation

$$\frac{\mathrm{d}^2\psi}{\mathrm{d}z^2} + Q^2(z)(1+d)\psi = 0, \tag{2.87}$$

where

$$d = \frac{1}{2Q^3}\frac{\mathrm{d}^2 Q}{\mathrm{d}z^2} - \frac{3}{4Q^4}\left(\frac{\mathrm{d}Q}{\mathrm{d}z}\right)^2. \tag{2.88}$$

If $d \ll 1$, then (2.87) approaches (2.85), *i.e.*, the Taylor–Goldstein equation. Obviously, Q cannot equal zero since then $d \to \infty$. We shall see in Chapter 6 when $Q = 0$ wave reflection will occur. Einaudi and Hines, 1970 make the assumption that $Q^2 = 0$ is the condition for significant reflection under the WKB method. If, however, Q becomes large and $\mathrm{d}Q/\mathrm{d}z$ becomes small, then d becomes small. If Q is a vertical wavenumber, then large Q implies small vertical wavelength, and if $\mathrm{d}Q/\mathrm{d}z$ is small, then the vertical scale of the background variables is small. Thus, if the WKB method is to apply, then the wave must appear to be propagating in a medium which is changing slowly relative to the vertical wavelength of the wave.

The WKB method is commonly used in wave studies on almost all atmospheric scales. In the stable planetary boundary layer and lower troposphere background wind speed and stratification can change significantly over a typical vertical wavelength. In such cases, one would not expect the WKB method to be applicable; however, studies by, for example, Kim and Mahrt (1992), Grisogono (1994), Teixeira, Miranda, and Valente (2004), and Teixeira and Miranda (2009) have been successfully used in the lower troposphere and boundary layer. In many applications, the WKB approximation coincides with hydrostatic gravity waves, and then the Taylor–Goldstein equation takes the form

$$\frac{d^2\hat{w}}{dz^2} + \frac{N^2}{(c-u_0)^2}\hat{w} = 0. \tag{2.89}$$

2.5 ENERGETICS

A characteristic of all waves is the ability to transport energy. Indeed, if not for this property the study of wave phenomena would be little more than a mathematical curiosity. However, on all atmospheric scales of motion waves exist and transport energy. In this section, we introduce the concepts of wave kinetic and potential energy.

2.5.1 WAVE ENERGY

We begin by multiplying (2.6) by u_1 and (2.7) by w_1 and adding the two equation to get

$$\frac{D}{Dt}\left[\frac{1}{2}\rho_0(u_1^2 + w_1^2)\right] + u_1\frac{\partial p_1}{\partial x} + w_1\frac{\partial p_1}{\partial z} + \rho_1 g w_1 = -\rho_0 u_1 w_1 \frac{du_0}{dz}, \tag{2.90}$$

where the total derivative is

$$\frac{D}{Dt} = \frac{\partial}{\partial t} + u_0\frac{\partial}{\partial x}, \tag{2.91}$$

and we have replaced P_1 by p_1 to have a consistent notation. We can write

$$u_1\frac{\partial p_1}{\partial x} + w_1\frac{\partial p_1}{\partial z} = \frac{\partial}{\partial x}(u_1 p_1) + \frac{\partial}{\partial z}(w_1 p_1) - p_1\left(\frac{\partial u_1}{\partial x} + \frac{\partial w_1}{\partial z}\right), \tag{2.92}$$

and using this and the continuity equation (2.8) in (2.90) gives, after some rearrangement of terms, the equation for the total perturbation energy per unit volume for the case of constant u_0

$$\frac{D}{Dt}\left[\frac{1}{2}\rho_0\left(u_1^2 + w_1^2\right)\right] + \rho_1 g w_1 = -\frac{\partial}{\partial x}(u_1 p_1) - \frac{\partial}{\partial z}(w_1 p_1). \tag{2.93}$$

If we define ζ_1 as the vertical displacement of an air parcel from its equilibrium position, then $w_1 = \partial\zeta_1/\partial t$ and from (2.9) the perturbation density, ρ_1, is

$$\rho_1 = -\frac{d\rho_0}{dz}\zeta_1. \tag{2.94}$$

Note that because $\frac{d\rho_0}{dz} < 0$, an upward displacement of the air results in a positive density perturbation, *etc.* Using (1.67) and (2.94) in (2.9) results in

$$\rho_1 g w_1 = \frac{D}{Dt}\left(\frac{1}{2}\rho_0 N^2 \zeta_1^2\right), \tag{2.95}$$

where we have assumed that ρ_0 and N are constants in time and uniform in the horizontal direction. Then (2.93) becomes

$$\frac{D}{Dt}\left[\frac{1}{2}\rho_0\left(u_1^2 + w_1^2\right) + \frac{1}{2}\rho_0 N^2 \zeta_1^2\right] = -\frac{\partial}{\partial x}(u_1 p_1) - \frac{\partial}{\partial z}(w_1 p_1). \tag{2.96}$$

The first term in the brackets on the left-hand-side of (2.96) is the *perturbation kinetic energy*, and the second term is the *perturbation potential energy*. To verify this latter statement, consider the gravitational potential energy (PE) gained by a fluid parcel when it is displaced vertically a distance ζ_1 by the wave, *i.e.*,

$$\text{PE} = -\int_0^{\zeta_1} F_b \mathrm{d}z, \tag{2.97}$$

where F_b is the buoyant force per unit volume exerted on the air parcel. From (1.45) and using (1.67), the buoyant force per unit volume is

$$F_b = g\frac{d\rho_0}{dz}z. \tag{2.98}$$

Using (2.98) in (2.97) and integrating gives the desired result, *i.e.*,

$$\text{PE} = -\frac{1}{2}g\frac{d\rho_0}{dz}\zeta_1^2 = \frac{1}{2}\rho_0 N^2 \zeta_1^2. \tag{2.99}$$

The equation for the total perturbation energy per unit volume is

$$\frac{DE}{Dt} + \frac{\partial}{\partial x}(u_1 p_1) + \frac{\partial}{\partial z}(w_1 p_1) = 0, \tag{2.100}$$

where

$$E = \frac{1}{2}\rho_0(u_1^2 + w_1^2 + N^2\zeta_1^2). \tag{2.101}$$

The second and third terms on the left-hand-side of (2.100) represent the divergences of the fluxes of wave energy in the horizontal and vertical directions respectively.

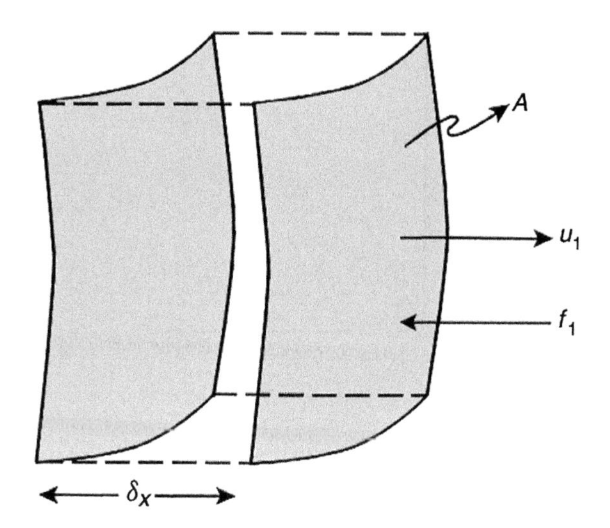

FIGURE 2.7 Horizontal displacement of a unit surface area by the horizontal wave-perturbation velocity u_1.

An examination of the units of the terms $u_1 p_1$ and $w_1 p_1$ in (2.100) shows that these represent fluxes of wave energy in the x- and z-directions respectively. To see this, consider Fig. 2.7 which illustrates the horizontal displacement δx of a surface of unit area A by the wave-perturbation velocity u_1 against a force f_1 due to the pressure perturbation. Then,

$$u_1 p_1 \approx \frac{f_1 \delta x}{A \delta t} = \frac{\delta W}{A \delta t}, \tag{2.102}$$

where δW is the differential work done by the wave; then $u_1 p_1$ is the work done per unit area per unit time, but this is the horizontal flux of wave energy in the x-direction. The same analysis applies to the vertical term, $w_1 p_1$.

In the absence of a background wind, (2.100) becomes

$$\frac{\partial E}{\partial t} + \frac{\partial}{\partial x}(u_1 p_1) + \frac{\partial}{\partial z}(w_1 p_1) = 0, \tag{2.103}$$

and we see that the wave energy per unit volume is a conserved quantity, *i.e.*, the total rate of change of E is equal to the local divergence of the energy flux.

Because the wave energy is periodic in space and time, it is more useful to discuss average values rather than local values. Because we are considering linear waves, an average over a wavelength or over a wave period is sufficient. Considering for the moment all three spatial dimensions, we can write

$$\overline{E} = \frac{1}{2}\rho_0 \overline{(u_1^2 + v_1^2 + w_1^2)} + \frac{1}{2}\rho_0 N^2 \overline{\zeta_1^2}, \tag{2.104}$$

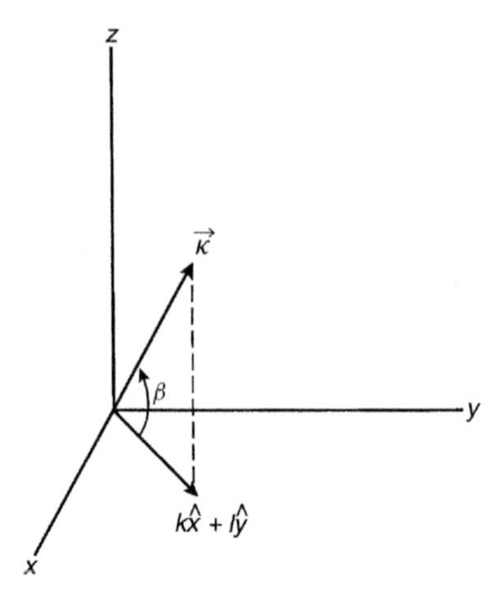

FIGURE 2.8 Projection of the three-dimensional wave vector onto the horizontal plane.

where the over bars indicate time averaging. Now let us assume there is a gravity wave with amplitude a_1, *i.e.*,

$$w_1 = a_1 \cos(\phi), \tag{2.105}$$

where $\phi = kx + ly + mz - \omega t$. The polarization equations (2.22) and (2.25) plus an additional one for the y-direction are

$$u_1 = -\frac{k}{k^2 + l^2} m a_1 \cos\phi, \tag{2.106}$$

$$v_1 = -\frac{l}{k^2 + l^2} m a_1 \cos\phi, \tag{2.107}$$

where l is the wavenumber in the y-direction. Averaging the wave kinetic energy, KE, over 2π radians and using (2.105)–(2.107) gives

$$\overline{\mathrm{KE}} = \frac{1}{2}\rho_0 \overline{(u_1^2 + v_1^2 + w_1^2)} = \frac{1}{4}\rho_0 \left(\frac{k^2 + l^2 + m^2}{k^2 + l^2}\right) a_1^2 = \frac{1}{4}\rho_0 \frac{a_1^2}{\cos^2\beta}, \tag{2.108}$$

where β is the angle the wave vector $\vec{\kappa}$ makes with the x–y plane as illustrated in Fig. 2.8. The vertical displacement, ζ_1, appearing in the expression for the potential energy in (2.99) is related to w_1 by

$$\zeta_1 = \int w_1 \mathrm{d}t = \frac{a_1}{\omega} \sin\phi. \tag{2.109}$$

Then using (2.109) in (2.99) and averaging over 2π and using (2.57) gives

$$\overline{\mathrm{PE}} = \frac{1}{4}\rho_0 \frac{a_1^2}{\cos^2\beta}. \tag{2.110}$$

We see that the average kinetic and potential energies are equal, but this should not come as a surprise. As with all simple harmonic motions, for example, a pendulum or a weighted spring, the kinetic and potential energies are equal but of opposite phase. When KE is at its maximum value, PE is at its minimum value, *etc.* When we calculate the average energies over a cycle, the results must be equal. Adding $\overline{\mathrm{KE}}$ and $\overline{\mathrm{PE}}$ gives the mean perturbation energy per unit volume as

$$\overline{E} = \frac{1}{2}\rho_0 \frac{a_1^2}{\cos^2\beta}. \tag{2.111}$$

The flux of wave energy across the surface bounding a volume of fluid is obtained by averaging the flux terms in (2.100) over a wavelength to get the *energy flux density vector*

$$\vec{F} = \overline{p_1 \vec{V}_1} = \frac{1}{2}\Re(p_1 \vec{V}_1^*), \tag{2.112}$$

where $\vec{V}_1 = u_1\hat{x} + v_1\hat{z}$.[3] Solving for p_1 using polarization equations (2.22) and using (2.105) and (2.106), the energy flux density vector is

$$\vec{F} = \frac{1}{2}\frac{\rho_0\omega m^2 a_1^2}{k^3}\hat{x} - \frac{1}{2}\frac{\rho_0\omega m a_1^2}{k^2}\hat{z}. \tag{2.113}$$

For the two-dimensional wave, $k = \kappa\cos\beta$. Using this and (2.57) in (2.113) gives

$$\vec{F} = \frac{1}{2}\rho_0\frac{a_1^2}{\cos^2\beta}\left[\frac{Nm^2}{(k^2+m^2)^{3/2}}\hat{x} - \frac{Nkm}{(k^2+m^2)^{3/2}}\hat{z}\right]. \tag{2.114}$$

Using (2.58), (2.59), and (2.111) in (2.114) gives,

$$\vec{F} = \overline{E}(u_g\hat{x} + w_g\hat{z}) = \overline{E}\vec{v}_g. \tag{2.115}$$

Equation (2.115) is an important result; it tells us that the flux of wave energy is in the direction of the group velocity. Using (2.115) we can write (2.103) as

$$\frac{\partial \overline{E}}{\partial t} + \nabla \bullet \overline{E}\vec{v}_g = 0. \tag{2.116}$$

In the absence of a background wind and assuming uniform thermal stratification k and m will be constants, and so we can write (2.116) as

$$\frac{\partial \overline{E}}{\partial t} + \vec{v}_g \bullet \nabla\overline{E} = 0. \tag{2.117}$$

Thus, we see that wave-perturbation energy is constant in the special case of calm winds and constant Brunt–Väisälä frequency.

In the general case, when the background flow is neither calm nor uniform, the equation for the mean wave energy is, after time averaging (2.100),

$$\frac{\mathrm{D}\overline{E}}{\mathrm{D}t} + \frac{\partial}{\partial x}(\overline{u_1 p_1}) + \frac{\partial}{\partial z}(\overline{w_1 p_1}) = -\rho_0\overline{u_1 w_1}\frac{\mathrm{d}u_0}{\mathrm{d}z}. \tag{2.118}$$

[3] We remind the reader that the average $\overline{p_1\vec{V}_1}$ is taken over real values; however, in $\frac{1}{2}\Re(p_1\vec{V}_1^*)$, p_1, and $\vec{V}$ are complex values.

The term on the right-hand-side of (2.118) is of interest. In the turbulence theory, the term $-\rho_0\overline{u'w'}\mathrm{d}u_0/\mathrm{d}z$, where u' and w' are turbulence quantities, represents the production of turbulence kinetic energy. This production is due to the work done by the Reynolds stress $(-\rho_0\overline{u'w'})$ against the mean rate of strain $(\mathrm{d}u_0/\mathrm{d}z)$. It is always positive. However, such an interpretation of the right-hand-side of (2.118) is not apparent. Indeed, if the wave stress $-\rho_0\overline{u_1w_1}$ and the mean wind shear $(\mathrm{d}u_0/\mathrm{d}z)$ are of opposite sign, then it would appear that the wave energy will increase; however, it is not clear where this increased energy originates. Since we have assumed *a priori* that the background flow is steady, the energy cannot originate there. If the wave stress and wind shear have the same sign, then it would appear that the wave energy will decrease, but then where does this energy go? The problem is that our analysis of wave energy is only to first order. If we went to second order, then terms such as $u_1\frac{\partial}{\partial z}(u_0u_1)$, $w_1\frac{\partial}{\partial z}(u_0u_1)$, *etc.* would appear, and these would more clearly define the energy transfer process.

2.5.2 WAVE-ACTIVITY CONSERVATION LAWS

2.5.2.1 Wave Action

In the previous section, we demonstrated that in the presence of a background wind the total average wave energy, $\overline{E}$ is not conserved. We now review the argument proposed by Bretherton (1966) which demonstrates that a form of the wave energy is conserved in the presence of a background wind. We begin by multiplying the linearized equation for horizontal momentum (2.6) by the vertical velocity perturbation, w'. We assume wave-like perturbations of the form $w' = w_1\exp[i(kx-\omega t)]$, and take the real parts of the terms to get

$$(\omega - u_0k)u_1w_1 = \frac{k}{\rho_0}w_1p_1. \tag{2.119}$$

Integrating (2.119) over a wave period, and making use of the definition of the intrinsic frequency (2.18) we have

$$\rho_0\overline{u_1w_1} = \frac{k}{\Omega}\overline{w_1p_1}. \tag{2.120}$$

From (2.112) and (2.115) we see that

$$\overline{w_1p_1} = w_g\overline{E} \tag{2.121}$$

and

$$\overline{u_1p_1} = u_g\overline{E}, \tag{2.122}$$

where w_g and u_g are the vertical and horizontal components respectively of the group velocity. Then, using (2.121) in (2.120) gives

$$\rho_0\overline{u_1w_1} = \frac{k}{\Omega}w_g\overline{E}. \tag{2.123}$$

Using (2.123), the average of the energy transfer term, *i.e.*, the right-hand-side of (2.118), can be written as

$$\rho_0 \overline{u_1 w_1} \frac{\mathrm{d}u_0}{\mathrm{d}z} = \frac{k}{\Omega} \frac{\mathrm{d}u_0}{\mathrm{d}z} w_g \overline{E} = -\frac{\overline{E}}{\Omega} \frac{\mathrm{D_g}\Omega}{\mathrm{D}t}, \tag{2.124}$$

where $\mathrm{D_g}/\mathrm{D}t = \partial/\partial t + \vec{v}_g \bullet \nabla$. The equation for the mean wave energy density (2.100) can now be written as

$$\frac{\partial(\overline{E})}{\partial t} + \nabla \bullet \vec{v}_g \overline{E} - \frac{\overline{E}}{\Omega} \frac{\mathrm{D_g}\Omega}{\mathrm{D}t} = 0, \tag{2.125}$$

or more simply as

$$\frac{\partial}{\partial t} \left(\frac{\overline{E}}{\Omega} \right) + \nabla \bullet \left(\vec{v}_g \frac{\overline{E}}{\Omega} \right) = 0. \tag{2.126}$$

Bretherton (1966) refers to $\overline{E}/\Omega$ as the *wave action*. Unlike wave energy, wave action is a conserved quantity.

2.5.2.2 Pseudo-Energy and Pseudo-Momentum

McIntyre's classic paper "On the 'wave momentum' myth" (McIntyre, 1981) elegantly shows the difference between "fluid" energy and momentum and "wave" energy and momentum. Flow and wave conservative quantities are not the same things. Conservation relations are based on the symmetry of canonical pairs such as energy and time or momentum and space. Thus, conservation of *energy* requires that laws governing a given dynamical system be independent of *time* including any external forces. In this case, what is conserved is the *pseudo-energy*[4]

$$\frac{\overline{E}}{\overline{\Omega}} \omega. \tag{2.127}$$

A different symmetry condition for waves propagating in a material medium requires the medium to be independent of time. In this case what is conserved is *pseudo-momentum*

$$\frac{\overline{E}}{\overline{\Omega}} \vec{K}. \tag{2.128}$$

McIntyre (1981) gives the an example which illustrates the "pseudo-momentum." Consider the laboratory experiment sketched in Fig. 2.9. A closed rectangular tank (or a wave guide) is unbounded in the horizontal direction and contains a stably stratified fluid between horizontal boundaries. An obstacle is dragged along the bottom of the tank at constant speed. At a certain speed U, gravity waves appear behind and move with the obstacle. The fluid remains at rest and the wave drag is balanced by the towing force. The resultant forces due to the wave are the same if:

[4] Some authors use *pseudoenergy*. We use pseudo-energy to empathize the adjective *pseudo*.

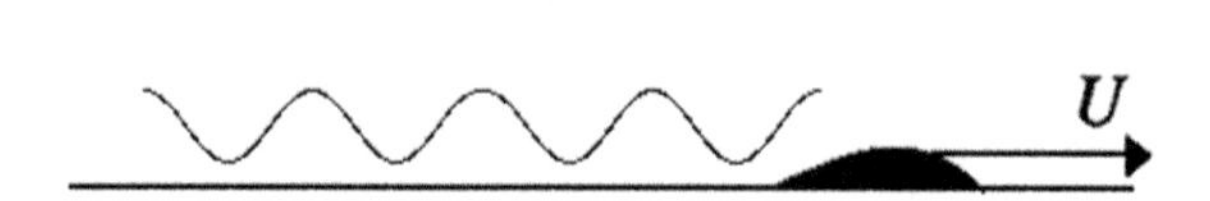

FIGURE 2.9 Sketch of a hypothetical experiment of flow over an obstacle.

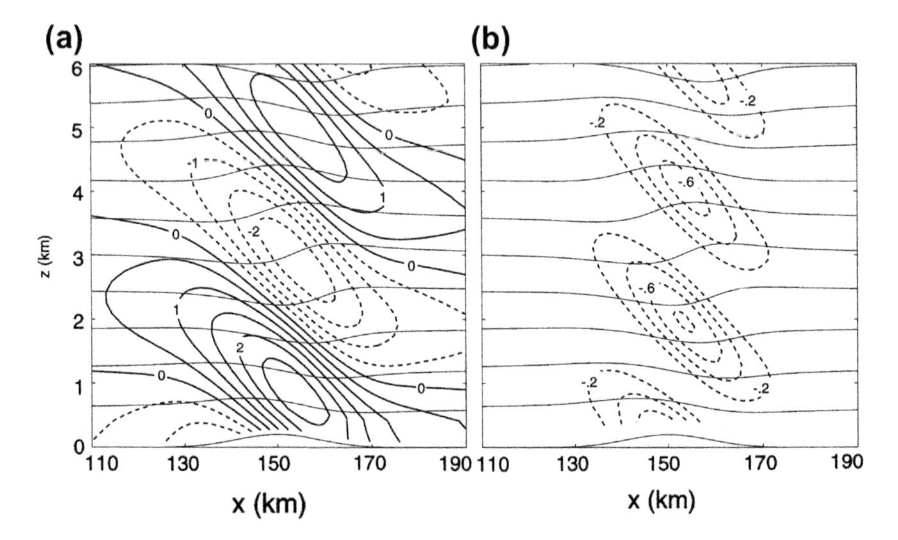

FIGURE 2.10 Illustration of the difference between flow momentum (a) and pseudomomentum (b). Values are in kg ms^{-1}. Taken from Durran (1995).

(a) the waves had momentum equal to their pseudo-momentum, $\overline{E}k/\Omega$, and
(b) the medium were absent.

Statements (a) and (b) constitute the McIntyre's pseudo-momentum rule. Figure 2.10 taken from Durran (1995) illustrates the difference between momentum and pseudo-momentum in the case of a mountain wave. Figure 2.10a shows contours of perturbation horizontal momentum, $\rho u - \bar{\rho}\bar{u}$, and Fig. 2.10b shows contours of pseudo-momentum. In the middle and upper atmospheres, the fluxes of pseudo-energy and pseudo-momentum are considered rather than the erroneous fluxes of wave energy and wave momentum. Further discussions of pseudo-momentum and pseudo-energy can be found for example in Andrews and McIntyre (1978), Dunkerton (1981), Held (1985), Tanaka (1996), Durran (1995), Lott (1998), and Fritts and Alexander (2003).

PROBLEMS

1. In an calm isothermal atmosphere with a temperature of 280 K, a gravity wave with wavelength 25 km and phase speed 15 ms^{-1} propagates upward at an angle of 48° to the horizontal. The amplitude of w_1 is 1 ms^{-1}. Calculate the amplitudes of u_1, θ_1, and p_1.
2. A constant density balloon is drifting with a speed of 5 ms^{-1}. A gravity wave passes by and the balloon oscillates up and down with a period of 10 min. (a) What is the frequency of the wave as seen from the ground? (b) What is the wavelength of the wave?
3. Show that $\left(\frac{\rho_s}{\rho_0(z)}\right)^{1/2} = \exp\left(\frac{z}{2H_s}\right)$.
4. Show that $\frac{1}{\rho_0}\frac{\partial \rho_0}{\partial z} = -H_s^{-1}$.
5. Show that when the gravity waves are in hydrostatic balance, the Taylor–Goldstein equations is:

$$\hat{w}'' + \left[\frac{N^2}{(c - u_0)^2}\right]\hat{w} = 0.$$

6. In a clam atmosphere, if N' in (1.50) equals 0.5ω, at what angle, β, to the horizontal is a gravity wave with frequency ω propagating?
7. A gravity wave at height 10 km above the ground surface has a horizontal wavelength of 25 km, a frequency of 0.03 s^{-1}, and amplitude $w_1 = 2\mathrm{ms}^{-1}$. If the atmosphere has a mean speed of 15 ms^{-1} and a Brunt–Väisälä frequency of 0.02 s^{-1}, what will be the wave amplitude at 1 km above he ground surface?
8. For a gravity wave with the ratio $\frac{p_1}{\rho_0 u_1} = -8$ and the background wind speed is 5 ms^{-1}, what is the phase speed of the wave? Is the wave moving with or against the background wind?
9. What parts of equation (2.54) correspond to upward and downward propagating wave energy?
10. (a) Show that $\left|\frac{w_g}{c_z}\right| \leqslant 1$. (b) What is the orientation of the phase fronts when $\left|\frac{w_g}{c_z}\right| = 1$?
11. If

$$w_1 = 5e^{z/2H_s}\exp\left[i\,(kx + mz - \omega t)\right]$$

and

$$u_1 = 20e^{z/2H_s}\exp\left[i\,(kx + mz - \omega t)\right]$$

and if the background wind speed is -3 ms^{-1} at height 12 km and -15 ms^{-1} at height 16 km and H_s is 8 km and the surface air density is 1 kg m^3, what is the value of $\mathrm{D}\bar{E}/\mathrm{D}t$ at height 14 km?

12. A gravity wave has group velocity $\vec{V}_g = 10\hat{x} - 20\hat{z}$. Make a graph showing the wave vector, the horizontal and vertical numbers, the wave fronts, and the direction of the phase speed.
13. The vertical velocity perturbation, $w_1(x, z, t)$, has an amplitude of 1 ms^{-1} at 800 hPa. What will be the wave amplitude at 100 hPa in an isothermal atmosphere?
14. Explain why does the WKB method not work when the profiles of wind speed and stratification changes over distances less than the vertical wavelength of an upward propagating gravity wave?
15. A gravity wave has intrinsic phase speed of 10 ms^{-1}, horizontal wavelength of 30 km, vertical wavelength of 10 km, and Brunt–Väisälä frequency 0.015 s^{-1}. Calculate the angle the energy flux vector makes with the horizontal axis, β.
16. Let the amplitude of the gravity wave, $|\tilde{w}|$ in Problem 9 be 1 ms^{-1}. Calculate the wave action of the wave.
17. Calculate the magnitude of pseudo-momentum vector from the wave in Problem 9.
18. Calculate the momentum and heat flux for a gravity wave with unite amplitude and horizontal wavelength of 10 km. Repeat these calculations but now the temperature wave and the horizontal velocity wave are 10° out of phase with w_1, *i.e.*, $\varphi = \theta_w + 10°$ where θ_w is the phase of w_1.
19. What is the phase relation between w_1 and u_1 such that (a) the wave vector lies in the first quadrant and (b) the wave vector lie in the second quadrant?
20. In what directions are the wave fronts for 18(a) and 18(b)?
21. Write the polarization equation for v_1 in terms of u_1.

3

MOUNTAIN WAVES

3.1 INTRODUCTION

In this chapter, we examine the very important topic of gravity waves generated when a stably stratified fluid passes over a quasi-stationary obstacle. Here *quasi-stationary* is taken to mean an obstacle that moves with a constant speed much less than the mean flow speed. Generically we shall call these *mountain waves*. Mountain waves are stationary, *i.e.*, they remain fixed to the generating obstacle. Mountain waves have been studied and modeled more than any other kind of gravity wave (see, for example, Smith, 1979; Baines, 1995). Since the publication of the first edition of this book in 2002, Google Scholar lists (as of December 2011) 15,900 articles with the phrase "mountain waves" in the body of the text.[1] The types and sizes of these obstacles include, for example, mountain ranges, broad valleys, large islands, isolated peaks, large thunderstorms, small ridges and valleys, and gently rolling hills. Areas of isolated surface temperature anomalies such as arctic leads or large cities can act as an equivalent mountain and generate gravity waves (Malkus and Stern, 1953; Tjernström and Mauritsen, 2009). Mountain waves play an essential role in the global circulation of the atmosphere. Mountain waves can produce destructive Bora-type winds that occur worldwide. We shall see that the essence of these complex flows are contained in the linear wave theory. The linearized equations of motion predict that the vertical displacement of a stably stratified flow almost always leads to the generation of gravity waves of some form. Of all the possible linear atmospheric waves, only the amplitudes of terrain-generated gravity waves can be unambiguously calculated. Large scale terrain

[1] Search Google within the fields of Physics, Astronomy, and Planetary Science.

features such as mountain ranges create waves that transport energy from the lower atmosphere up to the middle and upper atmosphere. It is now recognized that this deposition of energy and momentum is an essential component of synoptic and global scales of motion. Accordingly, these wave effects must be included in numerical models if accurate climate and weather forecasts are to be made.

The amplitudes of mountain waves are functions of the vertical and horizontal scales of the wave-generating obstacle and the local meteorological conditions. This is significant because the amplitudes of "ordinary" gravity waves, *i.e.*,

$$\hat{w}(k, z, t) = A\,\mathrm{e}^{i(kx+mz-\omega t-iz/2H_s)} + B\,\mathrm{e}^{i(kx-mz-\omega t-iz/2H_s)}, \tag{3.1}$$

where A and B are unknown constants. Mountain waves are stationary relative to the ground surface, and they do not experience dispersion, *i.e.*, all the waves launched by the mountain have the same phase speed. Under certain conditions, the amplitudes of mountain waves can grow as they move upward and eventually break similar to waves at an ocean coast. These breakdowns sometimes result in outbreaks of turbulence more commonly known as *clear air turbulence* (CAT) as shown in Fig. 3.1 taken from Lilly (1971).

To an observer, mountain waves are stationary, *i.e.*, $c = 0$ for all wavenumbers. However, because the wave is part of the flow its intrinsic phase speed cannot be zero. Thus, relative to the background flow the wave must be moving upwind at the same speed as the wind, *i.e.*, $c_\mathrm{I} = -u_0$. Remember that c_I is the intrinsic phase speed; it is the wave speed observed in a reference frame moving with the flow. These velocities are illustrated in Fig. 3.2. In all cases of mountain waves, it is the flow directed over the obstacle that generates the gravity wave. If the flow is over a mountain, then u_0 is the component of the background wind perpendicular to the ridge line of the mountain as illustrated in Fig. 3.3. If the background wind is parallel to the ridge line, then $u_0 = 0$ and waves will not be generated. In this chapter, we shall always take u_0 to be the component of the background wind directed over an obstacle. Because the wave is stationary relative to the ground, $\omega = 0$, and Ω is then

$$\Omega = \omega - ku_0 = -ku_0. \tag{3.2}$$

The intrinsic phase speed, $c_\mathrm{I} = \Omega/k$, and we see that

$$c_\mathrm{I} = \frac{\Omega}{k} = \frac{-ku_0}{k} = -u_0, \tag{3.3}$$

i.e., the wave speed relative to the flow is equal and opposite to the flow speed. Because the wave vector is always in the direction of c_I, the sign of the horizontal wavenumber must be given by

$$k = -|k|\mathrm{sgn}(u_0). \tag{3.4}$$

where $\mathrm{sgn}(u_0)$ signifies the sign of u_0. If the wave appears stationary, then it must be moving against the wind at the same speed as the wind. If the wind speed changes with height, then c_I must also change with height. This opens the question

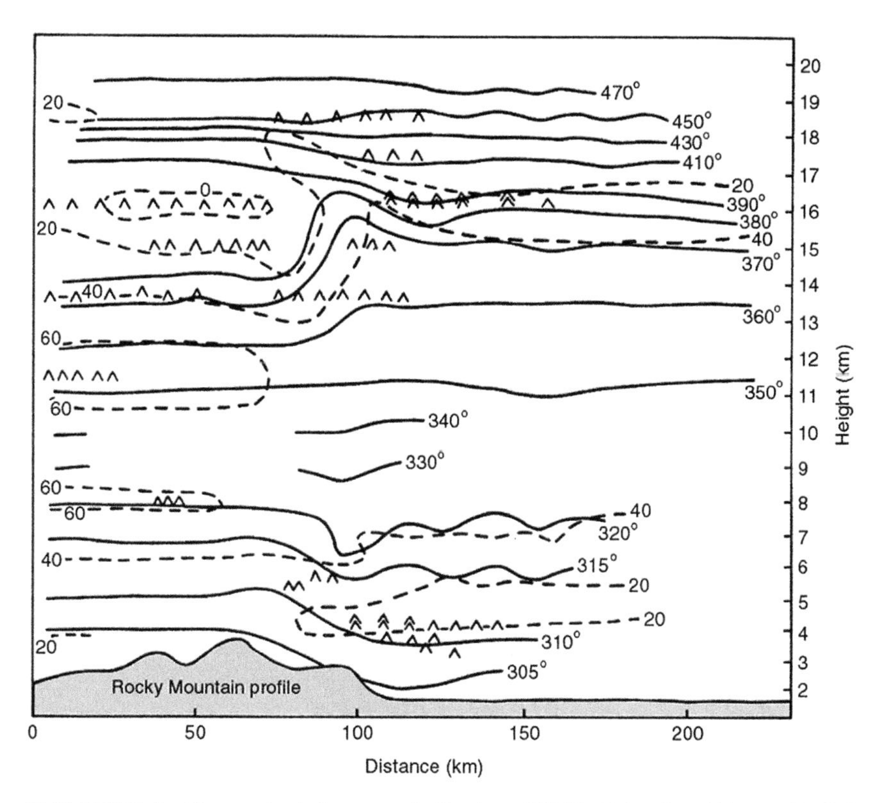

FIGURE 3.1 Waves and turbulence over the Continental Divide, Colorado on February 17, 1970: potential temperature (solid lines, K) and westerly wind component (dashed lines, m s^{-1}). ^ indicates light turbulence, and ^^ indicates severe turbulence. (Taken from Lilly, 1971.)

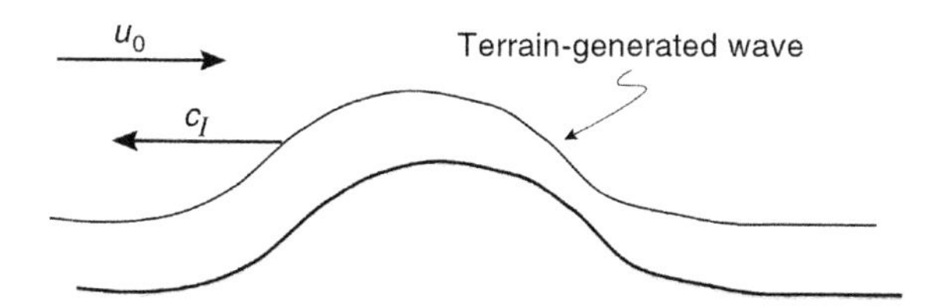

FIGURE 3.2 The intrinsic phase speed, c_I, of a terrain-generated wave is equal to the background wind speed, u_0, but in the opposite direction. To a stationary observer the wave is stationary.

of what happens to the wave if $u_0 \rightarrow 0$? Because mountain waves are stationary, there will be a critical level (see Section 5.4) where $u_0 = 0$.

Two types of terrain-generated gravity waves exist, *lee waves* and *mountain waves*; however, special flow conditions are required for lee waves to form. The lee wave extends down wind from the generating obstacle. The wave is trapped

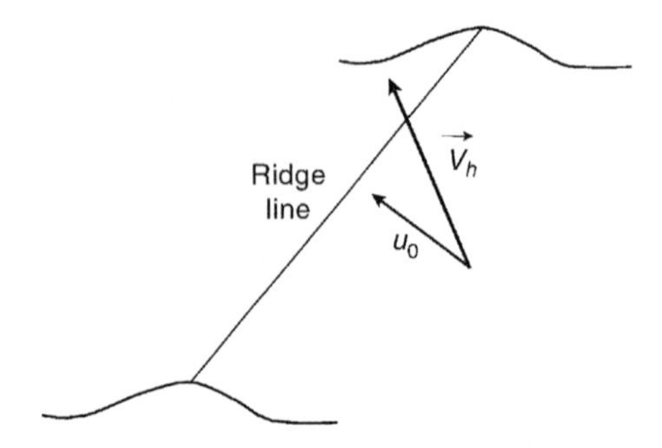

FIGURE 3.3 Gravity waves are generated only by the component of the background wind, u_0, directed perpendicular to the ridge line.

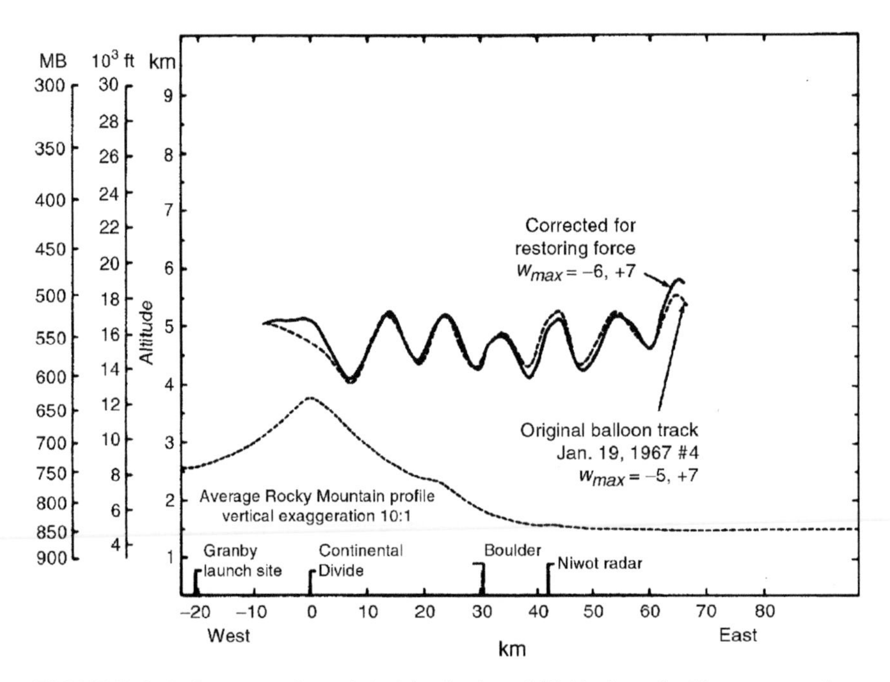

FIGURE 3.4 Lee waves downwind of the Continental Divide determined by constant volume balloon flights on January 19, 1967. Sometimes the crests of the waves are marked by standing clouds. (Taken from Vergeiner and Lilly, 1970)

between the ground surface and an upper level where wave reflection occurs, and is characterized by a nearly constant single horizontal wave length. Lee waves can extend many wave lengths downwind from an obstacle as illustrated in Fig. 3.4. Lee waves are at times marked by somewhat evenly spaced bands of clouds extending downwind from mountains and ridges. Lee waves will be discussed in Chapter 4.

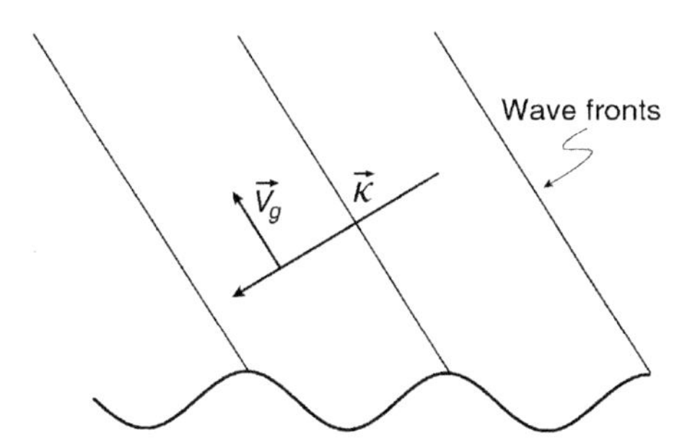

FIGURE 3.5 Wave fronts, wave vector $\vec{\kappa}$, and group velocity vector $\vec{V}_g$, over a surface corrugation with $u_0 > 0$. The transport of wave energy determines the tilt of the wave fronts.

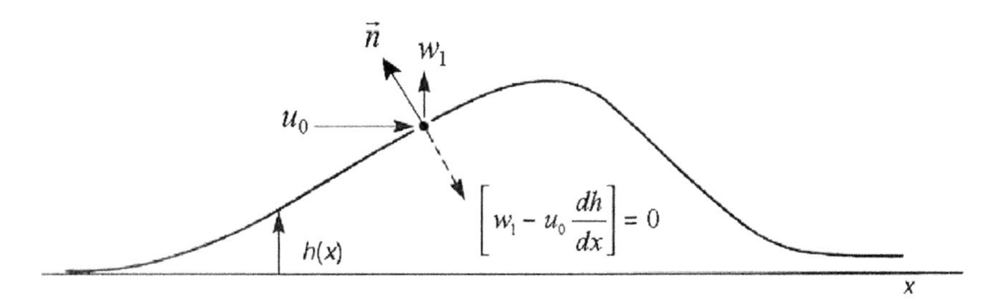

FIGURE 3.6 The flow component normal to a streamline must be zero. This provides the boundary condition at the ground surface.

3.2 UNIFORM FLOW OVER A SURFACE CORRUGATION

The simplest mountain wave problem is the case of uniform two-dimensional flow with constant Brunt–Väisälä frequency over a surface corrugation. From Chapter 2, we can tell much about such waves before interring the mathematics. Because the waves are generated at the ground, the waves must carry energy upward, thus $w_g > 0$, and because $w_g > 0$, the vertical phase speed must be downward, and therefore $m < 0$, *i.e.*, the wave fronts are moving downward. Let us take $u_0 > 0$, and then from (3.4) the horizontal wave vector points upstream, *i.e.*, $k < 0$. Because the horizontal group velocity is in the same direction as the horizontal wave vector, $u_g < 0$. Since wave fronts are parallel to the group velocity vector, the wave fronts tilt upstream as illustrated in Fig. 3.5.

Now let the surface height be given by

$$h(x) = H\Re(e^{\pm ik_s x}), \tag{3.5}$$

where H is the amplitude of corrugation, and

$$k_s = \frac{2\pi}{\lambda_s}, \tag{3.6}$$

where λ_s is the wavelength of the corrugation. Note that regardless of the sign of the exponential in (3.5) $\Re h(x) = H \cos(k_s x)$. So why do we have $\pm k$ in (3.5)? We know from (3.4) that the sign of the horizontal wavenumber is opposite to the sign of u_0. If $u_0 > 0$, then the wave must be moving in the $-x$-direction. Remember that while the wave is stationary to a fixed observer, it is moving relative to the flow. Thus for $u_0 > 0$ we chose

$$h(x) = H\,\mathrm{e}^{-ik_s x}. \tag{3.7}$$

Now for constant u_0 and N, the Taylor–Goldstein equation (2.36) along with (3.2) takes the form

$$\frac{\mathrm{d}^2 \hat{w}}{\mathrm{d}z^2} + \left[\frac{N^2}{u_0^2} - k^2\right]\hat{w} = 0. \tag{3.8}$$

We define the Fourier transforms as

$$\psi(x) = \frac{1}{2\pi}\int_{-\infty}^{+\infty} \hat{\psi}(k)\mathrm{e}^{ikx}\,\mathrm{d}k, \tag{3.9}$$

$$\hat{\psi}(k) = \int_{-\infty}^{+\infty} \psi(x)\mathrm{e}^{-ikx}\,\mathrm{d}x. \tag{3.10}$$

Since

$$m^2 = \frac{N^2}{u_0^2} - k^2 \tag{3.11}$$

the general solution of (3.8) is

$$\hat{w}(k,z) = A\,\mathrm{e}^{-imz} + B\,\mathrm{e}^{imz} \tag{3.12}$$

We must now select the proper boundary conditions. At the upper boundary we use the so-called *radiation condition*, *i.e.*, only upward propagating wave energy is allowed. Accordingly, above the upper boundary we assume that wave reflections do not occur. Thus, we must choose the A solution in (3.12) and drop the B solution at the top boundary. To evaluate A we must use the bottom boundary condition. Because the background flow is frictionless and irrotational, it can be represented by a stream function. In streamline flow, the velocity vector must always be tangent to a streamline, *i.e.*, there can be no velocity component normal to a streamline. Because the ground surface is a flow streamline we require,

$$\vec{V} \bullet \vec{n} = 0, \tag{3.13}$$

where $\vec{V} = (u_0 + u_1)\hat{x} + w_1\hat{z}$ is the flow velocity and $\vec{n}$ is the unit normal to the streamline. Figure 3.6 illustrates these vectors. The equation for the surface streamline is

$$\phi = z - h(x). \tag{3.14}$$

In the linear theory, $h(x)$ is assumed to be a small first-order perturbation. If this is not the case, then $h(x)$ would have to be written as $h(x,z)$, and the lower boundary condition would be nonlinear.[2] The unit vector normal to the streamline is given by

$$\vec{n} = \frac{\nabla\phi}{|\nabla\phi|} = \frac{-\frac{dh}{dx}\hat{x} + \hat{z}}{|\nabla\phi|}. \tag{3.15}$$

Using (3.15) in (3.13) gives

$$\vec{V} \bullet \vec{n} = \left[(u_0 + u_1)\hat{x} + w_1\hat{z}\right] \bullet \left(-\frac{dh}{dx}\hat{x} + \hat{z}\right) = 0. \tag{3.16}$$

Solving for w_1 give the *linear boundary condition* as

$$w_1(x, 0) = u_0\frac{dh}{dx}. \tag{3.17}$$

Note that since both u_1 and dh/dx are first-order terms, their products are dropped in (3.17). The Fourier transform of (3.17) is

$$\hat{w}(k, 0) = \int_{-\infty}^{+\infty} u_0\frac{dh}{dx}\, e^{-ikx}\, dx. \tag{3.18}$$

Using (3.7) in (3.18) gives

$$\hat{w}(k, 0) = -iu_0k_sH\int_{-\infty}^{+\infty} e^{-i(k+k_s)x}\, dx \tag{3.19}$$

for the bottom boundary condition.

The integral in (3.19) is the Fourier transform of the *sifting function, i.e.*,

$$\delta(k + k_s) = \frac{1}{2\pi}\int_{-\infty}^{+\infty} e^{-i(k+k_s)x}\, dx. \tag{3.20}$$

The sifting function has the property

$$f(y) = \int_{-\infty}^{+\infty} f(x)\delta(x - y)dx. \tag{3.21}$$

Recognizing the delta function in (3.20), the value of A in (3.12) becomes

$$A = \hat{w}(k, 0) = -i2\pi u_0k_sH\delta(k + k_s), \tag{3.22}$$

so that the complete solution is

$$\hat{w}(k, z) = -i2\pi u_0k_sH\, e^{-imz}\delta(k + k_s). \tag{3.23}$$

To get the solution in physical space, we must take the inverse Fourier transform, *i.e.*,

$$w_1(x, z) = -iu_0Hk_s\int_{-\infty}^{+\infty} e^{-imz}\, e^{ikx}\delta(k + k_s)dk. \tag{3.24}$$

[2] Smith (1977) discusses the effects of nonlinear boundary conditions on the wave structure. One effect of the nonlinearity is enhanced steeping of gravity waves with height which can lead to wave breakdown.

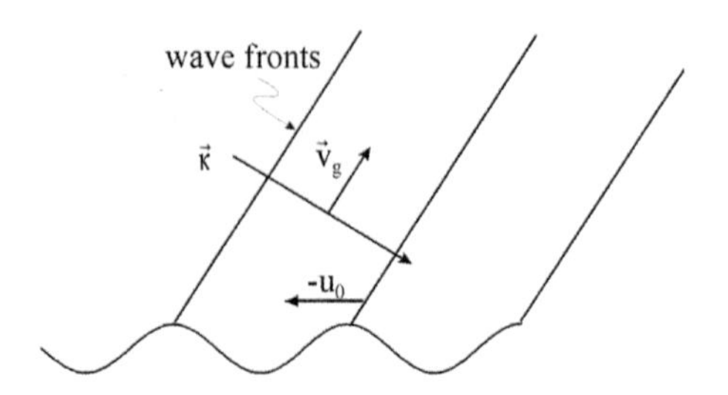

FIGURE 3.7 Same as Fig. 3.5, but with $u_0 < 0$.

The integration of (3.24) is straightforward, *i.e.*, we simply set $k = -k_s$ resulting in the complex solution

$$w_1(x, z) = -iu_0 H k_s \, e^{-i(k_s x + m_s z)}, \tag{3.25}$$

where the vertical wavenumber is

$$m_s = \left(\frac{N^2}{u_0^2} - k_s^2\right)^{1/2}. \tag{3.26}$$

The sifting property of the delta function selects only that wavenumber that is in resonance with the forcing wavenumber at the ground surface, k_s; all other waves destructively interfere. The solution in physical space is attained by taking the real part of (3.25). This gives

$$w_1(x, z) = -u_0 k_s H \sin(k_s x + m_s z) \quad \text{if } u_0 > 0. \tag{3.27}$$

Now let us see what happens if we select $u_0 < 0$, then $h(x) = H \, e^{ik_s x}$. In this case we get

$$w_1(x, z) = u_0 k_s H \sin(k_s x - m_s z) \quad \text{if } u_0 < 0. \tag{3.28}$$

The wave field (3.28) is plotted in Fig. 3.7. We see that the wave fronts still tilt upstream. Comparing Fig. 3.7 with Fig. 3.5 we see that one figure is the reflection of the other. In other words, if we simply rotated Fig. 3.5 180° about a vertical axis we would have Fig. 3.7. This can also be done by changing the signs of k_s and u_0 from plus to minus to get (3.27).

Let us use (3.27) to evaluate the other perturbation quantities using the linearized equations (2.6)–(2.9). The governing equations are then:

$$w_1(x, z) = -u_0 H k_s \sin(k_s x + m_s z), \tag{3.29}$$

$$u_1(x, z) = u_0 H m_s \sin(k_s x + m_s z), \tag{3.30}$$

$$p_1(x, z) = -\rho_0 u_0^2 H m_s \sin(k_s x + m_s z), \tag{3.31}$$

$$\theta_1(x, z) = H \frac{d\theta_0}{dz} \cos(k_s x + m_s z). \tag{3.32}$$

Fig. 3.8 illustrates the variations of wave perturbation pressure, wind speeds, and potential temperature along the surface corrugation for positive background wind.

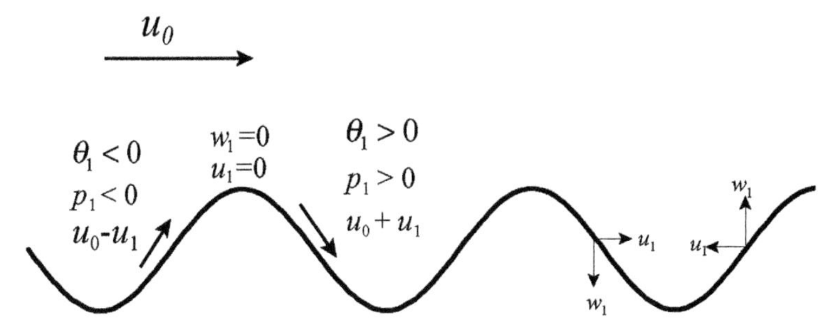

FIGURE 3.8 Variations of wave-perturbation quantities along a surface corrugation.

Along the windward side of the hill, the horizontal wind speed is reduced and the pressure is increased. Because of the stable stratification, the upward air current brings cooler air. On the lee side of the hill, the reverse is true. The pressure difference across the hill causes an acceleration of the downslope flow. Indeed, mountain waves often produce extreme downslope winds such as those recorded near Boulder, Colorado (see, for example, Lilly, 1978; Peltier and Calrk, 1979; Durran, 1986). Similar winds are known as Bora, Chinook and Föhn.

The perturbation continuity equation (2.8) allows us to define a first-order vertical *streamline displacement*, $\zeta_1(x, z)$, of the flow streamlines, *i.e.*,

$$w_1(x, z) = u_0 \frac{\partial \zeta_1(x, z)}{\partial x}. \tag{3.33}$$

Comparing (3.33) with (3.17) we see that $\zeta_1(x, 0)$ represents the surface topography. Fig. 3.9 shows the streamline displacements over two cycles of a surface corrugation for the case where $H = 50$ m, $\lambda_s = 2000$ m, $u_0 = 4$ ms^{-1}, and $N = 0.023$ s^{-1}. For these values, the vertical wavelength is $\lambda_z = 1320$ m. We see in Fig. 3.9 that the horizontal phase of the terrain shape is repeated at $z = \lambda_z$, as indicated by the thick line at that height. The inverse of the terrain shape, *i.e.*, where the wave is 180° out of phase with the surface terrain, is seen at $z = \lambda_z/2$. The up-wind tilt of the wave fronts is clearly seen in Fig. 3.9.

When

$$\frac{N}{u_0} < k_s, \tag{3.34}$$

the waves are evanescent and the solution is

$$w_1(x, z) = -u_0 H k_s \mathrm{e}^{-qz} \sin(k_s x), \tag{3.35}$$

where $q = k\sqrt{1 - N^2/\omega^2}$. Fig. 3.10 illustrates the wave streamlines for the evanescent case. Here u_0, N, and H are as in Fig. 3.9, but now $\lambda_s = 1000$ m. Note that now the wave amplitude decreases exponentially with height, and the wave fronts are vertical. The wave-perturbation quantities are symmetrically distributed with respect to the crests at the ground surface, and as we shall see in Chapter 6 a net or average wave stress is not exerted on the terrain. Evanescent waves

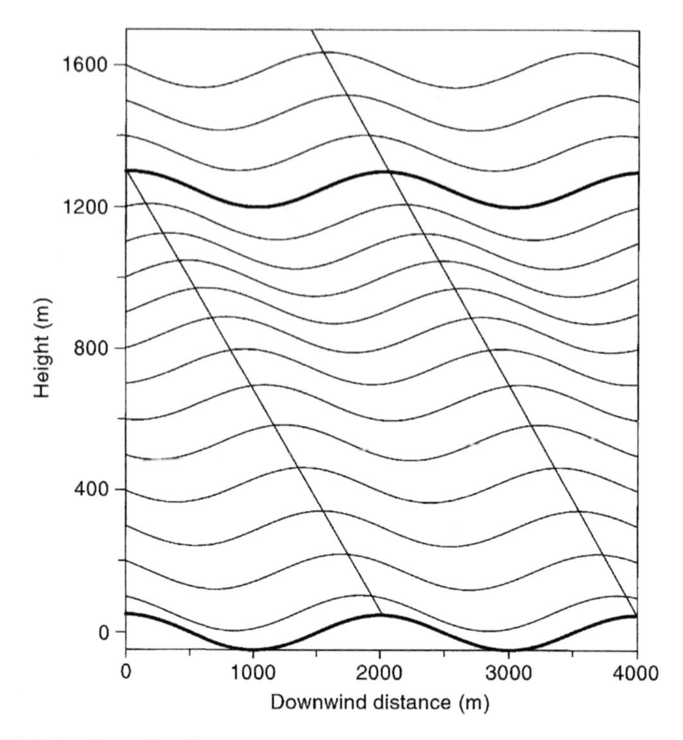

FIGURE 3.9 Streamline displacements over a surface corrugation with $H = 50$ m and wavelength $\lambda_x = 2000$ m. $N = 0.023\ \mathrm{s}^{-1}$ and $u_0 = 4\ \mathrm{ms}^{-1}$.

occur when u_0 is large, or N is small, or when λ_s is small. If u_0 is too large, then the frequency of the forced vertical oscillations of the air parcels as they pass over the surface corrugation will be greater than the resonant frequency of the atmosphere, N. As we have seen in Chapter 2, when a stably stratified flow is forced to oscillate at a frequency greater than its natural frequency, only evanescent waves are produced.

The surface corrugation, though simple, has been used in many studies of mountain waves. There are many regions of approximately evenly spaced parallel mountains. Fig. 3.11 is a photograph taken by Jeaneane Payne of quasi-parallel ridges in the Appalachian Mountains. However, the surface corrugation has been used to approximate a single mountain as for example in Kim and Mahrt (1992). This is further discussed in the next section. It must also be noted that the results discussed above pertain only to conditions of uniform wind and stratification. In the general case, the Taylor–Goldstein equation must be solved numerically using the radiation condition at the top of the model and the kinematic boundary condition (3.18) at the ground surface.

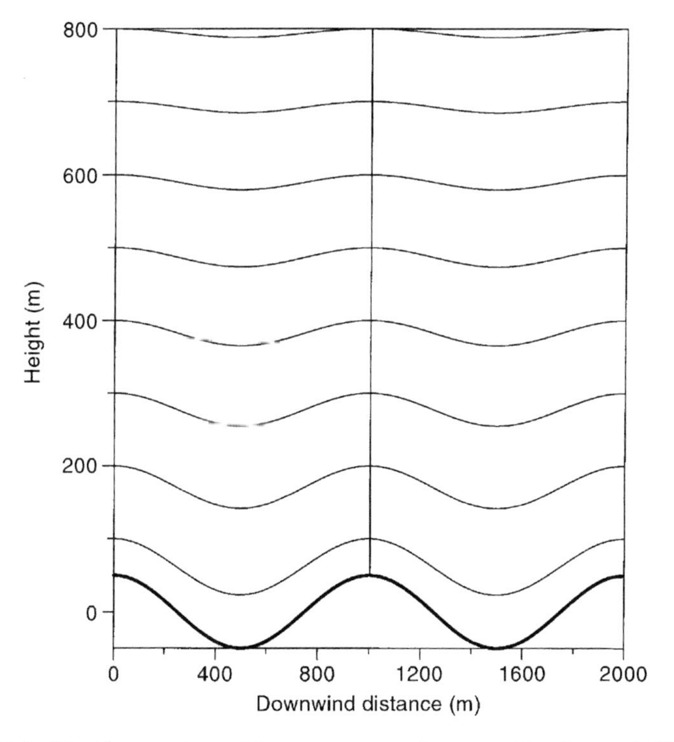

FIGURE 3.10 Evanescent mountain waves over a surface corrugation. Same as in Fig. 3.8, but now $\lambda_x = 1000$ m. Note that now the wave fronts are vertical.

3.3 THE TWO-DIMENSIONAL MOUNTAIN

We now consider the case of flow over an isolated two-dimensional mountain. The simplest mountain shape is one whose gross section is symmetric. The most often used shapes are the *bell-shaped* mountain[3] given by

$$h(x) = \frac{Hb^2}{x^2 + b^2} \tag{3.36}$$

and the *Gaussian-shaped* mountain given by

$$h(x) = H\,\mathrm{e}^{-x^2/b^2}, \tag{3.37}$$

where H is the maximum mountain height. In (3.36), b is the half-width of the bell-shaped mountain, and in (3.37) $0.83b$ is the half-width of the Gaussian-shaped mountain. The bell-shape cross section has been used by, for example, Queney (1948), Smith (1976), and Grisogono, Pryor, and Keislar (1993); the Gaussian-

[3] Also known as the Witch-of-Agnesi.

FIGURE 3.11 Quasi-parallel ridges in the Appalachian Mountains near Townsend, Tennessee. Image Credit: Jeaneane Payne, Image Builders, Knoxville, TN.

shape cross section has been used by, for example, Hines (1989), Nappo and Chimonas (1992), and Grisogono (1994). Here we shall use the Gaussian mountain.

The solution to (3.8) requires the Fourier transform of the bottom boundary condition (3.17), which is given by

$$\hat{w}(k,0) = iu_0k\hat{h}(k). \tag{3.38}$$

Assuming a horizontally infinite domain, the Fourier transform of (3.37) is

$$\hat{h}(k) = Hb\sqrt{\pi}\,\mathrm{e}^{-(kb/2)^2}. \tag{3.39}$$

Using (3.39) in (3.38) gives the linear lower boundary condition

$$\hat{w}(k,0) = iku_0Hb\sqrt{\pi}\,\mathrm{e}^{-k^2b^2/4}. \tag{3.40}$$

Note that when the background wind is not vertically uniform, then u_0 in (3.40) must be replaced by $u_0(z = 0)$. However, this can be problematical when modeling real flow situations since then $u_0(0) = 0$ and gravity waves will not be launched.

Unlike the waves over a surface corrugation which can be described by a single wave, the isolated mountain is represented by a spectrum of waves. The Taylor–Goldstein equation must be solved for each wavenumber. These waves will destructively interfere everywhere except above and down wind of the ridge. The waves will combine to form standing wave packets. These standing disturbances resemble those generated by water flowing over rocks in a briskly running stream. Unlike the surface corrugation case, we can neither draw nor visualize the wave vectors and the group velocity vectors. However, the fundamental physics remains unchanged. With this in mind, we proceed with our analysis.

For gravity waves in the troposphere where the atmospheric scale height can be neglected, the Taylor–Goldstein equation is

$$\hat{w}' + \left[\frac{N^2}{u_0^2} - \frac{u_0'}{u_0} - k^2\right]\hat{w} = 0, \tag{3.41}$$

where primes indicate vertical derivatives. For the case of constant flow and stratification, and imposing the radiation condition at the upper boundary, the solution of (3.41) for a particular wavenumber, k is

$$\hat{w}(k, z) = \hat{w}(k, 0)\mathrm{e}^{-imz}, \tag{3.42}$$

where

$$m = \left[\frac{N^2}{u_0^2} - k^2\right]^{1/2}. \tag{3.43}$$

The solution in physical space is obtained by summing the contributions of the waves given by (3.42), and this is achieved by taking the real part of the inverse Fourier transform of (3.42). Thus,

$$w_1(x, z) = \frac{1}{2\pi}\Re\int_{-\infty}^{+\infty} \hat{w}(k, 0)\mathrm{e}^{-imz}\,\mathrm{e}^{ikx}\,\mathrm{d}k. \tag{3.44}$$

Because the ridge is symmetric (3.44) can be written as

$$w_1(x, z) = \frac{1}{\pi}\Re\int_{0}^{+\infty} \hat{w}(k, 0)\mathrm{e}^{i(kx-mz)}\,\mathrm{d}k. \tag{3.45}$$

If $u_0 > 0$, we must use negative horizontal wavenumbers, so that

$$w_1(x, z) = \frac{1}{\pi}\Re\int_{0}^{+\infty} \hat{w}(-k, 0)\mathrm{e}^{-i(kx+mz)}\,\mathrm{d}k. \tag{3.46}$$

Using (3.40) in (3.46) gives

$$w_1(x, z) = -\frac{u_0 H b}{\sqrt{\pi}}\Re\int_{0}^{\infty} k\,\mathrm{e}^{-(kb/2)^2}\,\mathrm{e}^{-i(kx+mz)}\,\mathrm{d}k. \tag{3.47}$$

Taking the real part of (3.47) and distinguishing between propagating and evanescent waves, we can split the integral in (3.47) into two parts, *i.e.*,

$$\begin{aligned} w_1(x, z) &= -\frac{u_0 H}{\sqrt{\pi}}\left[\int_{0}^{k_c} bk\,\mathrm{e}^{-(kb/2)^2}\sin(kx + mz)\mathrm{d}k \right. \\ &\quad \left. - \int_{k_c}^{+\infty} bk\,\mathrm{e}^{-(kb/2)^2}\,\mathrm{e}^{-qz}\sin(kx)\mathrm{d}k\right], \end{aligned} \tag{3.48}$$

where q is given by (2.67), and $k_c = N/u_0$ is the *cut-off wavenumber* for propagating waves. Recall that for wavenumbers greater than k_c, m is complex. The first integral in (3.48) represents the contributions to the vertical velocity by the propagating waves, and the second integral represents the contributions by the evanescent waves. We can write (3.48) as

$$w_1(x, z) = \frac{1}{\sqrt{\pi}} \left[\int_0^{k_c} w_P(k) dk + \int_{k_c}^{\infty} w_E(k) dk \right], \tag{3.49}$$

where

$$w_P(k) = -u_0 H bk \, e^{-(kb/2)^2} \sin(kx + mz) \tag{3.50}$$

and

$$w_E(k) = -u_0 H bk \, e^{-(kb/2)^2} \, e^{-qz} \sin(kx). \tag{3.51}$$

Comparing (3.50) with (3.29) we see that $w_P(k)$ is the first-order vertical velocity perturbation over a sinusoidal surface with horizontal wavenumber k and amplitude H weighted by the factor $bk\, e^{-(kb/2)^2}$. Thus, we see that the wave over an isolated mountain is the sums of propagating waves each generated by a surface corrugation with wavenumber k and amplitude $H[bk\, e^{-(kb/2)^2}]$. Comparing (3.51) with (3.35) we see that w_E is a similar sum of evanescent waves. This is the essence of the linear theory. Each wave contributes independently to the perturbation variables.

From (3.50) and (3.51) we see that the structures of waves launched by a Gaussian-shaped mountain are strongly influenced by the term $kb \exp\left[-(kb/2)^2\right]$ which is plotted in Fig. 3.12. The maximum value occurs when $kb = \sqrt{2}$, *i.e.*, when the length of the excited wave is $\lambda_x \sim 4.4b$. This wavelength corresponds to about 98% of the width of the Gaussian-shaped mountain.[4] Thus, the most excited wave scales with the width of the mountain. From (3.41), we see that the wave is propagating if $L_s^2 > k^2$ where $L_s = N/u_0$ is the *Scorer* parameter. The wave structure over an isolated mountain is determined by the relative magnitudes of L_s and b. To see this, recall that for a wave to be propagating in a flow with constant wind speed and stratification $(L_s - k) > 0$. Now let us consider the case when $\lambda_x \sim 4b$. Then, $b \sim 0.25\lambda_x$, corresponds to the maximum width of the mountain. The most energetic wavenumber will be $1/b$, and the propagation condition is,

$$\frac{N^2}{u_0^2} - \frac{1}{b^2} > 0, \tag{3.52}$$

or

$$\frac{Nb}{u_0} > 1. \tag{3.53}$$

We can further simplify this relation by writing

$$\frac{N}{u_0/b} \approx \frac{T_c}{T_p} > 1, \tag{3.54}$$

[4] A similar result is obtained for the bell-shaped ridge.

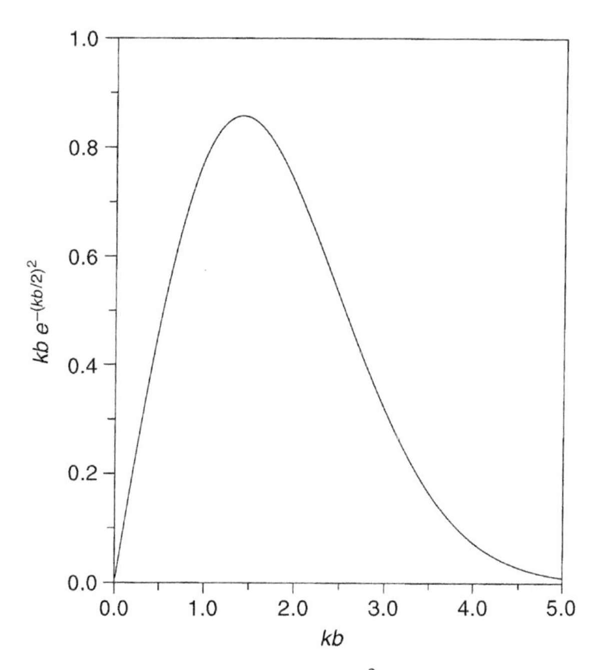

FIGURE 3.12 Response function, $kb\{\exp -(kb/2)^2\}$, plotted as a function of dimensionless wavenumber kb.

where T_c is the time required for an air parcel to pass over the mountain, and T_p is the period of the buoyancy oscillation of the air parcel. Thus, an air parcel must make at least one vertical oscillation while crossing the mountain if a wave is to be generated.

If $L_s b \leqslant 1$, then the waves will be mostly evanescent with amplitudes decreasing with horizontal and vertical distance from the mountain top. This case corresponds to combinations of narrow mountain width, weak stratification, and strong winds.[5] Fig. 3.13 shows the wave field for the case, $L_s b \approx 1$ with $u_0 = 10\ \text{ms}^{-1}$, $N = 0.01\ \text{s}^{-1}$, and $b = 1$ km. Lee waves stream downwind of the mountain with amplitudes decreasing with distance, but the rate of decrease is not great. Energy transport is vertical and downwind. The lee wave crests downwind of the mountain suggest the possibility of wave clouds often observed in mountainous regions. Fig. 3.14 shows the wave field when $L_s b > 1$ which corresponds to combinations of wide mountain, strong stratification, and weak winds. In this case, L_s is as in Fig. 3.12, but now $b = 10$ km. Keller (1994) proposes that for typical atmospheric conditions, smooth topography with half-width greater than about 10 km mountains will generate gravity waves that are hydrostatic. Hydrostatic gravity waves as described by (2.90) propagate upward directly above

[5] In almost all cases, the curvature term in (3.50) is not significant.

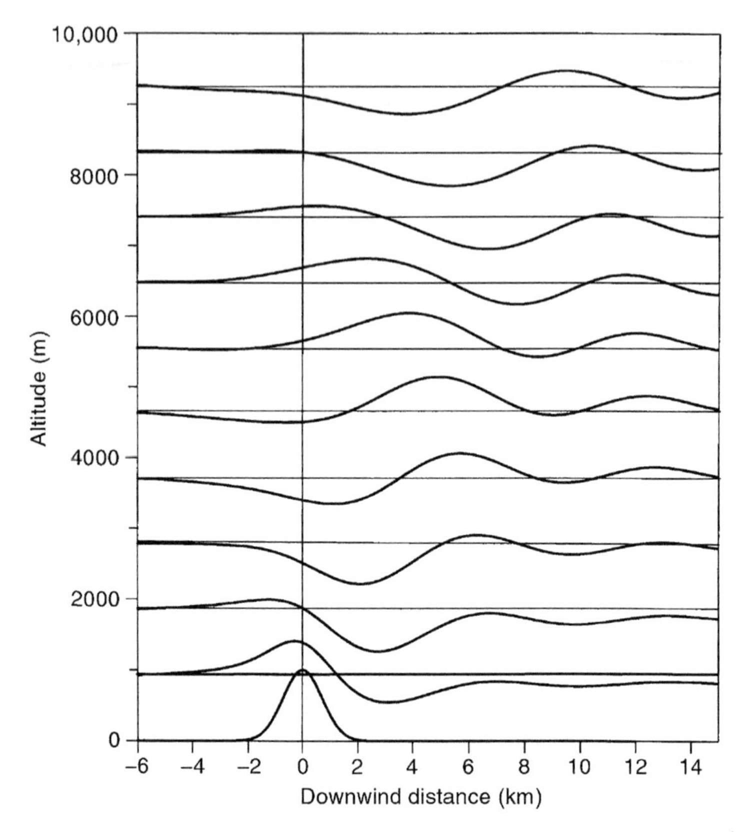

FIGURE 3.13 Wave field over a two-dimensional narrow mountain; $N = 0.01\ \mathrm{s}^{-1}$; $u_0 = 10\ \mathrm{ms}^{-1}$, and $b = 1$ km.

the mountain, and the energy transport is vertical. From Fig. 3.14, the vertical wavelength is seen to be a bit less than 6.5 km, which is the height where the mountain shape is reproduced. If we set $m = L_\mathrm{s}$ then $\lambda_z = 2\pi u_0/N$, and $\lambda_z \approx 6.3$ km, and we see that the vertical wavenumber is essentially independent of the horizontal wavenumber since the latter is very small. Indeed, this is also the reason for the vertical transport of wave energy.

It is clear that as the width of mountain increases, the wave field becomes increasingly hydrostatic. The time required for the flow to traverse the mountain shown in Fig. 3.14 is about 1 h. Imagine now a mountain range say with $b = 100$ km, *i.e.*, an effective width of about 400 km. An air parcel traveling at 10 ms^{-1} will require about 11 h to traverse the range. On this time scale, the Coriolis force due to the earth's rotation will be effective. These Coriolis accelerations are given by (Holton, 2004):

$$\frac{\mathrm{d}u}{\mathrm{d}t} = (2\Omega_\mathrm{E} \sin\phi)v, \tag{3.55}$$

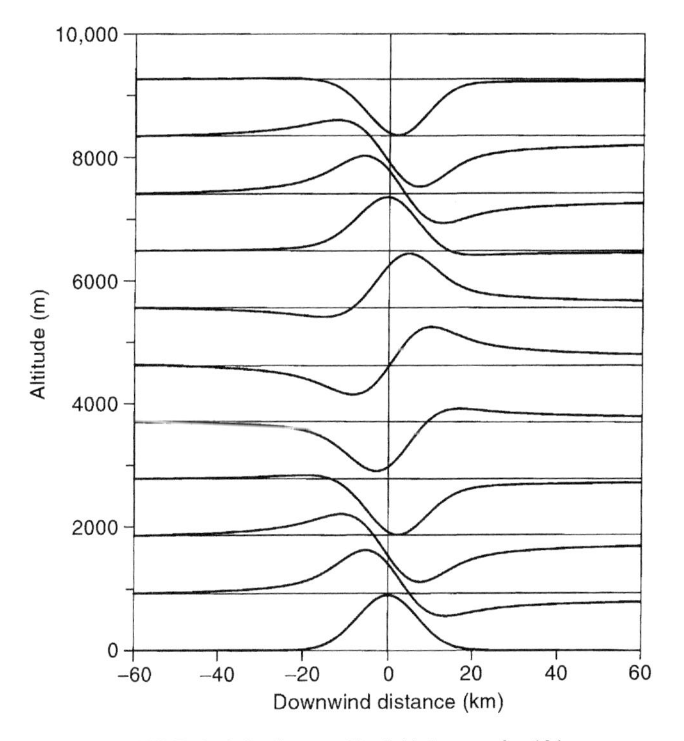

FIGURE 3.14 Same as Fig. 3.11, but now $b = 10$ km.

$$\frac{dv}{dt} = -(2\Omega_E \sin\phi)u, \tag{3.56}$$

$$\frac{dw}{dt} = (2\Omega_E \cos\phi)u, \tag{3.57}$$

where Ω_E is the angular velocity of the earth (7.292×10^{-5} s^{-1}) and ϕ is the latitude. Because of the Coriolis forces, the gravity wave problem is three-dimensional. Gill (1982) gives the appropriate equation:

$$\frac{\partial^2}{\partial t^2}\left(\frac{\partial^2 w_1}{\partial z^2} + \frac{\partial^2 w_1}{\partial x^2} + \frac{\partial^2 w_1}{\partial y^2}\right) + f^2 \frac{\partial^2 w_1}{\partial z^2} + N^2\left(\frac{\partial^2 w_1}{\partial x^2} + \frac{\partial^2 w_1}{\partial y^2}\right) = 0, \tag{3.58}$$

where f is the *Coriolis parameter* equal to $2\Omega_E \sin\phi$. The waves described by (3.58) are called *inertia-gravity waves* because inertial forces have an influence (see, for example, Blumen, 1965; Bretherton, 1969; Gill, 1982; Dunkerton, 1984; Sato, 1994; Allen and Vincent, 1995; Vincent, Allen, and Eckermann, 1997; Thomas, Worthington, and McDonald, 1999; Guest *et al.*, 2000). A characteristic of inertia-gravity waves is a spiraling velocity perturbation vector with height. In the Northern Hemisphere, the spiral is anticyclonic (clockwise) for upward propagating waves and cyclonic (counter clockwise) for downward propagating waves. The reverse holds for the southern hemisphere. This spiral is explained by Bretherton (1969): an upward displacement of an air parcel will result in a positive x displacement (3.57). This in turn will result in a negative y displacement (3.56), this will induce a negative x displacement (3.55), *etc.* The result is an oscillation similar to inertial oscillations. Observations by Thomas, Worthington, and McDonald (1999) show a case of anticyclonic rotation in the troposphere and cyclonic rotation in the stratosphere. This suggests that hydrostatic gravity were generated by the jet stream with downward propagating waves in the troposphere and upward propagating waves in the stratosphere.

If we assume wave solutions of the form $\hat{w}\,\mathrm{e}^{i(kx+ly-\Omega t)}$, and for constant wind and stratification, (3.58) becomes

$$\frac{\mathrm{d}^2\hat{w}}{\mathrm{d}z^2}+\left[\frac{N^2(k^2+l^2)-\Omega^2(k^2+l^2)}{\Omega^2-f^2}\right]\hat{w}=0, \tag{3.59}$$

where from (3.2)

$$\Omega=-u_0k-v_0l. \tag{3.60}$$

We can rotate the coordinate system so that the x-axis is aligned with the mean wind. Then (3.59) becomes

$$\frac{\mathrm{d}^2\hat{w}}{\mathrm{d}z^2}+\left[\frac{N^2k^2-\Omega^2k^2}{\Omega^2-f^2}\right]\hat{w}=0 \tag{3.61}$$

and $\Omega=-u_0k$. Using the Scorer parameter ($L_\mathrm{s}=N/u_0$), we can write (3.61) as

$$\frac{\mathrm{d}^2\hat{w}}{\mathrm{d}z^2}+\left[\frac{k^2u_0^2}{k^2u_0^2-f^2}\left(L_\mathrm{s}^2-k^2\right)\right]\hat{w}=0. \tag{3.62}$$

If $b\approx 100$ km, then the wavenumber of the most excited wave will be much less than L_s so that (3.62) becomes

$$\frac{\mathrm{d}^2\hat{w}}{\mathrm{d}z^2}+\left[\frac{k^2N^2}{u_0^2k^2-f^2}\right]\hat{w}=0. \tag{3.63}$$

Fig. 3.15, shows the wave field over a mountain with $b=100$ km, and background flow as in Fig. 3.12. The wave disturbances extend downwind of the mountain with slowly changing amplitudes and increasing wavelengths. The vertical wavenumber is $m=kN(u_0^2k^2-f^2)^{-1/2}$, and the angle, β, of the wave vector relative to the horizontal is

$$\beta=\tan^{-1}\left[\frac{N}{(u_0^2k^2-f^2)^{1/2}}\right]. \tag{3.64}$$

From (3.64), we see that if k is large, then $\beta\to 0°$, and as k decreases $\beta\to 90°$. If $k<f/u_0$, then m is imaginary, and the waves are evanescent. For the case shown in Fig. 3.15, the critical horizontal wavelength is about 6.3 km. Thus, we see in Fig. 3.15 that the vertical transport of energy directly above the mountain

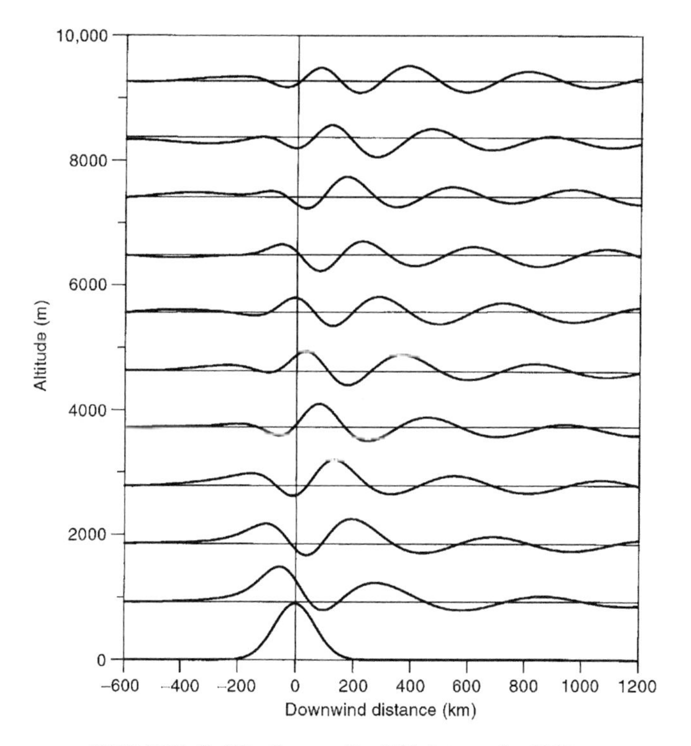

FIGURE 3.15 Same as Fig. 3.11, but now $b = 100$ km.

range is accomplished by the short horizontal wavelengths, and the downwind transport of energy is accomplished by the longer wavelengths. This explains in Fig. 3.15 the tendency for horizontal wavelengths of the downwind disturbances to decrease with increasing altitude. Perhaps one of the first *in situ* observations of a terrain-induced inertia-gravity wave, shown in Fig. 3.16, was presented by Dörnbrack *et al.* (2001). The wave was observed on January 26, 2000 using an aircraft-mounted lidar to measure cloud aerosols in the stratosphere over the Scandinavian mountain range. The Scandinavian mountain ridge has sufficient width (~300 km) to excite inertia-gravity waves. Comparison of Fig. 3.14 with Fig. 3.15 shows similar structures, and this again illustrates the utility of the linear theory.

Figures similar to Figs. 3.13–3.15 appear in the classic paper by Queney (1948). His results were made using a bell-shaped ridge (3.37). Comparing his results with Figs. 3.13–3.15 shows little difference, and this suggests that the linear theory for terrain-generated waves is not overly sensitive to the shape geometry.

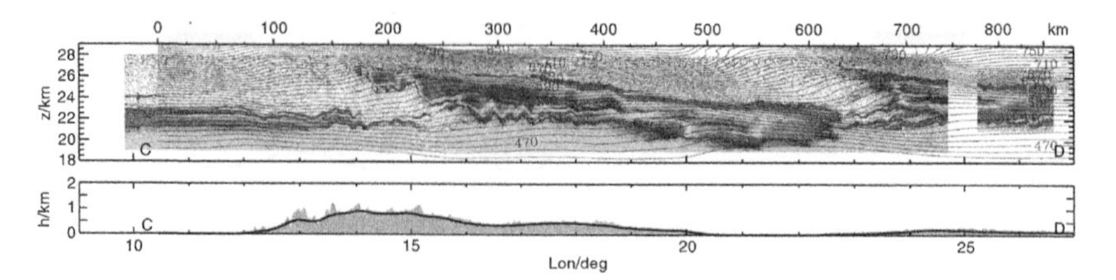

FIGURE 3.16 Top: Inertia-gravity wave observed over the Scandinavian mountain range on January 26, 2001. Bottom: Smoothed terrain profile beneath flight track. Taken from Dörnbrack *et al.* (2001).

3.4 THE THREE-DIMENSIONAL MOUNTAIN

Mountain waves over isolated 3-dimensional terrain features have been extensively studied (see, for example, Blumen and McGregor, 1976; Smith, 1980; Hines, 1988; Kim and Mahrt, 1992; Baines, 1995, and more recently Smith *et al.*, 2002; Jiang and Doyal, 2008, and references therein). The principal difference between flows over two-dimensional and three-dimensional mountains is that air parcels can pass around a three-dimensional mountain as well as over it. For large Froude numbers, *i.e.*, $F \equiv U/HN \gg 1$, where U is the upwind flow speed, H is the maximum mountain height, and N is the Brunt–Väisälä frequency, the flow tends to pass over the mountain, but for $F \ll 1$, the flow tends to pass around the mountain (see, for example, Smith, 1980; Snyder *et al.*, 1985). Because we have introduced the y-dimension into the problem, we must also include this dimension in the Eulier equations, (2.1)–(2.4), and Fourier transform of the terrain becomes two-dimensional, *i.e.*,

$$\hat{h}(k,l) = \frac{1}{4\pi^2}\int_{-\infty}^{\infty}\int_{-\infty}^{\infty} h(x,y)\exp[-i(kx+ly)]\,\mathrm{d}x\,\mathrm{d}y \tag{3.65}$$

with inverse transform given by

$$h(x,y) = \int_{-\infty}^{\infty}\int_{-\infty}^{\infty} \hat{h}(k,l)\exp[i(kx+ly)]\mathrm{d}k\,\mathrm{d}l. \tag{3.66}$$

The Taylor–Goldstein equation now takes the form

$$\frac{\mathrm{d}^2\hat{w}}{\mathrm{d}z^2} + \left[\frac{(k^2+l^2)N^2}{(ku_0+lv_0)^2} - \frac{ku_0'+lv_0'}{ku_0+lv_0} - (k^2+l^2)\right]\hat{w} = 0, \tag{3.67}$$

where l and v_0 is the wavenumber and background wind speed in the y-direction respectively. From the polarization equations, the horizontal perturbation velocities are

$$\hat{u}(k,l,z) = \frac{ik}{k^2+l^2}\left[\frac{l\hat{w}(lu_0'-kv_0')}{k(ku_0+lv_0)} + \frac{\mathrm{d}\hat{w}}{\mathrm{d}z}\right], \tag{3.68}$$

$$\hat{v}(k,l,z) = \frac{-il}{k^2+l^2}\left[\frac{k\hat{w}(lu_0'-kv_0')}{l(ku_0+lv_0)} - \frac{\mathrm{d}\hat{w}}{\mathrm{d}z}\right]. \tag{3.69}$$

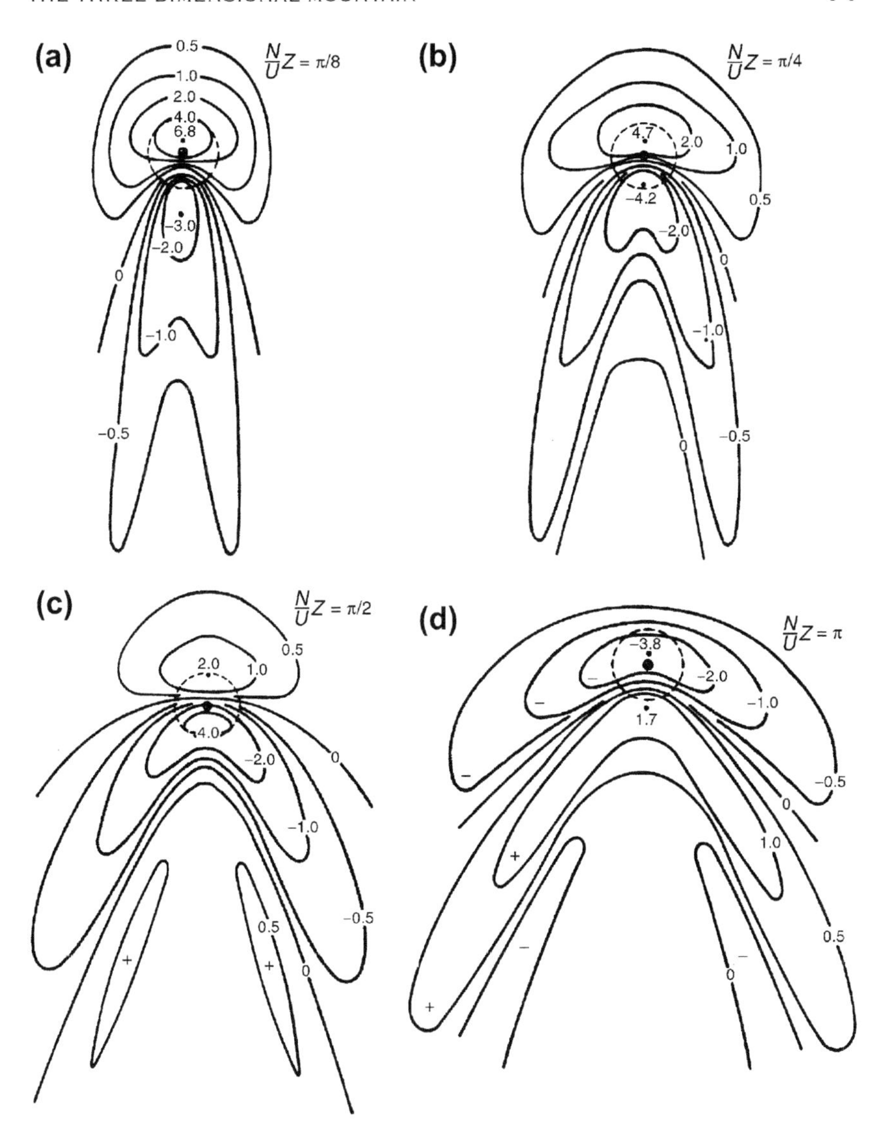

FIGURE 3.17 Streamline displacements over a three-dimensional symmetric bell-shaped mountain. (Taken from Smith (1980).)

Note that by setting $l = 0$ in (3.68) and (3.69), the solutions for the two-dimensional ridge is obtained. For a constant background wind, (3.67) reduces to

$$\frac{\mathrm{d}^2\hat{w}}{\mathrm{d}z^2} + \left[\frac{(\kappa_H^2)N^2}{(ku_0 - lv_0)^2} - \kappa_H^2\right]\hat{w} = 0, \tag{3.70}$$

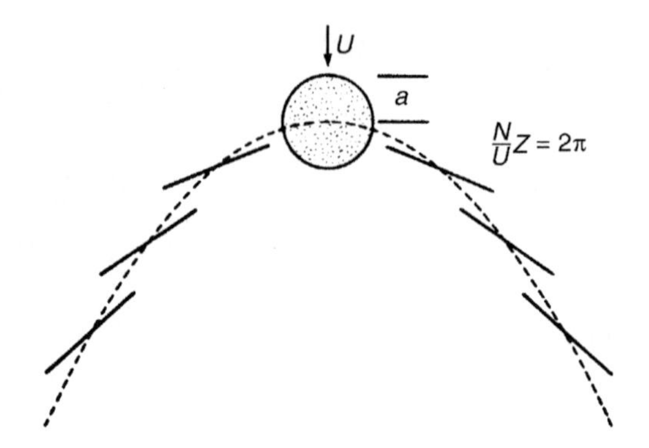

FIGURE 3.18 The parabola given by (3.89) along which the wave energy and stress above a three-dimensional hill is concentrated. (Taken from Smith (1980).)

where $\kappa_H = \sqrt{k^2 + l^2}$, and (3.68) and (3.69) reduce to

$$\hat{u}(k, l, z) = \frac{ik}{\kappa_H^2} \frac{\mathrm{d}\hat{w}}{\mathrm{d}z}, \tag{3.71}$$

$$\hat{v}(k, l, z) = \frac{il}{\kappa_H^2} \frac{\mathrm{d}\hat{w}}{\mathrm{d}z}. \tag{3.72}$$

The bottom boundary condition is

$$w_1(x, y, 0) = u_0 \frac{\partial h(x, y)}{\partial x} + v_0 \frac{\partial h(x, y)}{\partial y}, \tag{3.73}$$

with Fourier transform

$$\hat{w}(k, l, 0) = i(ku_0 + lv_0)\hat{h}(k, l). \tag{3.74}$$

The most widely used mountain shapes are the three-dimensional Gaussian-shaped mountain,

$$h(x, y) = H\,\mathrm{e}^{-[(x/a)^2 + (y/b)^2]}, \tag{3.75}$$

where a and b are length scales in the x- and y-directions respectively, and the three-dimensional bell-shaped mountain,

$$h(x, y) = \frac{H}{\left(\frac{x^2}{b^2} + \frac{y^2}{b^2} + 1\right)^{3/2}}. \tag{3.76}$$

If we assume horizontally-symmetric hills, *i.e.*, $a = b$, then the two-dimensional Fourier transform of (3.75) is

$$\hat{h}(k, l) = Ha^2\pi\,\mathrm{e}^{-\kappa_H^2 a^2/4}, \tag{3.77}$$

and for (3.76) it is

$$\hat{h}(k, l) = \frac{1}{2\pi} ha^2\,\mathrm{e}^{-a\kappa_H}. \tag{3.78}$$

Fig. 3.17 taken from Smith (1980), shows plane views of flow displacements over a three-dimensional, symmetric bell-shaped mountain. Horizontal distances have been scaled by a, and vertical distances have been scaled by N/u_0 where $N = 0.01\ \text{s}^{-1}$ and $u_0 = 10\ \text{ms}^{-1}$. Near the ground surface, the flow displacements are positive upwind of the mountain, and negative downwind. This pattern changes with increasing height so that eventually descending motion is upwind of the mountain, and ascending motion is downwind. We see also a broadening and weakening with height of the downstream displacement pattern. The results shown in Fig. 3.17 were obtained under the assumption that the wave perturbations are in *hydrostatic equilibrium* which is valid when $Na/u_0 \gg 1$ (Smith, 1980). Under the hydrostatic assumption, and taking the background wind to be in the x-direction, *i.e.*, $v_0 = 0$, (3.70) takes the form

$$\frac{\mathrm{d}^2\hat{w}}{\mathrm{d}z^2} + \frac{\kappa_H^2 N^2}{(ku_0)^2}\hat{w} = 0. \tag{3.79}$$

Because the mountain is the source of the waves, wave energy must radiate horizontally and vertically away from the mountain. It is this radiation of wave energy that results in the broadening of the displacement field shown in Fig. 3.17. The transport of wave energy is done by the group velocities. Using the definition of the intrinsic frequency, (3.2), the vertical wavenumber for the plane-wave solution of (3.79) is

$$m^2 = \frac{N^2\kappa_H^2}{\Omega^2}, \tag{3.80}$$

and the dispersion relation is

$$\Omega = \pm\frac{N\kappa_H}{m}. \tag{3.81}$$

Choosing the negative branch of (3.81), and noting that $\Omega = -ku_0$ the group velocity in the x-direction is

$$u_\mathrm{g} = u_0\frac{l^2}{\kappa_H^2}. \tag{3.82}$$

In a similar manner, we find that

$$v_\mathrm{g} = \frac{\partial\Omega}{\partial l} = -u_0\frac{kl}{\kappa_H^2} \tag{3.83}$$

and

$$w_\mathrm{g} = \frac{\partial\Omega}{\partial m} = \frac{u_0^2k^2}{N\kappa_H}. \tag{3.84}$$

Smith (1980) points out that in the terrain-attached reference frame, wave energy moves away from the mountain along straight lines defined by

$$\frac{x}{z} = \frac{u_\mathrm{g}}{w_\mathrm{g}}, \tag{3.85}$$

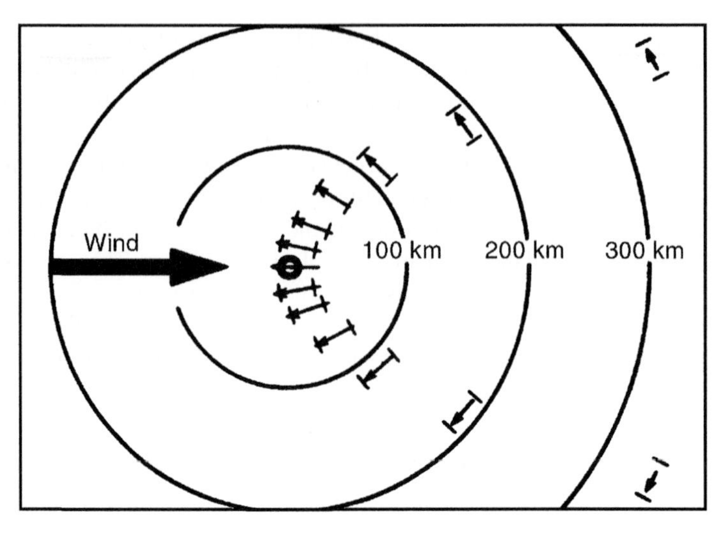

FIGURE 3.19 Direction and magnitude of wave stress above a three-dimensional mountain. (Taken from Hines (1988).)

$$\frac{y}{z} = \frac{v_g}{w_g}, \tag{3.86}$$

$$\frac{y}{x} = \frac{v_g}{u_g}. \tag{3.87}$$

Using (3.87) with (3.82) and (3.83) gives

$$\frac{x}{y} = -\frac{k}{l}, \tag{3.88}$$

which describes a family of straight lines (phase lines) along which energy propagates horizontally away from the mountain which is the origin of the coordinates. Using (3.86) and (3.87) gives

$$y^2 = \frac{N}{u_0 \kappa_H} zx, \tag{3.89}$$

which describes a parabola along which the wave energy and stress are concentrated as illustrated in Fig. 3.18. As z increases, the parabola widens. Figure 3.19, taken from Hines (1988) is similar to Fig. 3.18, but shows the direction and relative magnitude of the wave stress over a three-dimensional mountain. As will be shown in Chapter 6, the wave stresses τ_x and τ_y are

$$\tau_x = -\rho_0 \overline{u_1 w_1}, \tag{3.90}$$

$$\tau_y = -\rho_0 \overline{v_1 w_1}, \tag{3.91}$$

where the overline indicates an average over a wavelength. We see that the direction of the wave stress is always tangent to circles centered on the mountain.

3.5 NONOROGRAPHIC GRAVITY WAVES

This chapter has focused on the generation of gravity waves by vertical displacements of flow streamlines due to perturbation of topography. We have seen that these types of waves can propagate into the stratosphere and higher atmosphere. However, *nonorographic* gravity waves have strong effects in the middle and upper atmosphere, and have a major impact on the global circulation. Individual sources of nonorographic gravity waves are seldom recognized in the real atmosphere. Perhaps the best we can do is to explain these waves as not originating over mountains. For example, a major source of nonorographic waves are the tropical oceans. Other sources of stratospheric gravity waves is convection in the troposphere and the accompanying release of latent heat and the effects of storm systems. Indeed, the field is so wide reaching as to include the study of convectively-generated gravity waves in the lower atmosphere of Venus (see, for example, Baker, Schubert, and Jones, 2000 and references there in). We cannot treat here in any detail the subject of convectively-generated gravity waves; instead we present a brief literature review.

Smith and Lin (1982) investigated the relative strengths of gravity waves generated by terrain and orographic rain. They included adiabatic heating rate, $\dot{H}$, expressed in units of energy per unit mass, due to condensation so that the energy equation takes the form

$$c_v \frac{DT}{Dt} + p \frac{D\alpha}{Dt} = \dot{H}, \tag{3.92}$$

where T is the sensible temperature and α is the specific volume $1/\rho$. For constant background wind, the equation for the vertical velocity perturbation becomes

$$\frac{\mathrm{d}^2 w_1}{\mathrm{d}x^2} + \frac{\mathrm{d}^2 w_1}{\mathrm{d}z^2} + \frac{N^2(z)}{u_0} w_1 = \frac{g\dot{H}}{c_p T_0(z) u_0^2}. \tag{3.93}$$

Smith and Lin (1982) concluded that for typical wind speeds and rainfall rates the amplitudes of thermally-generated gravity waves will equal or exceed the amplitudes of terrain-generated gravity waves. Lin and Chun (1991) examined the effects of diabatic cooling due to evaporation of falling precipitation on a stably stratified shear flow. The cooling was confined to a sub-cloud layer, and their analysis included both a linear analytical model and a nonlinear numerical model. Their results showed in part that nonlinearity acts to reduce the amplitudes of wave disturbances above the cloud layer. The theoretical results of Smith and Lin (1982) were supported by Fritts and Nastrom (1992) who examined representative case studies within the Global Atmospheric Sampling Program (GASP). Their results suggest that convection as a source of gravity waves is at least as important as topographic forcing, and possible the most important source of gravity waves in the Tropics and Southern Hemisphere.

Fovell, Durran, and Holton (1992), studied gravity waves generated by moving two-dimensional mesoscale storms, *i.e.*, *squall lines* using a nonlinear numerical model. There results show a preference for excited waves to propagate against the direction of the storm's motion. In the absence of a wind relative to the storm, gravity waves in the stratosphere are excited by mechanical forcing due to oscillatory updrafts, a result first proposed by Pierce and Coroniti (1960), and apparently over looked. Alexander, Holton, and Durran (1995) used a fully-compressible, nonlinear, numerical, two-dimensional model of a midlatitude squall line to study the link between vertically propagating gravity waves and the wave-forcing mechanism. Fig. 3.20 shows a results of their simulation. Gravity waves radiating away from the storm's center are most striking, and the induced vertical velocities ranged from 5 to 20 ms^{-1}. They also related the peak in the frequency and vertical wavelength of the gravity wave spectrum with the updraft oscillation frequency and the vertical scale of the tropospheric heating respectively. Chun, Song, and Baik (1999) used linear theory and the Advanced Regional Prediction System (ARPS) (Xue *et al.*, 1995) to study gravity waves generated by a multi-cell convective system. Their results showed that updrafts, *i.e.*, forced internal gravity waves, at the head of the storm's outflow or density current can generate consecutive convective cells that move downstream and develop as a main convective cell.

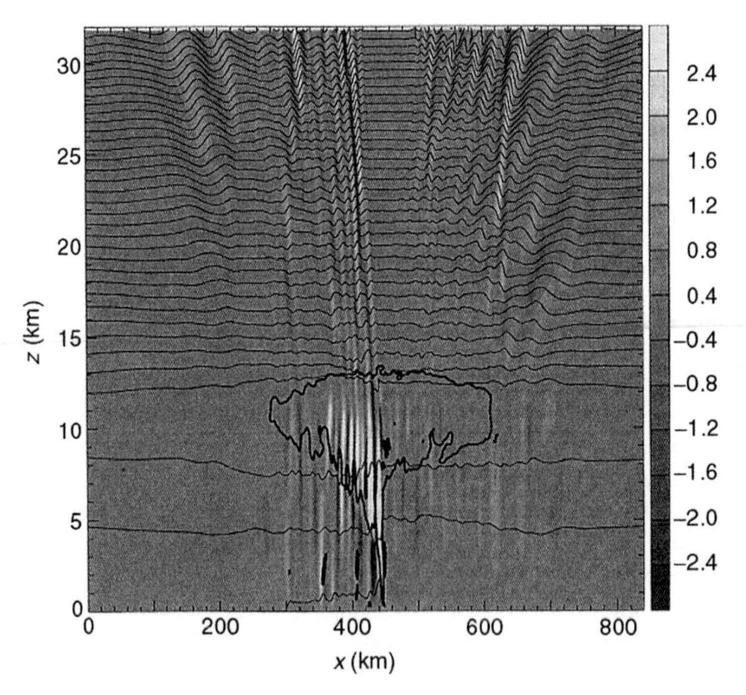

FIGURE 3.20 Isentropes (thin lines) at 4 h of a squall line numerical simulation. Thick line shows the cloud outline. Shading represents contours of vertical velocity. The vertical velocities range from 20 to −5 ms^{-1}. (Taken from Alexander, Holton, and Durran (1995).)

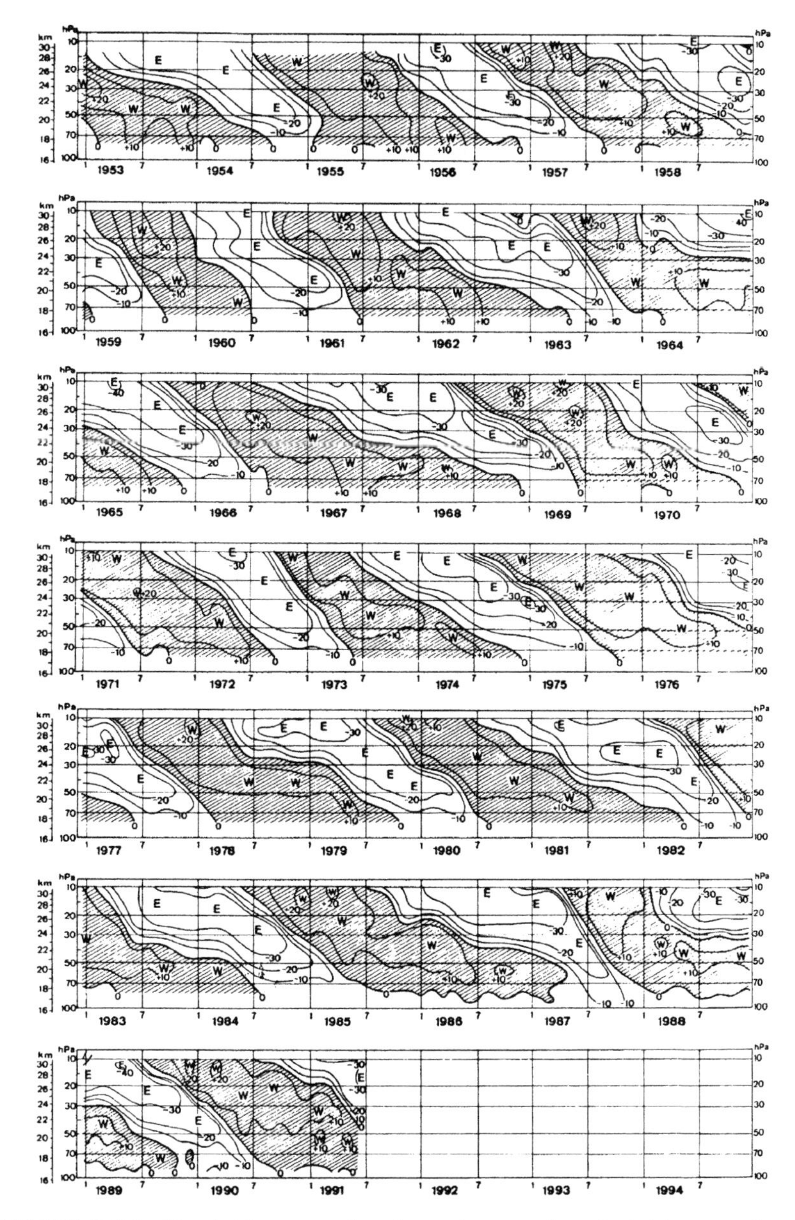

FIGURE 3.21 Time-height cross section of monthly-averaged zonal wind departures from the climatological average for that month at equatorial stations. The alternating downward propagating regimes of westerly (W) and easterly (E) winds (m s^{-1}) form the quasi-biennial oscillation. (Taken from Holton, 1992).

Thus, it appears that internal gravity waves play a major role in the initiation and maintenance of multi-cell convective systems.

Piani *et al.* (2000) extended modeling of convective gravity waves to three dimensions, and examined the role of these waves in the dynamics of the *quasi-biennial oscillation* (QBO). The quasi-biennial oscillation (see, for example, Lindzen and Holton, 1968; Holton and Lindzen, 1972; Holton, 2004) consists of zonally[6] symmetric easterly and westerly wind regimes which alternate regularly with a period varying from about 24 to 30 months. These wind regimes first appear above 30 km, but propagate downward at a rate of about 1 km/month. The downward propagation occurs without loss of amplitude between 30 and about 23 km; however, there is rapid attenuation of the wind regime below 23 km. The QBO is symmetric about the equator with a maximum amplitude of about 20 ms^{-1}, and a half-width of about 12° latitude. Fig. 3.21 illustrates the oscillating wind regimes of the QBO. It is now accepted that the QBO is the result of upward propagating convectively-generated gravity waves which interact with critical layers (see Chapter 5). At a critical layer, the wave is dissipated (absorbed), resulting in a divergence of wave stress, and a deceleration or acceleration of the wind. Research into the links between the QBO and convective gravity waves continues to be an active field of research as typified by, for example, the papers by Alexander, Beres, and Pfister (2000), Vincent and Alexander (2000), and Alexander and Vincent (2000). Fritts and Alexander (2003) give a review of QBO research.

PROBLEMS

1. Waves launched by a surface corrugation with $\lambda_s = 1000$ m and amplitude 10 m propagate upward in an atmosphere with $N = 0.012\ \text{s}^{-1}$ and wind speed where $u(0) = 10\ \text{ms}^{-1}$. The background wind speed is increasing at the rate of 0.01 ms^{-1} per 10 m. Assume the flow is nearly constant. At what height will the dominant wave become evanescent?
2. Given the expressions:

$$w(x) = a\,\mathrm{e}^{i(kx+mz-\omega t)},$$
$$u(x) = b\,\mathrm{e}^{i(kx+mz-\omega t)},$$

show that the wave stress is given by:

$$-\rho_0\overline{uw} = -\frac{1}{2}\Re\left(\rho_0 u w^*\right).$$

3. Derive an expression for the wave stress over a bell-shaped ridge with height

$$h(x) = \frac{Hb^2}{x^2 + b^2}.$$

[6] In dynamic meteorology, the *zonal wind* is the east-ward component of the wind.

4. An inertia-gravity wave is generated by a constant wind of 10 ms^{-1} flowing over a terrain obstacle at 43°N latitude. What wavelength is required for a critical level to exist? How wide is the terrain feature?
5. Plot the wave spectra for a surface corrugation with amplitude H and horizontal wavelength λ, and a Gaussian ridge with height given by

$$h(x) = H\,\mathrm{e}^{-(x^2/b^2)}.$$

6. Show that for gravity waves launched by a constant wind and stratification over a surface corrugation of wavenumber k and amplitude H, the equation:

$$i\Omega\tilde{\rho} + \tilde{w}\frac{\rho_0}{g}N^2 = 0$$

reduces to:

$$\theta_1 = H\frac{\partial\theta_0}{\partial z}\cos(kx + mz).$$

7. Use Bernoulli's equation to show that the air pressure on the upwind side of a hill is greater than the air pressure on the downwind side of a hill.
8. Calculate the wave stress over a Gaussian-shaped mountain with maximum height 1000 m, scale width, b, of 500 m, a constant wind of 15 ms^{-1}, and $N = 0.02s^{-1}$.
9. Calculate the amplitude of the most energetic wave in Problem 9.
10. Calculate the amplitude and wavelength of a surface corrugation that gives the same wave stress as in Problem 10.
11. Using Program MOUNTAIN_C.FOR at EURL/Ch3/CODES calculate the wave stress over the mountain in Problem 9 for $UB = 10$ ms^{-1} and $UT = 15$, 10, and -15 ms^{-1}.
12. Calculate and plot the normalized power spectrum in Problem 9.
13. Assume a two-dimensional mountain as in Problem 9. Let the background wind speed change linearly from 3 ms^{-1} at the ground surface and 18 ms^{-1} at 5 km. At what height does the dominant mountain wave become evanescent?
14. Consider the inertia-gravity wave (3.61). As $\Omega^2 \to f^2$ what angle β does the wave vector approach relative to the horizontal plane?
15. For flow over a three-dimensional mountain, what is the relation between $\hat{u}$ and $\hat{v}$?
16. A nonorographic gravity wave propagates upward in an atmosphere with constant wind speed and stratification. Use (3.93) to write an equation for the vertical wavenumber, m.
17. Consider a flow in the x-direction over a symmetric three-dimensional bell-shaped mountain. What is the net wave stress in the y-direction?
18. For the inertia-gravity wave, the critical level is a function of what parameter?
19. At latitude 43°N a constant wind flows over a two-dimensional mountain. If at some initial time $t = 0$ the flow is perpendicular to the axis of the mountain, how long will it take for the wind to blow parallel to the mountain?

4

DUCTED GRAVITY WAVES

4.1 INTRODUCTION

If one observes a nearly coherent wave which persists for several oscillation and has a near constant phase speed, then it is likely a ducted wave, *i.e.*, a horizontally traveling wave trapped between two levels. Ducted waves can be observed as periodic oscillations of atmospheric pressure, horizontal and vertical wind speed, wind direction, temperature, and water vapor (see, for example, the case studies described by Gossard and Munk (1954), Gossard and Hooke (1975), Koch and Golus (1988), Fritts *et al.* (2003), Nappo, Miller, and Hiscox (2008), Tjernstrom and Mauritsen (2009)). Figure 4.1 shows pressure waves observed at the ground surface; Fig. 4.2, taken from Fritts *et al.* (2003) shows wave-like structures of, vertical velocity at several heights above the ground surface, and Fig. 4.3 (courtesy of Dale Durran) shows three-dimensional ship waves launched by the Sandwich Islands. If these waves persist over time spans ranging from several minutes to a few hours or as shown in Fig. 4.3 and extend horizontally over hundreds of kilometers, then it is reasonable to assume that the waves are horizontally propagating, or stationary as in Fig. 4.3. In many cases, but not always, the bottom layer of the duct is the ground surface. These kinds of waves have been called *cellular* (Martyn, 1950; Hines, 1965) or standing cellular (Turner, 1973) to denote waves that propagate horizontally and present a standing pattern in the vertical direction as illustrated in Fig. 4.4. This cellular pattern appears stationary when viewed by

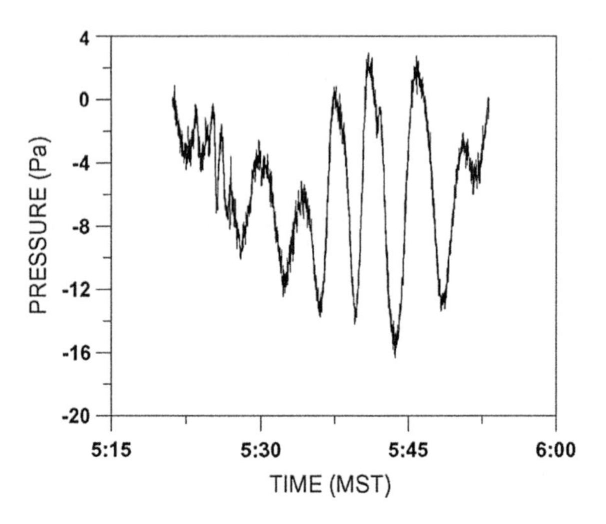

FIGURE 4.1 Typical signal for a ducted gravity wave.

an observer moving with the wave. However, to a stationary observer the wave oscillates at frequency $\omega = kc$, where k is the horizontal wavenumber and c is the wave phase speed.

As illustrated in Fig. 4.5, wave ducting requires two horizontal reflecting surfaces or layers. For complete reflection at the bottom surface, which we take to be at $z = 0$, $w_1(0) = 0$ which is a *node point.* For complete reflection at the upper surface, which we take to be at height $z = H_r$, $w_1(H_r) = 0$. If the bottom surface is solid, for example the ground, then there is complete reflection. If the upper surface is not solid, then there is partial reflection and refraction of the transmitted wave. Depending on the height of the reflecting level and the vertical wavelength of the wave, the upward and downward waves might interfere constructively or destructively. If the waves constructively interfere, then the wave becomes trapped between the ground surface and the reflecting level. The trapped wave is said to be *ducted.* We can think of the duct as being a *wave guide*. Studies by, for example, Bosart and Cussen (1973), Uccellini (1975), Balachandran (1980), Stobie, Einaudi, and Uccellini (1997), Koppel, Bosart, and Keyser (2000), Lac, Lafore, and Redelsperger (2002), and Alexander, Richter, and Sutherland (2006) show that thunderstorms often initiate ducted gravity waves which can propagate long distances ($\sim$1000 km), and initiate new thunderstorms along their paths. Chimonas and Nappo (1987) used gravity wave theory to argue that a long-lived wind gust observed by Doviak and Ge (1984) could be modeled as a ducted thunderstorm *bow wave*. Monserrat and Thorpe (1996) used ducting theory to explain a long-lived gravity wave event observed on Mallorca (Baleric Islands), and Rees *et al.* (2000) suggest that most of the high-frequency waves observed in the atmospheric boundary layer over the Brunt Ice Shelf, Antarctica

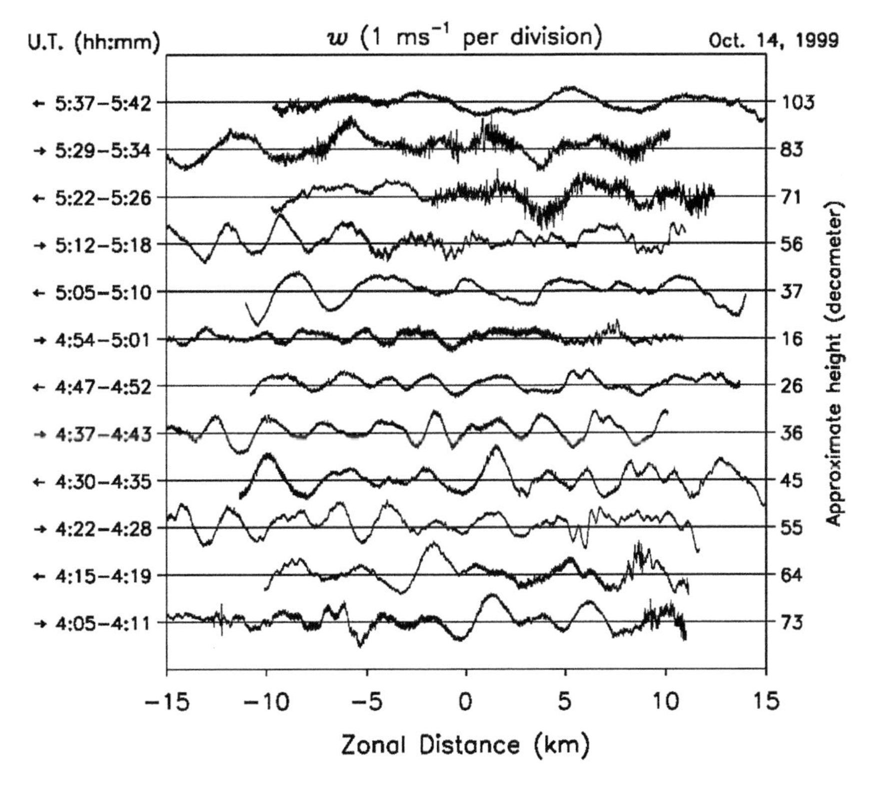

FIGURE 4.2 Ducted gravity waves at various heights. Taken from Fritts *et al.* (2003).

correspond to trapped gravity waves. Mentes and Kaymaz (2007) did a statistical analyses of ducting conditions over Istanbul, Turkey, and concluded that surface duct characteristics show clear seasonal differences, and that surface ducts in a stable atmosphere are found to be stronger than those in an unstable atmosphere. Haak *et al.* (2010) used four mesoscale forecast models to identify the most important components of forecasting systems that contribute to ducting in a coastal environment. Coleman, Knupp, and Herzmann (2010) used time-laps photography and radar to analyze a dramatic undular bore over Tama, Iowa, USA. The amazing time-lapse video can be seen at EURL\Ch4\L1. These few examples illustrate that gravity wave ducting is and continues to be a lively research topic.

4.2 WAVE REFLECTION AND REFRACTION AT AN ELEVATED LAYER

Under ordinary conditions, an upward propagating gravity wave may encounter a level where there is a discontinuity in vertical wavenumber, m, for example at

FIGURE 4.3 Lee waves off the Sandwich Islands. Courtesy of Dale Duran.

the tropopause or at the top of the stable PBL. When this happens, wave *reflection occurs*. If the wave is not perfectly reflected, then *transmission* occurs. If m is different on the other side of the level, then the wave vector will be different and *refraction* occurs. Figure 4.6 illustrates these mechanisms. The incident and reflected wave vectors make equal angles relative to the horizontal plane, and the transmitted wave vector makes a different angle to the horizontal plane. That the

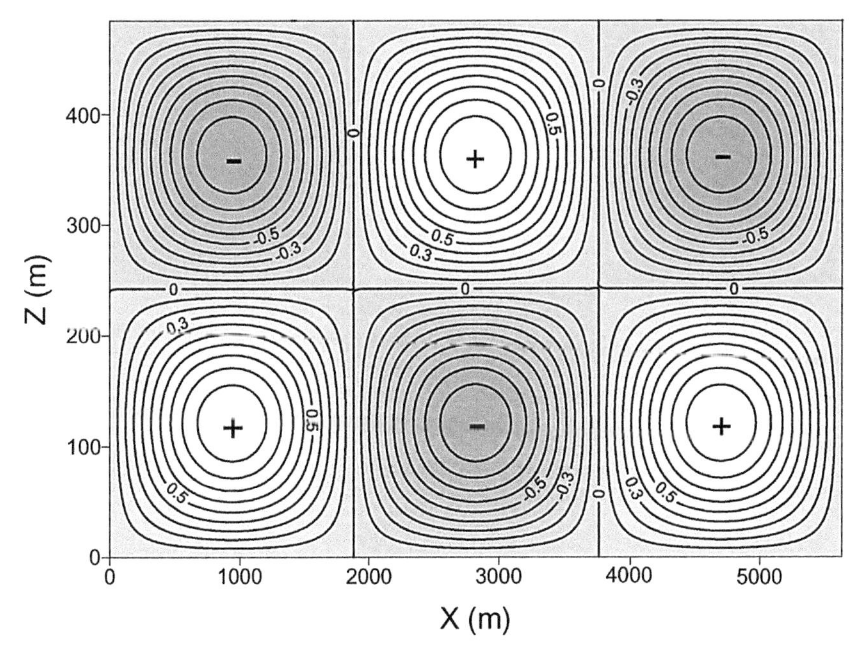

FIGURE 4.4 Cellular second mode of a ducted gravity wave.

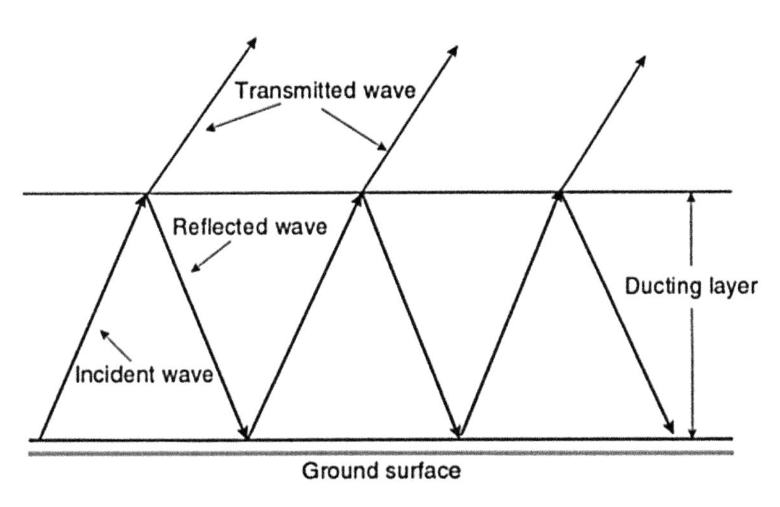

FIGURE 4.5 Illustration of wave reflection and transmission between the ground surface and some upper level. If the incident and reflected waves are in phase, then wave ducting can occur.

angles, of the incident and reflected waves relative to the horizontal plane are equal is not due to any special properties of the waves. Because the incident and reflected

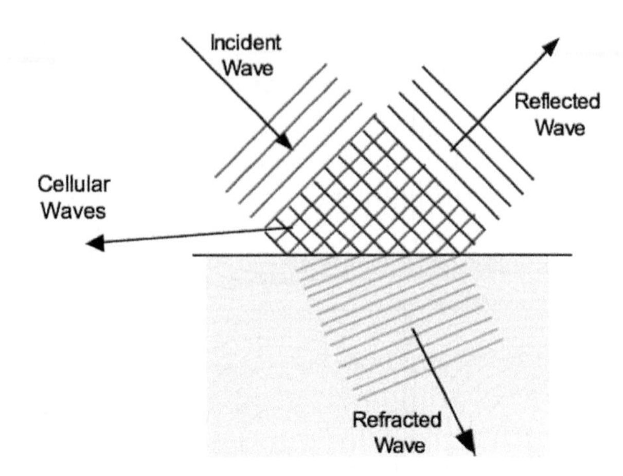

FIGURE 4.6 Reflection and refraction at a boundary between two immiscible fluids. The incident and reflected waves can form a standing wave.

waves have the same frequency these angles are necessarily equal due to (2.41). Note also that the incident and reflected wave fronts can interact in such a way as to create cellular or standing waves. Such a cellular structure is shown in Fig. 4.6. In optics, the characteristics of a ray of light are functions of the index of refraction of the medium. When considering gravity waves, the analogous index of refraction is m. If the reflection is partial, then some of the wave energy is transmitted through the level but with reduced amplitude. The transmitted wave can be either propagating (m real) or evanescent (m complex). If the reflected wave propagates to the ground surface, it is reflected upward.

We first consider wave reflection at an upper level of discontinuity in m. (A video of wave reflection and refraction is at EURL\Ch4\WAVE-R&R) In the general case, where the Brunt–Väisälä frequency and background wind speed are changing continuously with height the wave field can be very complicated, and not easily analyzed. We consider here the much simpler case of a two layer flow with constant but different stratification in each layer as illustrated in Fig. 4.7. If a background wind is present we shall assume it to be constant and equal at all heights, *i.e.*, there is no vertical wind shear. We also assume that the background densities are the same in both regions. The bottom layer, layer 1, extends from the ground surface (assumed flat and uniform) to the height of the interface, H_r. The upper layer, layer 2, is vertically unbounded. At the interface, the Brunt–Väisälä frequency changes discontinuously from N_1 below to N_2 above. In the bottom layer there is an upward propagating wave $\hat{w}_{1,u}$, and a downward propagating reflected wave $\hat{w}_{1,d}$. In the upper layer, we have only an upward propagating wave $\hat{w}_2$. If Ω is the same on either side of the interface, then, from (2.57), we see that

$$N_1 \cos\beta_1 = N_2 \cos\beta_2, \tag{4.1}$$

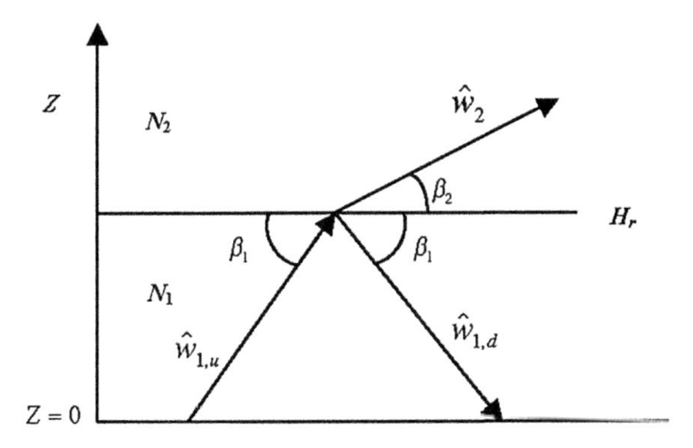

FIGURE 4.7 Wave reflection at height H_r where there exists a discontinuity in stratification.

which is similar to Snell's law in optics. Note that β is the angle the wave vector makes with the horizontal, not with the interface. Thus, (4.1) would hold even if the interface is sloping.

Hydrodynamics requires two conditions at the interface. The first condition, called the *dynamic* boundary condition, requires that the atmospheric pressure be continuous across the interface, *i.e.*,

$$p_{0,1}(H_r) + p_1(H_r) = p_{0,2}(H_r) + p_2(H_r), \tag{4.2}$$

where $p_{0,1}$ and p_1 are the background and perturbation pressures in layer 1, *etc.* If (4.2) is not met, then infinite vertical accelerations at the interface are possible. This is because at the interface $\partial p/\partial z \approx (p_2 - p_1)/\left[(H_r + \delta) - (H_r - \delta)\right]$ where δ is a small distance. Because the pressures are constant but different on either side of the interface $\partial p/\partial z \rightarrow \infty$ as $\delta \rightarrow 0$. The second condition, called the *kinematic* boundary condition requires that the two layers remain in contact with each other. If this condition is not satisfied, then voids or cavitation regions at the interface are possible as illustrated in Fig. 4.8. The kinematic boundary condition is satisfied if the vertical mass fluxes in each layer are equal at the interface, *i.e.*,

$$\rho_{0,1}(H_r)w_1(H_r) = \rho_{0,2}(H_r)w_2(H_r), \tag{4.3}$$

where $\rho_{0,1}$ is the background density in layer 1, and w_1 is the perturbation vertical velocity in layer 1, *etc.* To apply these boundary conditions, we need to know the amplitudes of the waves in each layer; however generally these are not known. Gill (1982) defines the ratio

$$Z = \frac{p_1}{\rho_0 w_1} \tag{4.4}$$

as the *impedance* which is defined in acoustics as the ratio of the force per unit area to the volume displacement of a given surface across which sound is being transmitted. Impedance can be thought of as the alternating-current analog of resistance in direct current.

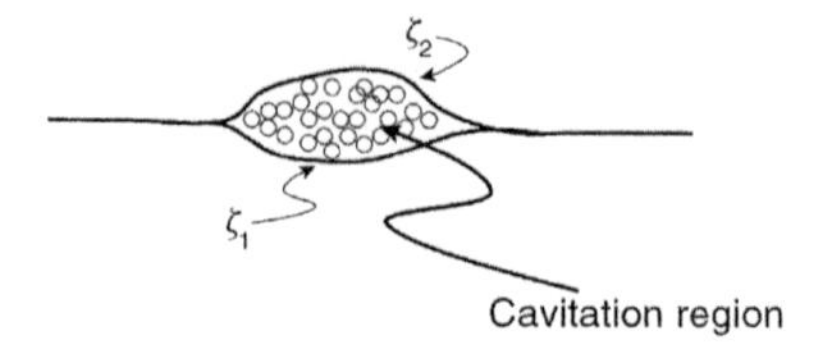

FIGURE 4.8 Cavitations or voids between two material surfaces. If the fluid is water, then the circles represent bubbles. If the fluid is moist air, then the circles represent condensation droplets.

The dynamic and kinematic conditions require that $Z_1(H_r) = Z_2(H_r)$. As Gill (1982) indicates, there are four possible forms of solutions to our problem since the solutions can be propagating or evanescent in regions 1 and 2. We first examine the case when the transmitted wave is propagating, and we assume *a priori* that Ω and k do not change in the reflection-transmission process. In layer 2, we have only upward propagating energy, and so the wave fronts must be moving downward. Thus, for $z > H_r$

$$w_2 = C\mathrm{e}^{-im_2(z-H_r)}\mathrm{e}^{-i(kx-\Omega t)}. \tag{4.5}$$

Using (2.22), (2.24), and (4.5) the pressure perturbation is

$$p_2 = \frac{1}{k^2}\rho_0\Omega C m_2\,\mathrm{e}^{-im_2(z-H_r)}\,\mathrm{e}^{-i(kx-\Omega t)}. \tag{4.6}$$

The impedance in the upper layer is then

$$Z_2 = \frac{\Omega m_2}{k^2}. \tag{4.7}$$

In the bottom layer, we have both upward and downward (reflected) energy propagation so that

$$w_1 = A\mathrm{e}^{im_1(H_r-z)} + B\mathrm{e}^{-im_1(H_r-z)}, \tag{4.8}$$

where A and B are the amplitudes of the upward and downward moving wave fronts respectively. We now define a *reflection coefficient, r*, as the ratio of the amplitude of the reflected wave to the incident wave, *i.e.*,

$$r = \frac{B}{A}, \tag{4.9}$$

so that

$$w_1 = A\left[\mathrm{e}^{im_1(H_r-z)} + r\mathrm{e}^{-im_1(H_r-z)}\right]\mathrm{e}^{-i(kx-\Omega t)}. \tag{4.10}$$

Note that since A and B can be complex, r also can be complex. The perturbation pressure in layer 1 is

$$p_1 = \frac{\rho_0\Omega A m_1}{k^2}\left[\mathrm{e}^{im_1(H_r-z)} - r\mathrm{e}^{-im_1(H_r-z)}\right]\mathrm{e}^{-i(kx-\Omega t)}. \tag{4.11}$$

Using (4.10) and (4.11), the impedance in the bottom layer is

$$Z_1 = \frac{\Omega m_1}{k^2}\left[\frac{e^{im_1(H_r-z)} - re^{-im_1(H_r-z)}}{e^{im_1(H_r-z)} + re^{-im_1(H_r-z)}}\right]. \tag{4.12}$$

Setting $Z_1(H_r) = Z_2(H_r)$ gives

$$\frac{m_2}{m_1} = \frac{1-r}{1+r}. \tag{4.13}$$

Solving (4.13) for r gives

$$r = \frac{m_1 - m_2}{m_1 + m_2}. \tag{4.14}$$

Note that from (2.51) the vertical wavenumbers of the upward and downward moving waves in layer 1 must be equal. Because k does not change in the reflection process, the angle of the wave vector relative to the horizontal plane is the same for both waves. However, because of (4.14) the amplitudes of the waves need not be equal.

If $r = 0$, then $m_1 = m_2$ and an interface between regions 1 and 2 does not exist. The wave is upward moving with constant amplitude in both regions. If $r = 1$, then from (4.14) $\Re m_2 = 0$, and the transmitted wave cannot propagate in the upper layer, but it is evanescent, *i.e.*, $m_2 = -iq_2$, where q is defined by (2.67). Taking the real parts of (4.5) and (4.10) for $r = 1$ we now have:

$$w_2 = Ce^{-q_2(z-H_r)}\cos(kx - \Omega t), \tag{4.15}$$

$$w_1 = 2A\cos[m_1(H_r - z)]\cos(kx - \Omega t). \tag{4.16}$$

Clearly, the kinematic boundary condition is satisfied at $z = H_r$ if $C = 2A$. This corresponds to a maximum w_1, and below H_r maxima are found at elevations where $m_1(H_r - z) = n\pi/2, n = 0, 2, 4, 6, \ldots$. The maximum at H_r gives maximum transfer of wave energy into the upper layer. However, because energy cannot propagate in the upper layer since $\Re m_2 = 0$, it is reflected back into the bottom layer. This is perfect reflection. Above H_r, wave amplitude decreases exponentially and move horizontally with phase speed Ω/k as does the interface. (See video at EURL\Ch4\VIDEO-1). For $0 < r < 1$, propagating waves will be excited in the upper layer but with less than maximum amplitude. Setting $r = 1 - b$, where $0 < b < 1$, (4.10) can be written as

$$w_1 = A\left[e^{im_1(H_r-z)} + (1-b)e^{-im_1(H_r-z)}\right]e^{-i(kx-\Omega t)}. \tag{4.17}$$

Expanding (4.17) and taking the real part gives

$$w_1 = (2-b)A\cos[m_1(H_r - z)]\cos(kx - \Omega t), \tag{4.18}$$

$$= (1+r)A\cos[m_1(H_r - z)]\cos(kx - \Omega t). \tag{4.19}$$

The amplitude of w_1 is diminished by the loss of energy into the upper layer, and the amplitude of w_2 is increased accordingly.

When $r = -1, \Re m_1 = 0$ and there are no propagating waves in the lower layer; however, the evanescent wave amplitude can initiate propagating waves in the upper medium at $z - H_r$. In quantum mechanics this process if called *tunneling*.

4.3 WAVE TRAPPING, ENERGY FLUX, AND WAVE RESONANCE

In the previous section, we did not consider the effects of the ground surface. In this section, we will look more closely at how waves in the lower layer are affected when wave trapping occurs. It is useful to consider a continuous source of wave energy caused by a steady flow over some terrain disturbance $h(x)$. Applying the boundary condition

$$w_1(0) = A(e^{im_1 H_r} + re^{-im_1 H_r}), \tag{4.20}$$

(4.10) becomes

$$w_1 = w_1(0)\left[\frac{e^{im_1(H_r - z)} + re^{-im_1(H_r - z)}}{e^{im_1 H_r} + re^{-im_1 H_r}} e^{-i(kx)}\right], \tag{4.21}$$

where

$$m_1 = \left[\frac{N_1^2}{u_0^2} - k^2\right]^{1/2}. \tag{4.22}$$

Note that if $r = 0$ in (4.21), then w_1 becomes similar to the wave-perturbation velocity field above a terrain disturbance as described in Chapter 3. Now let $r = e^{2im_1 H_r}$. If $m_1 H_r$ is an even multiple of $\pi/2$, then $r = 1$, and (4.21) becomes

$$w_1 = w_1(0)\frac{\cos[m_1(H_r - z)]}{\cos(m_1 H_r)} \cos kx. \tag{4.23}$$

Since wave energy cannot propagate through the interface, it is trapped in the lower layer. Also, the upward and downward momentum fluxes are of opposite sign and so cancel each other. Thus, the vertical energy flux within the duct is zero. To show this, consider the wave stress. For this we need the equation for u_1. Using the continuity (2.16) and (4.21) we get

$$u_1 = -\frac{u_0}{k} w_1(0)\left[\frac{e^{im_1(H_r - z)} - re^{-im_1(H_r - z)}}{e^{im_1 H_r} + re^{-im_1 H_r}} e^{-i(kx)}\right]. \tag{4.24}$$

For $r = 1$ (4.24) becomes

$$u_1 = \frac{i w_1(0) m_1 u_1(0)}{k}\left[\frac{\sin[m_1(H_r - z)]}{\cos(m_1 H_r)}\right]\cos(kx). \tag{4.25}$$

In Chapter 6, we shall see that the vertical flux of horizontal momentum, $\tau(z)$, is given by

$$\tau(z) = -\frac{1}{\ell}\int_{-\ell/2}^{\ell/2} \rho_0 u_1 w_1 \, \mathrm{d}x, \tag{4.26}$$

where ℓ is some horizontal scaling parameter usually taken to be the wave length. Using (4.23) and (4.25), (4.26) evaluates to zero, and hence there is no net flux of momentum out of the duct.

Now consider the case when $2m_1 H_r$ is an odd multiple of π, then $r = -1$, $m_1 = iq_1$, and waves in the lower layer are evanescent. However, as we have seen in Section 4.1, propagating waves will be excited in the upper layer because of tunneling. An example of this behavior is illustrated in Fig. 4.9, which shows the streamline displacements calculated by a numerical model over a surface corrugation with $u_0 = 2$ ms^{-1}, $N_1 \approx 0$ s^{-1}, $N_2 = 0.02$ s^{-1}, $\lambda_s = 1000$ m, amplitude $H = 25$ m, and $H_r = 200$ m. Below H_r, the wave fronts are vertically oriented with decreasing amplitude. The slight upstream tilt of the evanescent waves represents a small yet finite wave stress transport because $N_1 \neq 0$[1] The conditions shown in Fig. 4.9 are representative of a well mixed convective layer caped by a stable layer as may occur in the morning hours of a developing convective boundary layer.

Because the source of energy is continuous and because there is no loss of energy through the top of the layer, when $r = 1$ the wave amplitude in the bottom layer can grow without bound resulting in wave resonance. Using (4.14) we see that

$$r = -e^{i2m_1 H_r} = \frac{m_1 - iq_2}{m_1 + iq_2}. \tag{4.27}$$

If we rationalize the denominator in (4.27), and then take the square root, we get

$$ie^{im_1 H_r} = \frac{m_1 - iq_2}{(m_1^2 + q_2^2)^{1/2}}. \tag{4.28}$$

Separating (4.28) into real and imaginary parts leads to

$$\sin(m_1 H_r) = \frac{-m_1}{(m_1^2 + q_2^2)^{1/2}},$$

$$\cos(m_1 H_r) = \frac{-q}{(m_1^2 + q_2^2)^{1/2}},$$

and hence

$$\tan(m_1 H_r) = \frac{m_1}{q_2}. \tag{4.29}$$

For a given combination of background variables N, and H_r, there are unique *eigenvalues* (u_0, k) that are solutions to (4.29).

[1] This is because the numerical solution is unstable if $N_1 = 0$.

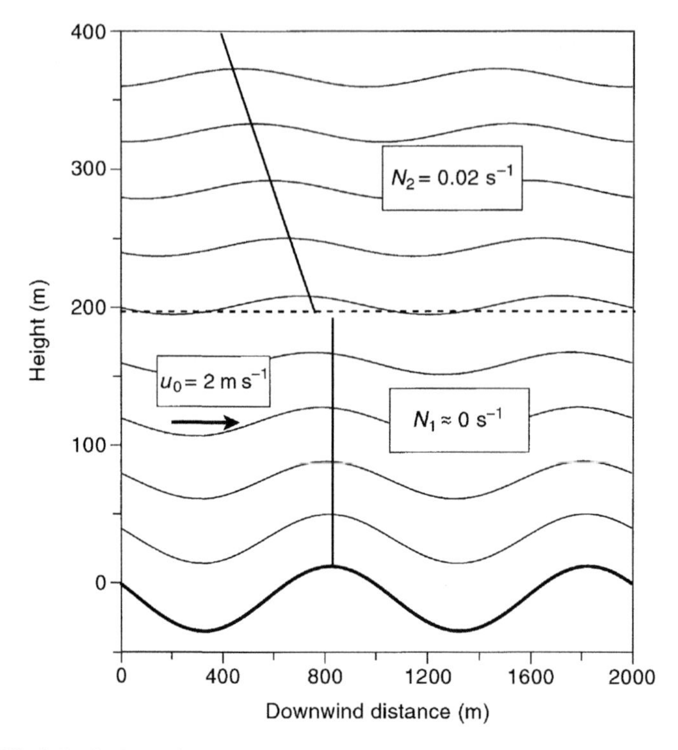

FIGURE 4.9 Leakage of wave energy through an evanescent region. An example of tunneling.

Examining Fig. 4.4, we can propose a solution of the form

$$w_1 = A \sin(m_1 z) \cos(kx - \Omega t) \tag{4.30}$$

with boundary conditions $w_1(0) = 0$ and $w_1(H_r) = 0$. These boundary conditions will be satisfied if $m_1 H_r = n\pi$, where n is an integer number. This implies that if H_r is an integer number of half wavelengths, $\lambda_z/2$, the reflecting level is a node. There is complete reflection at H_r with no wave motion of any kind above. Note that $r = e^{i2m_1 H_r} = e^{in2\pi} = 1$. Figure 4.10 shows the fundamental and first and second harmonics of the standing waves.

4.4 REFLECTION AT THE GROUND SURFACE

Wave reflection at an upper level was considered in the previous section. However, wave ducting also requires reflection at the ground surface, which is taken up in this section. This problem is simpler than reflection at an elevated level because we do not consider a transmitted wave. The reflection will be complete in the sense that the total energy of the incident and reflected wave is conserved.

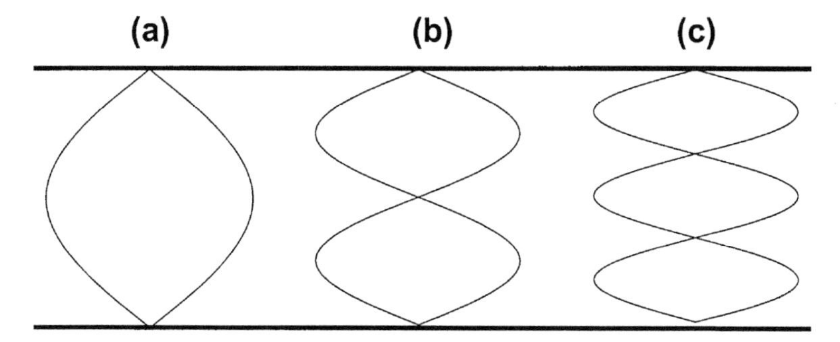

FIGURE 4.10 Standing waves or modes occur when a string is fixed at both ends; (a) fundamental or gravest mode; (b) first harmonic; (c) second harmonic.

In the linear theory, it is often assumed that the ground surface is horizontal, but in the real world horizontal surfaces of significant extent are rare. Gravity wave reflections from sloping surfaces are also important in oceanography (see, for example, Wunsch (1968), Phillips (1977), Thorpe (1987)).

To examine the effects of surface slope on wave reflection, we let the ground surface be inclined an angle γ to the horizontal as illustrated in Fig. 4.11. Because the reflected energy flux has the same value as the incident energy flux, it is conceptually easier to do the analysis in terms of group velocity rather than wavenumbers. Then, as shown in Fig. 4.11, the incident wave has downward directed group velocity $\vec{v}_{g,i}$ with upward directed wave vector $\vec{\kappa}_{\rm i}$, and the reflected wave has upward directed group velocity $\vec{v}_{g,r}$ and downward directed wave vector $\vec{\kappa}_{\rm r}$. Let $w_{\rm i}$ represent the vertical velocity perturbation with amplitude $A_{\rm i}$ associated with the incident wave, and let $w_{\rm r}$ represent the vertical velocity perturbation with amplitude $A_{\rm r}$ associated with the reflected wave. We also assume the background stratification and wind is constant, and accordingly the frequencies of the incident and reflected waves are equal. Then,

$$w_{\rm i}(x, z, t) = A_{\rm i}{\rm e}^{i(k_{\rm i}x+m_{\rm i}z-\omega t)} \tag{4.31}$$

and

$$w_{\rm r}(x, z, t) = A_{\rm r}{\rm e}^{i(k_{\rm r}x+m_{\rm r}z-\omega t)}, \tag{4.32}$$

where subscripts i and r refer to the incident and reflected waves respectively. The vertical wavenumber can be written as

$$m^2 = k^2\Delta^2, \tag{4.33}$$

where

$$\Delta^2 = \frac{N^2}{\Omega^2} - 1. \tag{4.34}$$

Then the vertical wavenumbers for the incident and reflected waves are

$$m_i^2 = k_i^2\Delta^2 \tag{4.35}$$

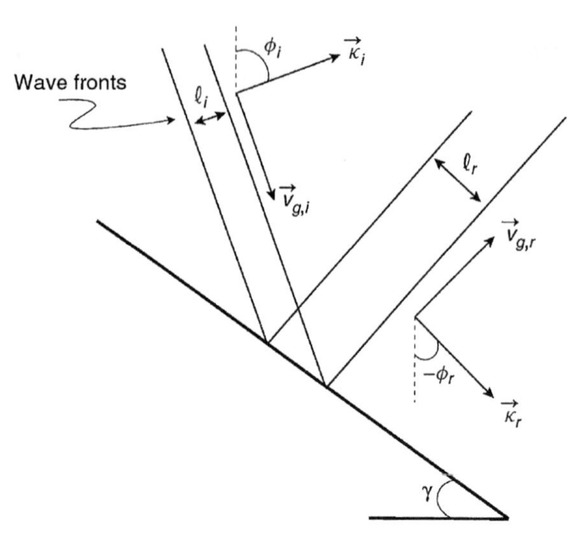

FIGURE 4.11 Wave reflection at a sloping ground surface.

and

$$m_{\mathrm{r}}^2 = k_{\mathrm{r}}^2 \Delta^2, \tag{4.36}$$

respectively. Since Δ is a constant, $\left|\frac{m_{\mathrm{i}}}{k_{\mathrm{i}}}\right| = \left|\frac{m_{\mathrm{r}}}{k_{\mathrm{r}}}\right|$ and we see that the magnitudes of the angles the wave vectors make with the horizontal or vertical directions are equal regardless of the inclination of the reflecting surface. This is a surprising result since intuition would suggest that the inclination of the surface would have a strong effect as, for example, reflection of a beam of light from a mirror. But the propagation of gravity waves is determined by N and ω which are constants. This is no less paradoxical than the fact that the speed of sound is independent of the speed of the sound source. However, we note that the vertical and horizontal wavenumbers of the incident and reflected waves are not equal, but rather their ratios are equal. Thus, wave reflection from the sloping surface results in change in wavelength, and a corresponding transfer of wave energy from one wavenumber to another.

As shown in Fig. 4.11, the angle the incident wave vector makes with the vertical direction is

$$\tan \phi_{\mathrm{i}} = \frac{k_{\mathrm{i}}}{m_{\mathrm{i}}} = \frac{1}{\Delta}. \tag{4.37}$$

The process of reflection implicitly assumes a reversal of direction of some kind. Either the horizontal or vertical wavenumber must change sign on reflection, but unless we know the inclination angle of the reflecting surface we cannot say which wavenumber changes sign. Accordingly, we can write

$$\tan\phi_{\mathrm{r}} = \frac{k_{\mathrm{r}}}{m_{\mathrm{r}}} = -\frac{1}{\Delta}. \tag{4.38}$$

From the linearized continuity equation (2.8) we have

$$ku + mw = 0, \tag{4.39}$$

so that

$$u_{\mathrm{i}} = -\frac{m_{\mathrm{i}}}{k_{\mathrm{i}}} w_{\mathrm{i}} = -\Delta w_{\mathrm{i}} \tag{4.40}$$

and

$$u_{\mathrm{r}} = -\frac{m_{\mathrm{r}}}{k_{\mathrm{r}}} w_{\mathrm{r}} = \Delta w_{\mathrm{r}}. \tag{4.41}$$

Using (4.31) and (4.32) along with (4.40) and (4.41) the perturbation velocities are

$$\vec{V}_{\mathrm{i}} = A_{\mathrm{i}}(-\Delta\hat{x} + \hat{z})e^{i(k_{\mathrm{i}}x + m_{\mathrm{i}}z - \omega t)} \tag{4.42}$$

$$\vec{V}_{\mathrm{r}} = A_{\mathrm{r}}(\Delta\hat{x} + \hat{z})e^{i(k_{\mathrm{r}}x + m_{\mathrm{r}}z - \omega t)}. \tag{4.43}$$

The velocity perturbation is the sum of the incident and reflected velocity perturbations so that

$$\vec{V} = \vec{V}_{\mathrm{i}} + \vec{V}_{\mathrm{r}}, \tag{4.44}$$

$$= A_{\mathrm{i}}(-\Delta\hat{x} + \hat{z})e^{i(k_{\mathrm{i}}x + m_{\mathrm{i}}z - \omega t)} + A_{\mathrm{r}}(\Delta\hat{x} + \hat{z})e^{i(k_{\mathrm{r}}x + m_{\mathrm{r}}z - \omega t)}. \tag{4.45}$$

Now let the elevation of the ground surface be given by $z_{\mathrm{g}} = -\alpha x$, where $\alpha = \tan\gamma$. At $z = z_{\mathrm{g}}$ we require

$$\vec{V} \bullet \vec{n} = 0, \tag{4.46}$$

where $\vec{n}$ is the outward drawn normal vector to the ground surface. For the configuration shown in Fig. 4.11, $z_{\mathrm{g}} = -\alpha x$ with x negative to the left of an arbitrary origin. The equation for the surface elevation is then

$$\phi_s = z + \alpha x \tag{4.47}$$

and

$$\vec{n} = \frac{\nabla\phi_s}{|\nabla\phi_s|} = \frac{\alpha\hat{x} + \hat{z}}{(\alpha^2 + 1)^{1/2}}. \tag{4.48}$$

Then at $z = z_{\mathrm{g}}$ we have from (4.46)

$$\left[A_1(-\Delta\hat{x} + \hat{z})e^{i(k_{\mathrm{i}}x - m_{\mathrm{i}}\alpha x - \omega t)} + A_{\mathrm{r}}(\Delta\hat{x} + \hat{z})e^{i(k_{\mathrm{r}}x - m_{\mathrm{r}}\alpha x - \omega t)}\right] \bullet (\alpha\hat{x} + \hat{z}) = 0. \tag{4.49}$$

Now if (4.49) is to hold for all times, then we require

$$k_{\mathrm{i}}x - m_{\mathrm{i}}\alpha x = k_{\mathrm{r}}x - m_{\mathrm{r}}\alpha x. \tag{4.50}$$

Solving (4.50) for k_r and using (4.37) and (4.38) we get

$$k_r = k_i \frac{1 - \alpha\Delta}{1 + \alpha\Delta}, \tag{4.51}$$

and we see that when the ground surface is sloped, the horizontal wavenumbers of the incident and reflected wave are not equal. Using (4.50) in (4.49) leads to

$$A_r = -A_i \frac{(1 - \alpha\Delta)}{(1 + \alpha\Delta)}. \tag{4.52}$$

If $\alpha\Delta < 1$ the reflection is forward, the vertical velocities w_i and w_r have opposite phase, and the reflected wavelength ℓ_r is greater than the incident wavelength ℓ_i as shown in Fig. 4.11. If $\alpha\Delta > 1$, then the reverse is true. If slope angle γ equals 0 or $\pi/2$, corresponding to a horizontal or vertical surface respectively, there is now change in the wave on reflection. A critical slope angle exists where $\gamma = -\phi_i$, *i.e.*, where $\alpha\Delta = 1$. In this case, the incident wave is reflected back on itself, and the waves constructively interfere resulting in a wave resonance.

4.5 WAVE DUCTS

We have seen that an internal gravity wave can be reflected at some level where m changes abruptly. If the distance between the ground surface and the reflecting level is a multiple of the vertical wavelength, then reflected and incident waves can constructively interfere to form a standing or cellular wave. When this happens the wave is trapped or ducted, and the ducting layer is a tuned wave guide.

4.5.1 THE PURE TEMPERATURE DUCT

The simplest wave duct is the temperature duct caused by a discontinuity in the temperature lapse rate. Figure 4.12 illustrates the problem. Below height H_d, N is constant and positive; above H_d, we set N equal to zero. We assume calm background winds; however, the analysis is unchanged if we have a constant background wind. We shall look for those waves that can exit in this duct. At the ground surface, $\hat{w}(k, 0) = 0$, and above H_d we require the wave amplitude to be bounded. The wave solutions to (2.50) are then

$$w_1(k, z) = A \sin\left[k\left(\frac{N^2}{\omega^2} - 1\right)^{1/2} z\right], \qquad 0 \leqslant z \leqslant H_d, \tag{4.53}$$

$$w_1(k, z) = B e^{k(z - H_d)} + C e^{-k(z - H_d)}, \qquad z > H_d. \tag{4.54}$$

From (4.54), we see that for decreasing wave amplitudes above H_d we must find those *eigenvalues* ω and k in (4.53) and (4.54) which result in $B \to 0$. At the interface, $z = H_d$, the dynamic and kinematic boundary conditions require that p and w be continuous. From the polarization (2.22) and (2.24), we see that

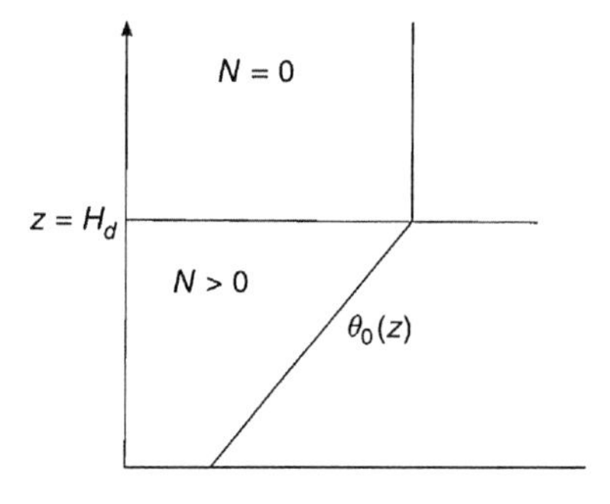

FIGURE 4.12 Schematic of a pure temperature duct of depth $H_{\rm d}$

$p_1 \propto \partial w_1/\partial z$ so that we require both w_1 and its derivative be continuous at $z = H_{\rm d}$. Thus,

$$A \sin\left[H_{\rm d}k\left(\frac{N^2}{\omega^2}-1\right)^{1/2}\right] = B + C, \tag{4.55}$$

$$Ak\left(\frac{N^2}{\omega^2}-1\right)^{1/2}\cos\left[H_{\rm d}k\left(\frac{N^2}{\omega^2}-1\right)^{1/2}\right] = kB - kC. \tag{4.56}$$

Now multiply (4.55) by k and add to (4.56) to get

$$\left(\frac{N^2}{\omega^2}-1\right)^{1/2}\cos\left[H_{\rm d}k\left(\frac{N^2}{\omega^2}-1\right)^{1/2}\right] + \sin\left[H_{\rm d}k\left(\frac{N^2}{\omega^2}-1\right)^{1/2}\right] = \frac{2B}{A}. \tag{4.57}$$

Then $B = 0$ only when

$$\left(\frac{N^2}{\omega^2}-1\right)^{1/2}\cos\left[H_{\rm d}k\left(\frac{N^2}{\omega^2}-1\right)^{1/2}\right] + \sin\left[H_{\rm d}k\left(\frac{N^2}{\omega^2}-1\right)^{1/2}\right] = 0. \tag{4.58}$$

Now let $k_* = H_{\rm d}k$ be a non-dimensional wavenumber, and let $\omega_* = \left(\frac{N^2}{\omega^2}-1\right)^{1/2}$ be a non-dimensional frequency. Then (4.58) becomes

$$\tan(k_*\omega_*) = -\omega_*. \tag{4.59}$$

Note that (4.59) is identical in form to (4.29). Indeed, if $N_2 = 0$ and $H_d m$ is an odd multiple of $\pi/2$, then (4.59) and (4.29) are identical.

The zeros of (4.59) are easily obtained numerically by iteration, and these eigenvalues (k_*, ω_*) are plotted in Fig. 4.13 which shows the dimensionless dispersion relation for the temperature duct. The slope of the curve in Fig. 4.13 is

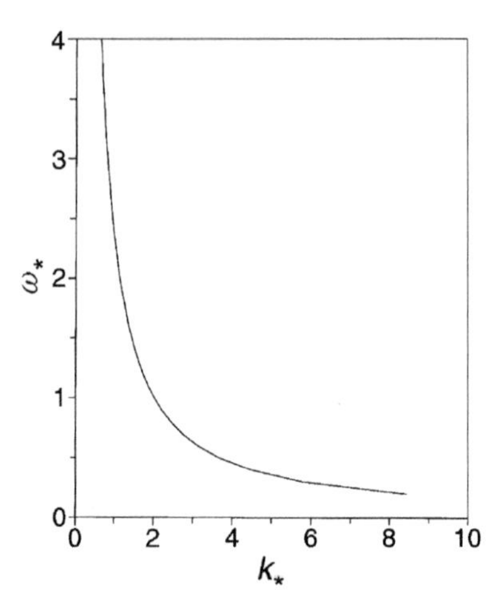

FIGURE 4.13 Dimensionless eigenvalues for the pure temperature duct with constant stratification.

the dimensionless horizontal group velocity, *i.e.*,

$$u_{g*} = \frac{\partial \omega_*}{\partial k_*} = \frac{N^2 \left(\frac{N^2}{\omega^2} - 1\right)^{-1/2}}{\omega^3 H_d} \frac{\partial \omega}{\partial k}, \tag{4.60}$$

where we have used the definitions of ω_* and k_*. If we let $\theta = k_* \omega_*$, we can define the functions $Y_1 = -\theta / k_*$ and $Y_2 = \tan(\theta)$, and the solutions to (4.59) occur where $Y_1 = Y_2$. Figure 4.14 shows plots of Y_1 and Y_2 as functions of θ for the case $k_* = 0.5$. The values of θ at the intersection of these curves are the solutions to (4.59), we see that there is a family of solutions because Y_2 is periodic with period π. Each solution corresponds to a vertical mode, *i.e.*, $Y_{2,0}$ is the *fundamental mode* and $Y_{2,1}$ is the *first harmonic*, *etc.* If we choose $A = 1$, $H_d = 300$ m and $N^2 = 0.003\ \mathrm{s}^{-2}$, then $k = 0.00167\ \mathrm{m}^{-1}$, $\omega_0 = 0.0144\ \mathrm{s}^{-1}$, and $\omega_1 = \omega_0/2$. The vertical velocities of the fundamental mode and first harmonic for this case are plotted in Fig. 4.15. The fundamental mode does not change sign between the ground surface and H_d, and the first harmonic changes sign once in this layer. Higher modes change sign accordingly, *i.e.*, two sign changes for the second harmonic, *etc.* Note also in Fig. 4.15, that the one-fourth of the vertical wavelength of the fundamental mode is less than the depth of the temperature duct. Above the ducting region the amplitudes of the modes decay exponentially, and there is no vertical transport of energy or stress. Figure 4.16 plots the phase and group velocities and frequency as functions of wavenumber for the fundamental mode for this problem. These curves were obtained by numerical solutions of the Taylor-Goldstein equation using techniques described in Chapter 11. Because $N = 0$ above the ducting region, the limits on phase speed are $0 < c < N/k$. If c is greater than N/k, the wave becomes evanescent. For the above values

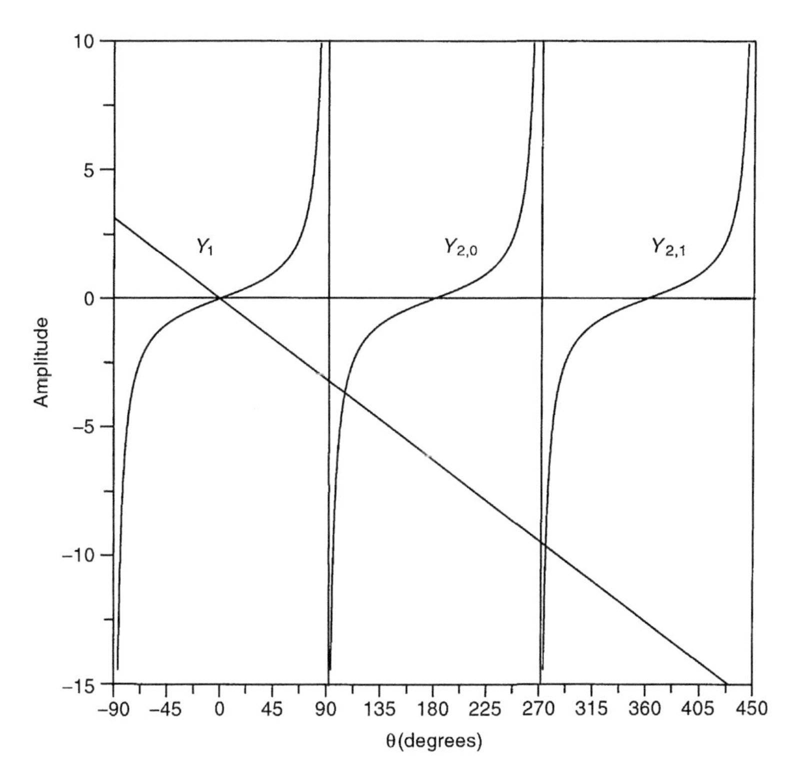

FIGURE 4.14 Graphical solutions to the pure temperature duct. $H_d = 300$ m; $N = 0.055\ \mathrm{s}^{-1}$.

of N and k, c must be less than about $32.8\,\mathrm{m\,s^{-1}}$. For $k = 0.00167\ \mathrm{m}^{-1}$ the phase velocities of the fundamental mode and first harmonic are calculated to be $c_0 = 8.6\,\mathrm{m\,s^{-1}}$ and $c_1 = 3.4\,\mathrm{m\,s^{-1}}$ respectively. The group velocities are calculated to be $u_{g,0} = 7.0\,\mathrm{m\,s^{-1}}$ and $u_{g,1} = 3.3\,\mathrm{m\,s^{-1}}$. The modes do not propagate along the duct with the same speed even though the modes have the same horizontal wavenumber. Thus, in any real situation one seldom sees more than one mode, and that mode is usually the fundamental or fastest traveling mode.

4.5.2 THE PURE WIND DUCT

We now examine the *wind duct* which is created by a wind jet in the background wind in the direction of wave propagation. This type of duct was first described by Chimonas and Hines (1986). It is a very common type of wave duct throughout the atmosphere. Consider the Taylor–Goldstein equation

$$\frac{\partial^2 \hat{w}}{\partial z^2} + Q^2 \hat{w} = 0, \tag{4.61}$$

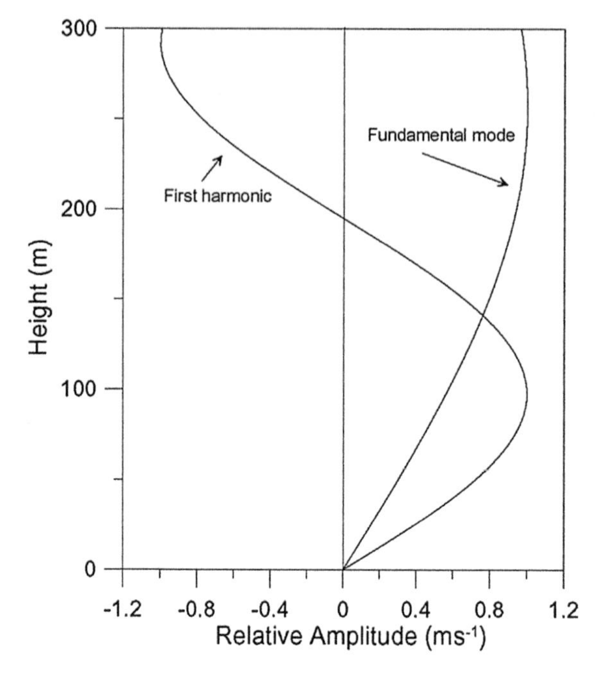

FIGURE 4.15 Fundamental mode and first harmonic for the solutions to the pure temperature duct shown in Fig. 4.12.

where

$$Q^2 = \frac{N^2}{(c - u_0)^2} + \frac{u_0''}{(c - u_0)} - k^2. \tag{4.62}$$

Figure 4.17 illustrates the wind duct. A velocity jet with maximum speed $u_{0,\max}$, defines a ducting region where $Q^2 > 0$. On either side of this region, $Q^2 < 0$ and the gravity waves are evanescent. Far from the jet, in the evanescent regions,

$$\frac{N^2}{(c - u_{0,\min})^2} < k^2. \tag{4.63}$$

A ducted wave will not form if a critical level is present. To avoid critical levels in our analysis, we consider only waves with phase speeds either less than $u_{0,\min}$ or greater than $u_{0,\max}$. Because the wind speed is not constant we must use numerical techniques. Let us assume that above the top level of the numerical model, z_{top}, N is constant and $u_0 = u_{0,\min}$. In the evanescent region above $z$top the wave solution is

$$w_1 = Ae^{q(z-z\text{top})} + Be^{-q(z-z\text{top})}, \tag{4.64}$$

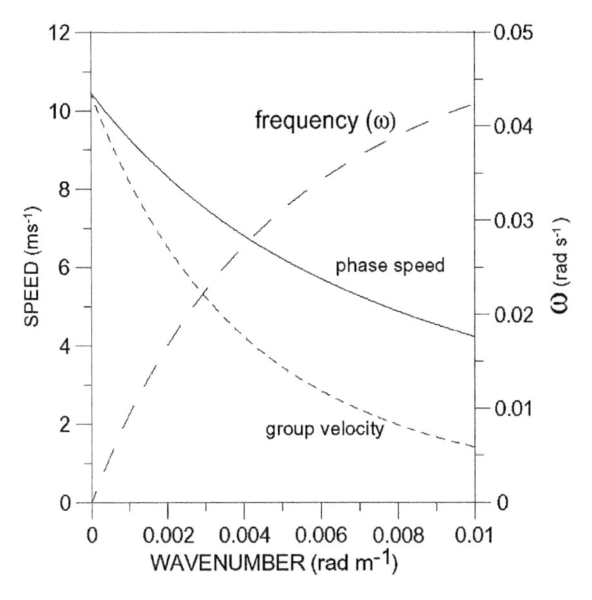

FIGURE 4.16 Phase speed, horizontal group velocity, and frequency as a function of wavenumber obtained from numerical solutions for the pure temperature duct in Fig. 4.12.

where

$$q = k^2 - \frac{N^2}{(c - u_{0,\min})^2}. \tag{4.65}$$

The kinematic and dynamic boundary conditions must be satisfied at $z = z_{\text{top}}$, and hence

$$w_1(k, c, z_{\text{top}}) = A + B, \tag{4.66}$$

and

$$\frac{\partial w_1(k, c, z_{\text{top}})}{\partial z} = qA - qB. \tag{4.67}$$

Solving for (4.66) and (4.67) for A gives

$$2A = w_1(k, c, z_top) + \frac{1}{q}\frac{\partial w_1(k, c, z_{\text{top}})}{\partial z}. \tag{4.68}$$

The numerical technique uses the so-called *shooting method*. We start at the bottom boundary where we assume $w_1(k, c, 0) = 0$ with a given value of k and an initial guess for the phase speed. We integrate (4.61) upward to z_{top} where we evaluate A using (4.68), and check for convergence, *i.e.*, $A < \epsilon$ where ϵ is a small number. If convergence is not reached, then we repeat the calculations starting at the bottom boundary, but with an updated value of phase speed. The process continues until $A \rightarrow \epsilon$. To determine all the characteristics of the duct, *i.e.*, phase and group

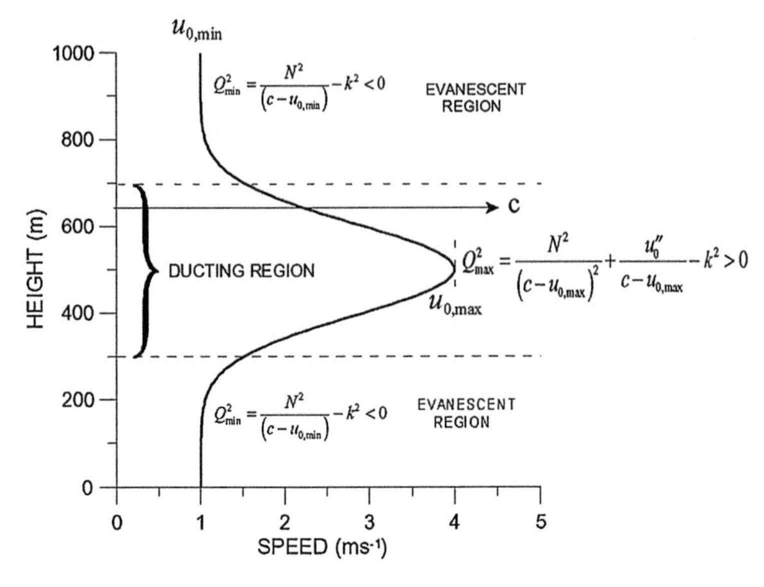

FIGURE 4.17 Schematic diagram of the wind duct.

velocities, the numerical procedure is again repeated but now with a new value of k.

As an example of the wind duct, consider the *low-level* jet shown in Fig. 4.18 along with the accompanying temperature profile. The analytical expressions for these curves are given by Mastrantonio, Einaudi, and Fua (1976), *i.e.*,

$$u_0(z) = \frac{u_s}{1 + \tanh(2)} \left[\frac{sz/H_j}{(s+1) + (z/H_j)^s} + \tanh\left(\frac{2z}{H_j}\right) \right] \tag{4.69}$$

and

$$T(z) = n \tanh(z/H_j) + T_0, \tag{4.70}$$

where u_s is the maximum wind speed at the jet; H_j is the height of the jet above the ground surface; s and n are shape parameters, and T_0 is the temperature at the ground surface. The maximum background wind speed is 5 m s^{-1} at 300 m above the ground surface. To avoid critical levels, we look for waves with phase speeds $c > u_s$. Note that at the speed jet we require

$$\frac{N^2 H_s}{(c - u_s)^2} > k^2. \tag{4.71}$$

We use the numerical methods described in Appendix A10.2 to calculate the phase speeds, horizontal group velocities and frequencies as functions of wavenumber. These results for the fundamental mode are plotted in Fig. 4.19. These curves have a markedly different appearance than those shown in Fig. 4.16. With decreasing

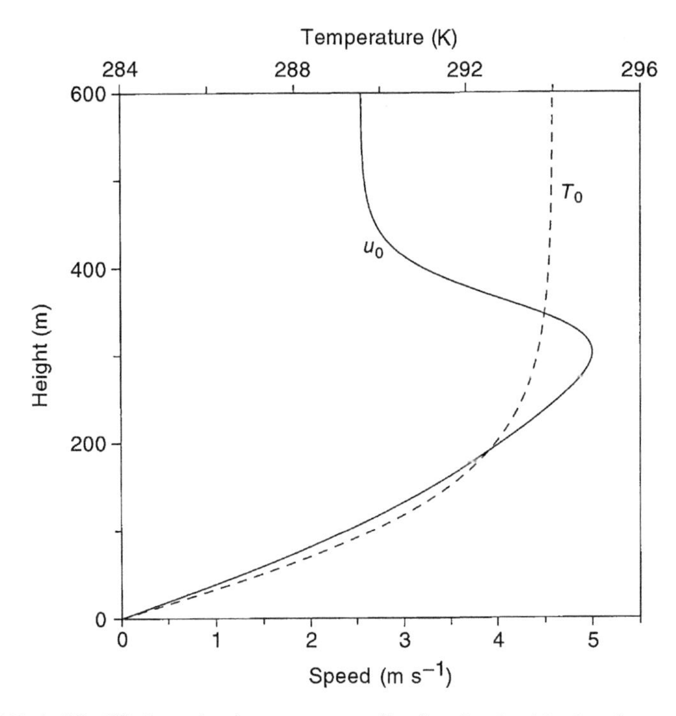

FIGURE 4.18 Wind speed and temperature profiles for a low-level jet, based on profiles from Mastrantonio et al. (1979).

wavenumber, the horizontal group velocity, $\partial\omega/\partial k$ initially increases slowly, but then rapidly decreases. Figure 4.20 shows the vertical variation of the perturbation velocities associated with the fundamental mode of the wind duct. Because the wave velocities are known only to within an undetermined constant, the values in Fig. 4.20 have been scaled so that the maximum magnitude of the horizontal velocity is $2\,\mathrm{m\,s^{-1}}$, which is a reasonable value in the boundary layer. The maximum value of u_1 occurs at the nose of the jet, and the secondary maximum at about 350 m occurs at the inflection point. The background wind is almost constant above 600 m (see Fig. 4.18); from Fig. 4.20 we see that u_1 and w_1 begin to decay exponentially at about 450 m. This marks the top of the ducting layer which extends to the ground surface.

4.5.3 WIND SPIRALS AND DUCTS

In the previous section, we said that the wind duct is very common. If there is a jet in the background wind speed, then there exists a possibility for a wave duct. Thus, the occurrence of wave ducts depends on the occurrence of speed jets, and whenever there is a turning of the background wind direction with height there is a possibility of a speed jet. The background wind speed seen by the wave is the

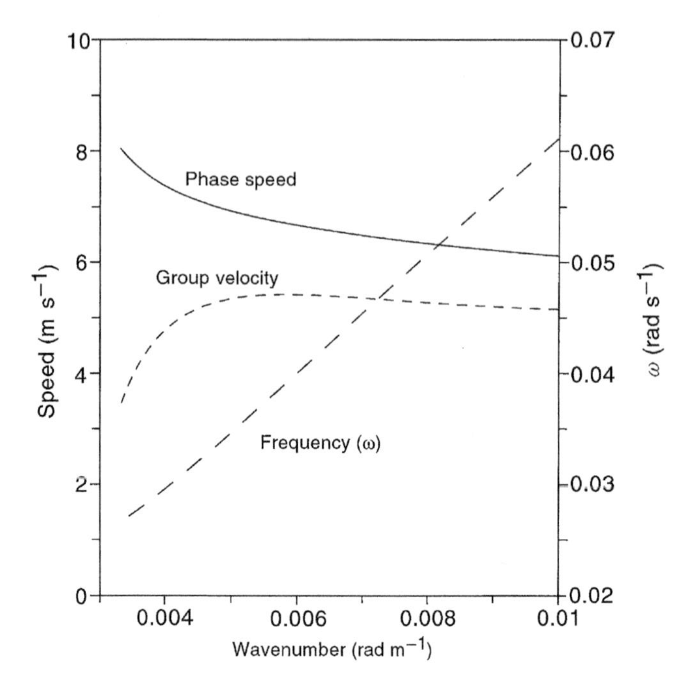

FIGURE 4.19 Phase speed, horizontal group velocity, and frequency as a function of wavenumber obtained from numerical solutions for a duct in a low-level jet defined by the profiles in Fig. 4.18.

projection of the horizontal wind vector onto the horizontal wave vector, *i.e.*,

$$u_0(z) = \vec{V}(z) \bullet \frac{\vec{k}}{|k|}. \tag{4.72}$$

In the lower troposphere and especially in the nighttime PBL, there can be significant turning of the wind with height. To illustrate, we consider the *Ekman wind spiral* which is derived in the Appendix A.9. Briefly, above the planetary boundary layer and away from the frictional force of the ground surface the wind tends to parallel the *isobars* or lines of constant atmospheric pressure. This type of flow, called *geostrophic wind*, is the result of a balance between the pressure gradient force which accelerates the fluid from high to low pressure and the Coriolis force which, in the Northern Hemisphere, accelerates the wind to the right of its direction of motion with a force that is proportional to wind speed. Thus, in the Northern Hemisphere the atmospheric flow is clockwise around a high pressure area and counter-clockwise around a low pressure area. At lower elevations, ground friction acts to slow the wind speed, and this decreases the magnitude of the Coriolis force. However, the pressure gradient force is unchanged, and as a result an imbalance of the forces occurs, and the winds are directed across the isobars towards lower pressure. The friction force grows as the ground surface is

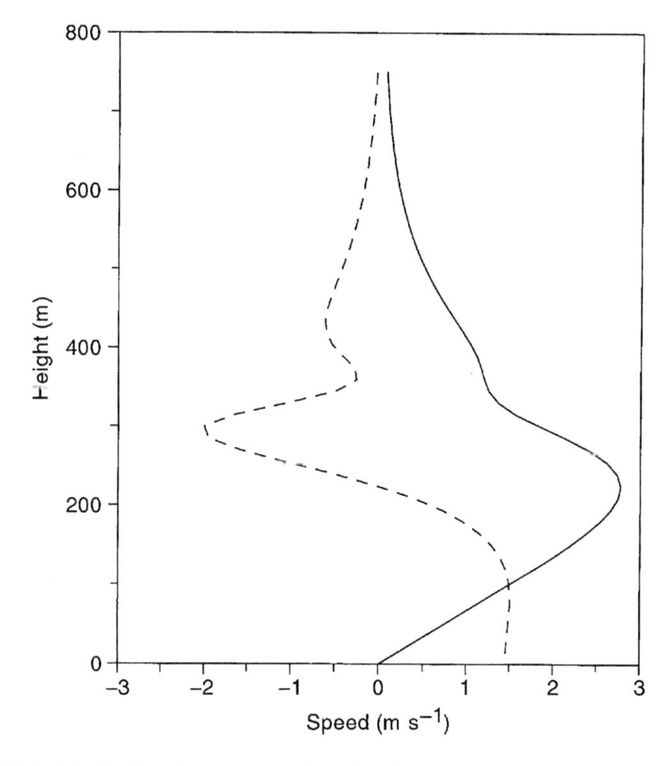

FIGURE 4.20 Profiles of wave perturbation velocities u_1 (dashed line) and w_1 (solid line) for the fundamental mode of the low-level jet shown in Fig. 4.18. The values were obtained from a numerical solution and were maximized so that the absolute magnitude of u_1 is 2 ms^{-1}.

approached, and as a result the turning of the wind toward lower pressure increases. The result of this interplay of forces is a flow that increases in speed and rotates to the right of its motion (in the Northern Hemisphere) with height forming a wind spiral. Ekman (1904) observed that the current on the ocean's surface was always to the right of the current at greater depth, and his analysis was later applied to the atmosphere. If our coordinate system is rotated so that the x-axis is in the direction to the geostrophic wind, U_g, then the orthogonal components of the background wind in the *Ekman layer* are

$$u_0(z) = U_g\left[1 - e^{-z/z_E}\cos(z/z_E)\right], \tag{4.73}$$

$$v_0(z) = U_g e^{-z/z_E}\sin(z/z_E), \tag{4.74}$$

where z_E is the depth of the Ekman layer, *i.e.*,

$$z_E = \sqrt{\frac{K}{f/2}}, \tag{4.75}$$

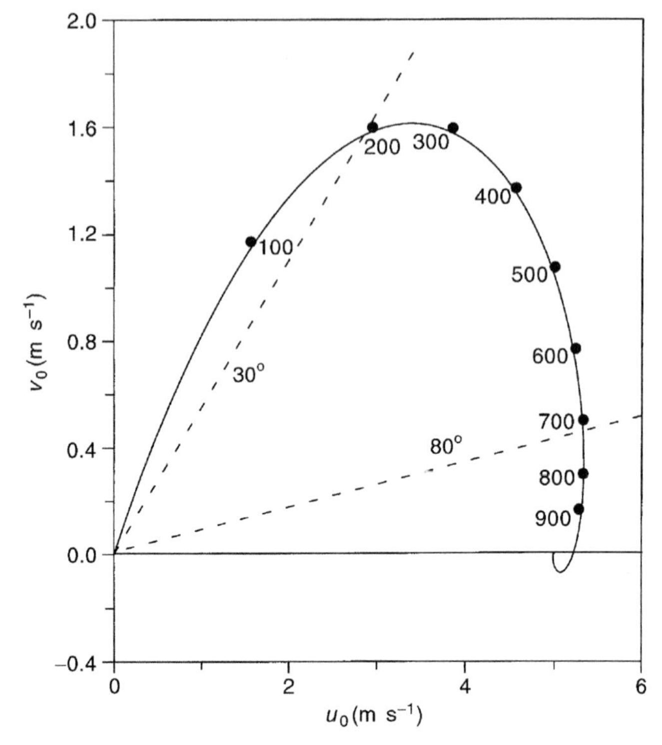

FIGURE 4.21 The Ekman spiral for the case $U_g = 5\ \mathrm{ms}^{-1}$, $f = 10^{-4}\ \mathrm{s}^{-1}$, and $K = 5\ \mathrm{m}^{2s^{-1}}$.

where K is the *eddy coefficient of friction* ($\mathrm{m}^2\mathrm{s}^{-1}$) and f is the *Coriolis parameter* (s^{-1}). When $z = \frac{\pi}{2} z_E$, $v_0 = 0$ and $u_0 = U_g$. For mid-latitudes, $f \approx 10^{-4}\mathrm{s}^{-1}$. Figure 4.21 shows the Ekman spiral for the case $U_g = 5\ \mathrm{ms}^{-1}$, $f = 10^{-4}\ \mathrm{s}^{-1}$, and K $= 5\ \mathrm{m}^{2s^{-1}}$; the dots along the curve mark the altitude. The vertical profiles along the vertical planes with orientations of 30° and 80° from north (shown in Fig. 4.21) are plotted in Fig. 4.22. Along the 30° plane a significant jet exists close to the ground; however, for the 80° plain only a week jet is seen. Thus, we see that whenever there is a turning of the wind with height, there is the possibility of a jet and ducted waves.

4.5.4 MOUNTAIN LEE WAVES

Mountain lee waves are ducted waves that extend long distances downwind from the mountain. We have see in Chapter 3 that a mountain generates a spectrum of vertically propagating gravity waves. For wave ducting to occur, downward wave reflection must occur, and this will happen at some height for some wave only if the vertical wavenumber is changing with height. When this change is abrupt, then substantial reflection can occur. Smith (1976) writes the solution of the Taylor-

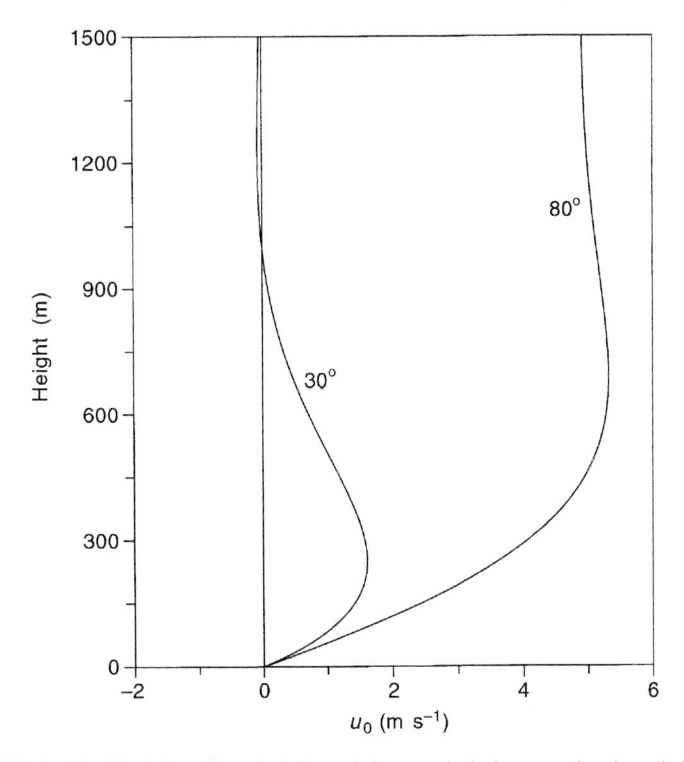

FIGURE 4.22 Vertical profiles of wind speed along vertical planes passing through the Ekman spiral shown in Fig. 4.15 at angles of 30° and 80° from north.

Goldstein equation for the mountain wave as

$$w(x,z) = \Re \int_{-\infty}^{\infty} iku_0 \hat{h}(k) \frac{\hat{w}(k,z)}{\hat{w}(k,0)} \mathrm{e}^{ikx} \mathrm{d}k. \tag{4.76}$$

For a particular value of the Scorer parameter, there will be a horizontal wavenumber $k = \pm k_n$, such that $\hat{w}(k,0) = 0$. The values of k_n result in pairs of singularities which are non-vanishing solutions to (4.76) as $x \to \infty$, *i.e.*, they form stationary ducted waves. Using the method of stationary phase (see Bender and Orszag 1999) Smith writes the asymptotic solution to (4.76) as

$$w(x,z) = \left(\frac{-2\pi u_0}{\overline{Z}_n^2} \right) \frac{\hat{w}(k_n,z)}{\overline{Z}_n} |\hat{h}(k_n)| \cos \left[k_n x - \tan^{-1} \frac{\hat{h}_{\mathrm{i}}(k_n)}{\hat{h}_{\mathrm{r}}(k_n)} \right], \tag{4.77}$$

where

$$\overline{Z}_n = \frac{\left[\int_0^\infty \hat{w}^2(k_n,z) \mathrm{d}z \right]^{\frac{1}{2}}}{\partial w / \partial z|_0}. \tag{4.78}$$

The parameter $\overline{Z_n}$ is a rough measure of the vertical distance between the ground surface and the altitude where most of the lee wave energy is located. Smith (1976) identifies the first term in parentheses in (4.77) as the sensitivity of the wave to excitation by the bottom topography. He argues that a thick neutral layer between the ground surface and the elevation where the wave energy is concentrated would lead to decreases lee wave amplitude. This is confirmed in (4.77) where the wave sensitivity is reduced by increasing $\overline{Z_n}$.

A somewhat mechanical approximation of lee wave calculations is to first assume a spectrum of horizontal wavenumbers with $k_{\text{max}} \sim 4b$, where b is the scale width of the mountain, the most energetic wavenumber. The phase speed on the lee wave is zero. Using known profiles of wind speed and temperature, we search modes that satisfy $w_1(x, 0) = 0$. If a mode cannot be found, then we must try a larger wavenumber, *etc.* There is as yet no verification of this method, but the approach seems novel.

It is clear that the presence of a lee wave depends on the atmospheric wind speed and temperature structures and the shape and size of the mountain. The preceding analysis was based on a two-dimensional isolated ridge. Extension to three dimensions, for example a Gaussian or bell-shaped hill, is straight forward but mathematically arduous. One reason for this is that lee waves cannot exist in a flow with constant wind speed and stratification. Thus, numerical evaluations of the wave equation are required. We note that lee waves from a two-dimensional mountain are manifest by the parallel cloud bands stretching downwind as seen in Fig. 4.23. Lee waves over three-dimensional objects often form *ship wave* clouds as shown in Fig. 4.3 or lenticular clouds as shown in Fig. 1.12:

PROBLEMS

1. A wave propagating at an angle β below the horizon is reflected off a surface which is inclined at angle γ above the horizontal. What angle does the reflected wave make with the reflecting surface?
2. Calculate the impedance, Z, for a wave with phase speed $5\,\text{m}\,\text{s}^{-1}$ and wavelength 10 km, where the background wind speed is constant at $8\ \text{ms}^{-1}$, the mean atmospheric temperature is 16 °C, and the atmospheric lapse rate is $7.5 \times 10^{-3}\ \text{kkm}^{-1}$.
3. Consider a wave-reflection level where the lapse rate changes from λ_1 to λ_2, the background wind speed changes from u_1 to u_2 and the horizontal wavelength is ℓ. Let and $\lambda_1 = 8\ \text{°C/km}, \lambda_2 = 5\ \text{°C/km}, u_1 = 3\ \text{ms}^{-1}, u_2 = 6\ \text{ms}^{-1}$ and phase speed $c = 15\ \text{ms}^{-1}$, where subscript 1 refers to the lower fluid and subscript 2 refers to the upper fluid. Find the value of ℓ such that the reflection coefficient, r = 0.75.
4. (a) Show that (4.26) is zero if r = 1. (b) Write the expression for (4.26) when $r = 0.90$.

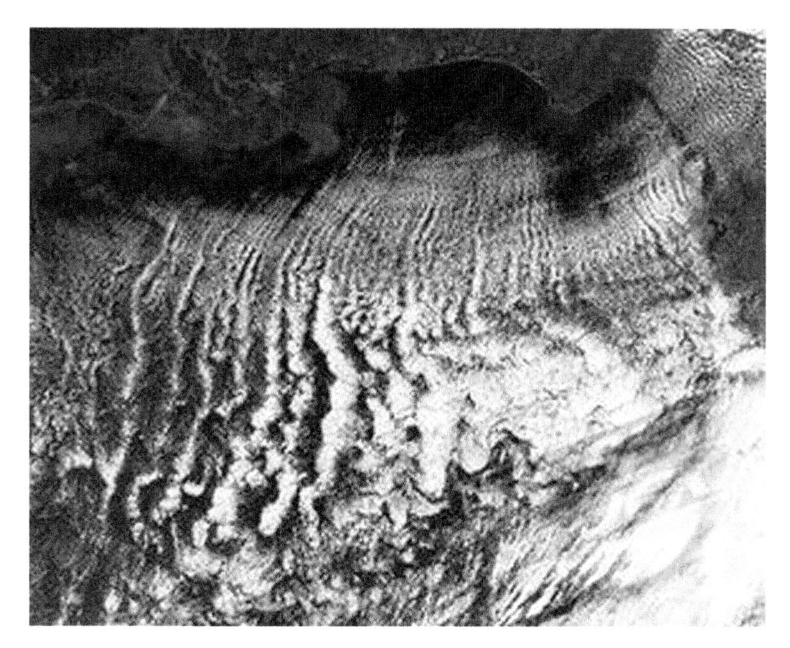

FIGURE 4.23 Mountain lee wave clouds. The clouds form at the crests of the standing waves.

5. Find the eigenvalues (c, k) for the fundamental and first harmonic waves in a pure temperature duct of depth 400 m and $N_1 = 0.025\ \mathrm{s}^{-1}$, and $N_2 = 0.0\ \mathrm{s}^{-1}$ for $z > 400$ m. Use *TEMP_DUCT_MODES.FOR* at *EURL\setminus CODES*.
6. Assume a Gaussian velocity jet given by:

$$u_0(z) = 1 + 10 \exp\left[-\frac{(z_{\max} - z)^2}{\zeta^2}\right],$$

where $z_{\max} = 1500$ m and $\xi = 200$ m. Let $N = 0.02\ \mathrm{s}^{-1}$, $c = 15\ \mathrm{ms}^{-1}$, and wavelength $\lambda_x = 10$ km. At what height above the ground surface does the wind-ducted wave switch from evanescent to propagating? What is the depth of the wind duct.
7. Using these profiles for N^2 and u_0 determine the ducting regions for modes with horizontal wavelengths of 1, 2, and 3 km; and corresponding horizontal phase speeds of 10.2, 13.0, and 15.8 ms^{-1}.

$$\begin{aligned} N^2 = &-0.0004465324411 + 1.847043985E - 005^{*}Z \\ &-1.384726947E - 007^{*}Z^2 + 4.006136214E - 010^{*}Z^3 \\ &-4.81009902E - 013^{*}Z^4 + 2.029982388E - 016^{*}Z^5, \end{aligned}$$

$$u_0 = -0.1121640968 + 0.07360626679^*Z - 0.0001948088975^*Z^2$$
$$-1.966057631E - 007^*Z^3 + 1.030246487E - 009^*Z^4$$
$$-7.854651945E - 013^*Z^5.$$

8. Consider a constant neutrally stable flow of 5 ms^{-1} over a surface corrugation with a horizontal wavelength 2 km and amplitude 400 m. What is the amplitude of the generated wave at 1000 m?
9. A gravity wave with intrinsic frequency $\Omega = 0.01\ \mathrm{s}^{-1}$ moving downward in a calm atmosphere with $N^2 = 0.02\ \mathrm{s}^{-1}$ is reflected off the side of a slope that is inclined 60° from the vertical direction. If the incident wave has a horizontal wavelength of 1000 m, what is the horizontal wavelength of the reflected wave?
10. Discuss the behavior of a gravity wave where $r = 1$ at a reflection level, but at the ground surface the amplitude of the reflected wave is 0.95 times the amplitude of the incident wave.

5

GRAVITY WAVE INSTABILITY AND TURBULENCE

5.1 INTRODUCTION

The major effect gravity waves have on the atmosphere is the transport of energy and momentum. Indeed, it is now believed that this transport maintains the general circulation and global climate. However, understanding the generation and propagation of gravity waves is only half of the problem. What remains is understanding how this energy and momentum are deposited in the atmosphere. One likely source is the breakdown of waves into turbulence. The turbulence then acts as a drag force on the local flow, and produces a vertical mixing of heat, momentum, water vapor, and trace gases. This hypothesis is supported by the fact that gravity waves and turbulence are often observed to exit simultaneously in almost all stably-stratified flows (see, for example, Turner, 1973; Caughey and Readings, 1975; Hunt, Kaimal, and Gaynor , 1985; Finnigan, 1988; Nappo and Johansson, 1999). Indeed, gravity wave instability appears to be the ultimate source of turbulence in the stable atmosphere (see, for example, Lindzen, 1981; Blumen *et al.*, 2001; Chimonas,

2002; Poulos *et al.*, 2001; Schöch *et al.*, 2004; Duck and Whiteway, 2005; Koch *et al.*, 2005; Yague *et al.*, 2007; Fritts *et al.*, 2009; Pham and Sarkar, 2010). We must make a distinction between *turbulence* and *instability*. It has been said that while we cannot define turbulence we know it when we see it. Turbulence is a subject unto itself with unique characteristics not related to linear gravity waves. While there is a theory of linear waves, there is not as yet a theory of turbulence. However, within the context of linear theory we can define the conditions for the onset of flow instability, which under certain conditions can evolve into a state of motion called turbulence (Drazin, 2002).

5.2 PARCEL EXCHANGE ANALYSIS OF FLOW STABILITY

We begin with a somewhat mechanistic examination of stability. The method is based on the stability analyses performed by Chandrasekhar (1961) and Ludlum (1967), and generalized by Hines (1971) to include the effects of stratification along slanting surfaces. The method looks at the total energy of two fluid parcels before and after a virtual vertical exchange of their positions. If the total energy of the system increases after the exchange, then work has been done on the system, and the system is stable. However, if the total energy decreases after the exchange, then work has been done by the system, and the system is unstable.

Figure 5.1 illustrates the flow under study. Fluid A has constant density ρ_A and uniform speed u_A; fluid B has constant density ρ_B and uniform speed u_B. Fluid parcels A and B are initially at heights z_A and z_B respectively above an arbitrary reference level. An impermeable interface lies between the two fluids. The initial potential energy, P_{I}, and kinetic energy, K_{I}, of the system are

$$P_{\mathrm{I}} = g(\rho_A z_A + \rho_B z_B) \tag{5.1}$$

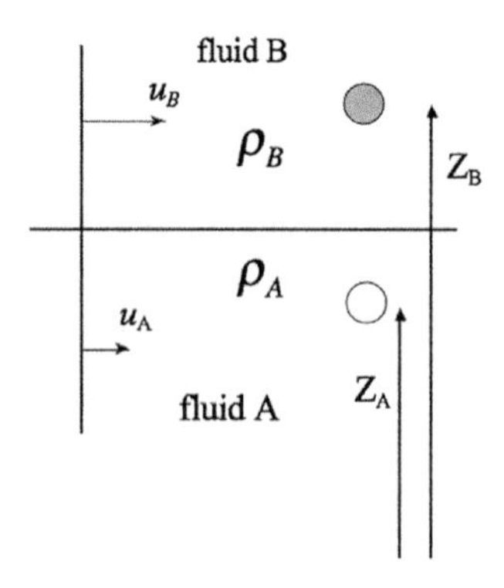

FIGURE 5.1 Fluid parcels in their initial configuration.

and

$$K_{\mathrm{I}} = \frac{1}{2}(\rho_A u_A^2 + \rho_B u_B^2). \tag{5.2}$$

We now exchange the positions of the parcels. The exchange process is adiabatic, and we require conservation of mass and momentum. Conservation of mass requires

$$\rho_A + \rho_B = \rho_A' + \rho_B', \tag{5.3}$$

where the primes indicate values after parcel exchange. Conservation of momentum gives

$$\rho_A u_A + \rho_B u_B = (\rho_A' + \rho_B')u_{\mathrm{f}} = (\rho_A + \rho_B)u_{\mathrm{f}}, \tag{5.4}$$

where u_{f} is an effective final speed. Solving (5.4) for u_{f} gives

$$u_{\mathrm{f}} = \frac{\rho_A u_A + \rho_B u_B}{\rho_A + \rho_B}, \tag{5.5}$$

which is the mass-weighted average velocity of the system. The final potential and kinetic energies are

$$P_{\mathrm{F}} = g(\rho_A' z_B + \rho_B' z_A) \tag{5.6}$$

and

$$K_{\mathrm{F}} = \frac{1}{2}(\rho_A + \rho_B)u_{\mathrm{f}}^2. \tag{5.7}$$

Using (5.5) in (5.7) gives

$$K_{\mathrm{F}} = \frac{1}{2}\frac{(\rho_A u_A + \rho_B u_B)^2}{\rho_A + \rho_B}. \tag{5.8}$$

The change in kinetic energy as a result of the exchange is

$$K_{\mathrm{F}} - K_{\mathrm{I}} = -\frac{\rho_A \rho_B}{2(\rho_A + \rho_B)}(u_A - u_B)^2. \tag{5.9}$$

If we assume that the density of the parcels is constant during the exchange, then the change in potential energy is

$$P_{\mathrm{F}} - P_{\mathrm{I}} = g(\rho_A - \rho_B)(z_B - z_A). \tag{5.10}$$

The work, W, done on the system to cause the change in parcel positions is,

$$\begin{aligned} W &= (K_{\mathrm{F}} + P_{\mathrm{F}}) - (K_{\mathrm{I}} + P_{\mathrm{I}}) \\ &= (P_{\mathrm{F}} - P_{\mathrm{I}}) + (K_{\mathrm{F}} - K_{\mathrm{I}}). \end{aligned} \tag{5.11}$$

Using (5.9) and (5.10) in (5.11) gives

$$W = g(\rho_A - \rho_B)(z_B - z_A) - \frac{1}{2}\frac{\rho_A \rho_B}{(\rho_A + \rho_B)}(u_A - u_B)^2. \tag{5.12}$$

If $W > 0$, then work has been done on the system, and the total energy of the system has increased. If $W < 0$, then work has been done by the system, and the total energy has decreased. Expressed more simply

$$W > 0 \Rightarrow \text{stable flow},$$
$$W < 0 \Rightarrow \text{unstable flow}.$$

The first term on the right-hand side of (5.12) is called the *buoyancy term.* It represents the change in the potential energy of the system, which can be either positive or negative depending on the relative sizes of ρ_A and ρ_B. If the buoyancy term is positive, it acts to make the system more stable. If the buoyancy term is negative, then it acts to make the system less stable. The second term on the right-hand side of (5.12) represents the change of kinetic energy of the system, which is always negative. The change in kinetic energy (5.9) always acts to decrease the total energy of the system. Thus, changes in kinetic energy act to create instability; it is called the *production term.*

Consider the case when $\rho_A = \rho_B = \rho$ and $u_A \neq u_B$. Then,

$$W = -\frac{1}{4}\rho(u_A - u_B)^2, \tag{5.13}$$

and we see that the system is always unstable. We refer to such a system as being *dynamically* unstable. If $\rho_A > \rho_B$ and $u_A = u_B$, then from (5.12) we see that if $z_B > z_A$ the system is *convectively* stable, but if $z_B < z_A$, the system is *convectively* unstable. In the latter case, fluid of greater density is forced to lie over fluid of lesser density.

We consider next the case of height-varying density and speed. Instead of exchanging two fluid parcels, we vertically move a fluid parcel from its initial location at point A to nearby location B. Expanding ρ_B and u_B to first order about the point A, we have

$$\rho_B = \rho_A + \left.\frac{\partial \rho}{\partial z}\right|_{z_A} (z_B - z_A), \tag{5.14}$$

$$u_B = u_A + \left.\frac{\partial u}{\partial z}\right|_{z_A} (z_B - z_A). \tag{5.15}$$

Substitution of (5.14) and (5.15) into (5.12) leads to

$$W = -g\frac{\partial \rho}{\partial z}(z_B - z_A)^2 - \frac{1}{2}\frac{\rho_A^2 + \rho_A\frac{\partial \rho}{\partial z}(z_B - z_A)}{2\rho_A + \frac{\partial \rho}{\partial z}(z_B - z_A)}\left[\frac{\partial u}{\partial z}(z_B - z_A)\right]^2. \tag{5.16}$$

Note that

$$\frac{\rho_A^2 + \rho_A\frac{\partial \rho}{\partial z}(z_B - z_A)}{2\rho_A + \frac{\partial \rho}{\partial z}(z_B - z_A)} \approx \frac{\rho_A}{2}, \tag{5.17}$$

so that (5.16) becomes

$$W = \rho_0 N^2 (z_B - z_A)^2 - \frac{\rho_0}{4}\left(\frac{\partial u}{\partial z}\right)^2 (z_B - z_A)^2, \tag{5.18}$$

where we have set $\rho_A = \rho_0$ and used (1.47). Equation (5.18) can be further simplified, *i.e.*,

$$W = \rho_0 (z_B - z_A)^2 \left(\frac{\partial u}{\partial z}\right)^2 (Ri - 0.25), \tag{5.19}$$

where

$$Ri = \frac{N^2}{\left(\frac{\partial u}{\partial z}\right)^2}. \tag{5.20}$$

Ri is called the *Richardson number* and is a fundamental parameter in turbulence theory (see, for example, Gossard and Hooke, 1975; Stull, 1988; Foken, 2008). It represents the ratio of the production or suppression of turbulence kinetic energy by buoyant forces to the mechanical production of turbulence kinetic energy by shear forces. We see that the criterion for stability is the value of Ri relative to 1/4. If $Ri > 1/4$, the flow is stable, *i.e.*, if the fluid is disturbed by a small vertical displacement, it will return to its initial configuration or harmonically oscillate about that state. The condition for *convective instability* is $Ri < 0$. *Dynamic instability* occurs over the range $0 \leqslant Ri \leqslant 0.25$.

5.3 WAVE INSTABILITY

5.3.1 INTRODUCTION

In this section, we examine the stability of stratified shear flows. The analysis is straightforward. A small linear wave-like disturbance, is introduced into the flow, and we examine the temporal evolution of the perturbation. If the amplitude of the disturbance remains bounded, then we say that the flow is stable. However, if the amplitude grows without limit, then an instability exists. Note that *linear stability analysis* can trace the evolution of a flow only to the onset of an instability. Subsequent flow development is not accessible to the linear theory. Also note that a stability analysis usually gives the conditions for the flow to be stable to small disturbances. The fact that a flow may be unstable to small disturbances is not a sufficient condition for instability. For example, in the linear analysis we usually assume a frictionless flow; however, all real fluids have some viscosity, and this can act to stabilize a flow (see, for example, Fritts and Geller, 1976; Thorpe, 1981; Hooke and Jones, 1986). The linear theory also neglects surface tension effects, and these may be strong enough to stabilize a shear flow between immiscible fluids. It is also possible that the disturbance wavenumber or phase speed required for instability is not present. Thus, we see that there are many reasons why a flow may be stable even though a linear stability analysis indicates it is unstable.

In the previous section we examined the stability of a stratified flow by evaluating the total energy of a two-particle system before and after an adiabatic exchange of fluid parcel positions. The analysis was kinematic. We did not look at how the instability developed. Indeed, the question of time dependence did not enter the analysis. It would therefore seem that the results are perhaps not very helpful; however, it did show that the source of flow instability is the velocity shear. We saw that the Richardson number provides a means for estimating flow stability, and that there is a range of values of Ri between 0 and 0.25 where the flow can be dynamically unstable. In this section, we will examine the dynamic stability of stratified flows, and examine the effects of this shear instability on the generation of gravity waves. We begin with the most simple flow situation, and proceed to more complicated flows.

5.3.2 KELVIN–HELMHOLTZ INSTABILITY

The simplest model of velocity shear is the Helmholtz profile illustrated in Fig. 5.2. We shall see that the Helmholtz profile is potentially unstable. The flow is constant but different in each of two semi-infinite planes, and the change in background speed occurs sharply across a thin immiscible interface. Thus, the vertical shear can be considered large or infinite. We wish to examine the stability of the interface to small wave perturbations. We assume the flow in each half plane is inviscid, incompressible, and irrotational, and neutrally stratified ($N = 0$). Further, we assume that the background pressures are in hydrostatic balance. Because the Brunt–Väisälä frequency is zero in each layer, we can anticipate that waves in each layer will be evanescent. The linearized Eulier equations are

$$\frac{\partial u_1}{\partial t} + u_0 \frac{\partial u_1}{\partial x} = -\frac{1}{\rho_0} \frac{\partial p_1}{\partial x}, \tag{5.21}$$

$$\frac{\partial w_1}{\partial t} + u_0 \frac{\partial w_1}{\partial x} = -\frac{1}{\rho_0} \frac{\partial p_1}{\partial z}, \tag{5.22}$$

$$\frac{\partial u_1}{\partial x} + \frac{\partial w_1}{\partial z} = 0. \tag{5.23}$$

We assume that as $z \rightarrow \pm\infty$, the perturbations go to zero. At the interface, $z = 0$, we apply the dynamic and kinematic boundary conditions. We assume wave solutions of the form

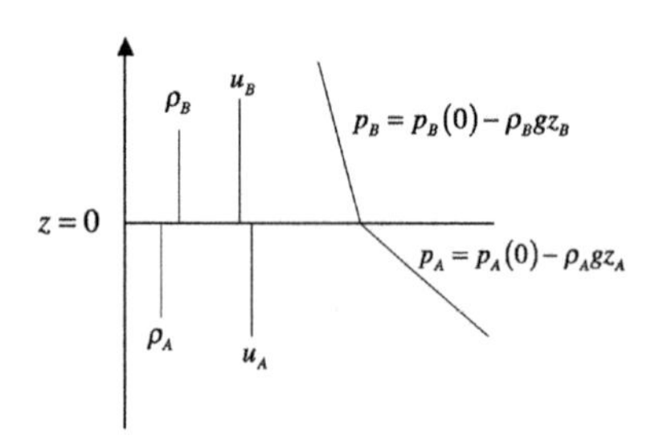

FIGURE 5.2 The Helmholtz profile. Similar to Fig. 5.1, but now he pressures in each fluid are considered.

$$u_1 = \tilde{u}(z)\mathrm{e}^{i(kx-\omega t)}, \tag{5.24}$$
$$w_1 = \tilde{w}(z)\mathrm{e}^{i(kx-\omega t)}, \tag{5.25}$$
$$p_1 = \tilde{p}(z)\mathrm{e}^{i(kx-\omega t)}, \tag{5.26}$$

so that the polarization equations in the upper fluid are

$$\Omega_B \tilde{u}_B = \frac{k\tilde{p}_B}{\rho_B}, \tag{5.27}$$
$$\Omega_B \tilde{w}_B = \frac{1}{\rho_B}\frac{\partial p_B}{\partial z} \tag{5.28}$$
$$\tilde{u}_B = \frac{i}{k}\frac{\partial \tilde{w}_B}{\partial z} \tag{5.29}$$

where Ω is the intrinsic frequency, and ρ_B is the density of fluid B. A similar set of equations are written for the lower fluid. The dynamic boundary condition requires that the pressure be continuous across the interface between two fluids. Thus, we need to solve (5.27)–(5.29) in terms of $\tilde{p}_B$. This leads to

$$\frac{\partial^2 \tilde{p}_B}{\partial z^2} - k^2 \tilde{p}_B = 0. \tag{5.30}$$

Then, using (5.30) the equations for the vertical variations in the upper fluid become

$$\tilde{p}_B = \tilde{p}_B(0)\mathrm{e}^{-kz}, \tag{5.31}$$
$$\tilde{u}_B = \frac{k\tilde{p}_B(0)}{\Omega_B \rho_B}\mathrm{e}^{-kz}, \tag{5.32}$$
$$\tilde{w}_B = \frac{ik\tilde{p}_B(0)}{\Omega_B \rho_B}\mathrm{e}^{-kz}. \tag{5.33}$$

In the lower fluid we have

$$\tilde{p}_A = \tilde{p}_A(0)\mathrm{e}^{kz}, \tag{5.34}$$
$$\tilde{u}_A = \frac{k\tilde{p}_A(0)}{\Omega_A \rho_A}\mathrm{e}^{kz}, \tag{5.35}$$
$$\tilde{w}_A = \frac{ik\tilde{p}_A(0)}{\Omega_A \rho_A}\mathrm{e}^{kz}. \tag{5.36}$$

The disturbed interface is illustrated in Fig. 5.3. The vertical wave displacements of the interface from its mean height are

$$\zeta_B = \tilde{\zeta}_B\,\mathrm{e}^{-kz+i(kx-\omega t)}, \tag{5.37}$$
$$\zeta_A = \tilde{\zeta}_A\,\mathrm{e}^{kz+i(kx-\omega t)}. \tag{5.38}$$

The interface displacements are related to the vertical velocity perturbations, *i.e.*,

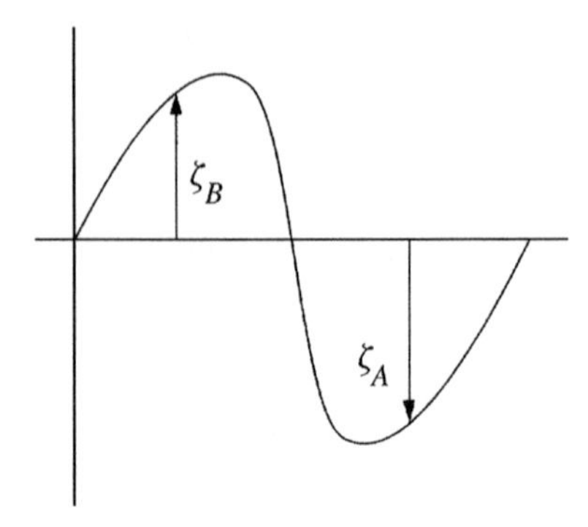

FIGURE 5.3 The disturbed interface between two immiscible fluids.

$$\frac{D\zeta_B}{Dt} = \frac{\partial \zeta_B}{\partial t} + u_B \frac{\partial \zeta_B}{\partial x} = w_{1,B}, \tag{5.39}$$

$$\frac{D\zeta_A}{Dt} = \frac{\partial \zeta_A}{\partial t} + u_A \frac{\partial \zeta_A}{\partial x} = w_{1,A}. \tag{5.40}$$

Using (5.25), (5.33), and (5.37) in (5.39) and solving for $\tilde{p}_B$ gives

$$\tilde{p}_B = -\frac{\Omega_B^2 \rho_B}{k} \tilde{\zeta}_B. \tag{5.41}$$

Likewise, in the lower layer we have

$$\tilde{p}_A = \frac{\Omega_A^2 \rho_A}{k} \tilde{\zeta}_A. \tag{5.42}$$

The dynamic boundary condition requires that the total pressure be continuous across the interface, *i.e.*,

$$p_{0,B} + p_{1,B} = p_{0,A} + p_{1,A}, \tag{5.43}$$

where $p_{0,A}$ and $p_{0,B}$ are the background pressures in lower and upper fluids respectively. The background pressures are given by the hydrostatic equation (1.43). Assuming that the mean interface height is zero and using (5.37) and (5.38) we get

$$p_{0,B} = p_{0,B}(0) - \rho_B g \zeta_B = \rho_B g \tilde{\zeta}_B e^{i(kx-\omega t)} \tag{5.44}$$

$$p_{0,A} = p_{0,A}(0) - \rho_A g \zeta_A = \rho_A g \zeta_A e^{i(kx-\omega t)}. \tag{5.45}$$

Note that the background pressures are neither steady nor horizontally uniform, but change with the height of the interface. This is a consequence of the hydrostatic condition. Using (5.41) and (5.42) the perturbation pressures are

$$p_{1,B} = -\frac{\Omega_B^2 \rho_B}{k} \tilde{\zeta}_B \, e^{i(kx-\omega t)}, \tag{5.46}$$

$$p_{1,A} = \frac{\Omega_A^2 \rho_A}{k} \tilde{\zeta}_A \, e^{i(kx-\omega t)}. \tag{5.47}$$

Using (5.44)–(5.47) in (5.43) and noting that the background pressures in each fluid must be equal at $z = 0$ gives

$$-\rho_B g\tilde{\zeta}_B - \frac{\Omega_B^2 \rho_B \tilde{\zeta}_B}{k} = -\rho_A g\tilde{\zeta}_A + \frac{\Omega_A^2 \rho_A \tilde{\zeta}_A}{k}. \tag{5.48}$$

The kinematic boundary condition requires the two fluids remain in contact at the interface; therefore $\tilde{\zeta}_B = \tilde{\zeta}_A$ and (5.48) becomes

$$(\rho_A - \rho_B)gk = \Omega_A^2 \rho_A + \Omega_B^2 \rho_B. \tag{5.49}$$

Using the definition of intrinsic frequency, $\Omega = \omega - kc$, (5.49) becomes

$$\frac{\omega}{k} = \frac{\rho_A u_A + \rho_B u_B}{\rho_A + \rho_B} \pm \left[\frac{g}{k}\frac{(\rho_A - \rho_B)}{(\rho_A + \rho_B)} - \frac{\rho_A \rho_B (u_B - u_A)^2}{(\rho_A + \rho_B)^2}\right]^{1/2}, \tag{5.50}$$

where u_B and u_A are the mean speeds in the upper and lower fluids respectively. Equation (5.50) is our final result; it represents the complex horizontal phase speeds of interface disturbances. The first term on the right-hand side represents the mass-weighted average horizontal speed across the interface. The first term in braces represents the buoyancy term, and the second term in braces represents the production term. As in Section 5.2, if $\rho_A > \rho_B$, the stratification is stable, and the buoyant force acts to suppress the growth of the disturbances. If $\rho_A < \rho_B$, then a convective instability results. The second term in braces is production term. In their effects, the buoyancy and production terms are similar to the potential and kinetic energy terms appearing in (5.12). If the buoyancy term is greater than the production term, ω is real and the interface displacements are given by

$$\zeta = \zeta(0)\mathrm{e}^{-kz}\cos(kx - \omega t), \quad z > 0, \tag{5.51}$$

$$\zeta = \zeta(0)\mathrm{e}^{kz}\cos(kx - \omega t), \quad z < 0. \tag{5.52}$$

Note that the size of the disturbances decrease exponentially away from the interface. These are evanescent waves as illustrated in Fig. 2.5. Although amplitude of the disturbance decreases exponentially with distance from the interface, the flow is stable. Thus, we can associate a real ω with a stable flow.

Now consider the case when $u_A = u_B = u_0$. Then, (5.50) becomes

$$\frac{\omega}{k} = u_0 \pm \left[\frac{g(\rho_A - \rho_B)}{k(\rho_A + \rho_B)}\right]^{1/2}. \tag{5.53}$$

Note that if $u_0 = 0$, then (5.53) becomes the phase speed (Turner, 1973),

$$c^2 = \frac{g}{k}\frac{(\rho_A - \rho_B)}{(\rho_A + \rho_B)}. \tag{5.54}$$

When $\rho_A > \rho_B$, c is real and the interface is convectively stable, but and if $\rho_A < \rho_B$, then c is complex, *i.e.*, $c = c_R + ic_I$ where c_R is the real phase speed and c_I is the imaginary phase speed. The interface is now convectively unstable. To see this, rewrite (5.51) in exponential form,

$$\zeta = \zeta(0)\mathrm{e}^{-kz}\,\mathrm{e}^{ik(x-ct)} = \zeta(0)\mathrm{e}^{-kz}\,\mathrm{e}^{ik[x-(c_R+ic_I)t]}. \tag{5.55}$$

Expanding the exponential factors in (5.55) gives

$$\zeta = \zeta(0)e^{k(c_I t - z)} \cos[k(x - c_R t)]. \tag{5.56}$$

We see that the perturbation ζ grows exponentially with time, and the growth rate is c_I. If $\rho_B \ll \rho_A$ and $u_0 = 0$, then (5.53) reduces to

$$\omega = \sqrt{kg}, \tag{5.57}$$

which is the dispersion relation for waves on deep water. When $\rho_A = \rho_B$ and if $u_A \neq u_B$, then

$$\frac{\omega}{k} = \frac{1}{2}(u_A + u_B) \pm i\frac{1}{2}|u_A - u_B|. \tag{5.58}$$

In this case the interface is always unstable because it will grow as $e^{0.5|u_A - u_B|kt}$.

We have said that the Helmholtz profile is potentially unstable, and this is confirmed in (5.50) where we see that for wavenumbers greater than some critical wavenumber, k_{crit}, where

$$k_{crit} = \frac{g}{\rho_A \rho_B} \frac{\rho_A^2 - \rho_B^2}{(u_A - u_B)^2}, \tag{5.59}$$

the production term will be greater than the buoyancy term, and ω will be complex. Now the flow is unstable and the growth rate is ω_I. Recall that in Section 5.1 we conjectured that instability would somehow lead to turbulence. However, we see now that instabilities of the Helmholtz profile take the form of exponential growth of the interface displacements as illustrated in Fig. 5.4 taken from Rosenhead (1931). The flow in each layer pushes the crests of the interface waves in their respective forward directions causing them to "roll up" and limit vertical growth. In nature, Kelvin–Helmholtz instabilities often manifest themselves as *billow clouds* as shown in Fig. 5.5. Kelvin–Helmholtz instabilities can be seen on radar as shown in Fig. 5.6 taken from Chimonas (1999), and on sodar as shown in Fig. 5.7 taken from Zamora (1983). Several videos of Kelvin–Helmholtz instability and billow clouds are found in EURL−Ch5−KH-WAVES. Note that billow clouds will appear only if condensation is possible. These waves rarely last more than a few minutes even if the flow is steady. This is because the growth of wave amplitude and the effects of velocity shear will cause the heavier (more dense) fluid in the lower layer to be displaced above the lighter (less dense) fluid of the upper layer. A process not unlike the parcel exchange discussed in Section 5.2. When this happens, a convective instability develops, and the waves breakdown resulting in turbulence. We see that the wave–turbulence process begins with a dynamic instability which leads to convective instability, and then turbulence. It is often assumed that turbulence occurs when $Ri < 0.25$. However, this is not accurate because dynamic stability is not the same thing as turbulence. Indeed, as we shall see in the next sections that a linear analysis of shear flows shows that a sufficient condition for stability against infinitesimal disturbances is that $Ri > 0.25$. Einaudi and Lalas (1974) discovered new properties of Kelvin–Helmholtz waves and included the effects of condensation.

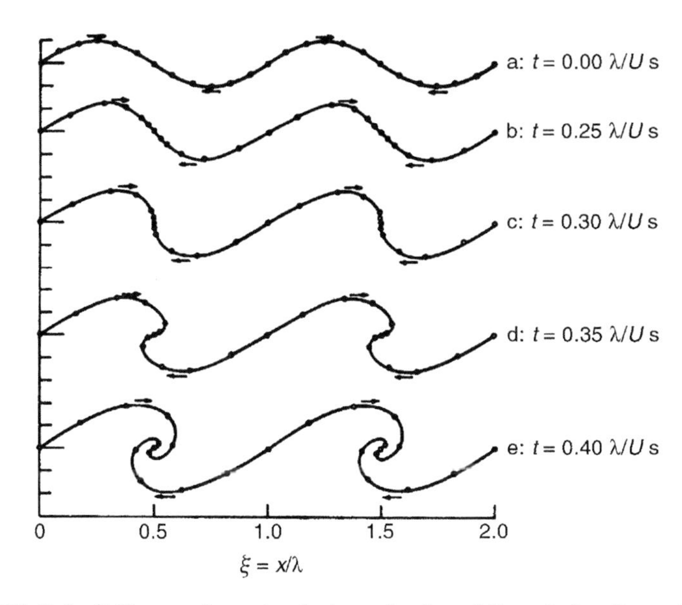

FIGURE 5.4 Rolling up of a vortex sheet as a function of dimensionless time and distance. (Taken from Rosenhead (1931).)

FIGURE 5.5 Billow clouds above Frösön, Sweden. (Photo by Roger Gyllenhammar.)

5.3.3 THE STABILITY OF SHEAR FLOWS

We have seen in the previous section that flow stability is controlled by the imaginary component of the wave frequency or equivalently the imaginary component of the phase speed. Linear stability analysis determines under what flow conditions the phase speeds of small disturbances remains real. In the

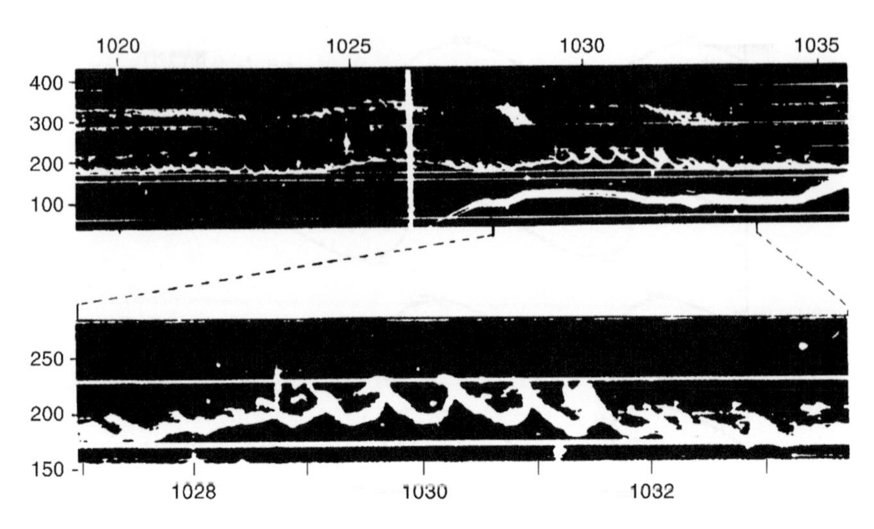

FIGURE 5.6 Kelvin–Helmholtz waves observed with an FM/CW radar on June 25, 1970 at the Navy Research Facility, San Diego, California by E.E. Gossard, J.R. Richter, and D.R. Jensen. The horizontal axis is time (PST) and the vertical axis is height in meters. (Taken from Chimonas (1999).)

previous section we saw that for an infinite shear layer, for example as shown in Fig. 5.2, the phase speed of small disturbances remains real if the wavenumber of the disturbance is less than some critical wavenumber. For wavenumbers greater than the critical value, the flow can become unstable. In real flows, this instability can lead to turbulence which can mix the two fluids producing a third layer in which the velocity shear is now finite. In this section we examine the stability of finite shear layers.

5.3.3.1 Inflection Point Instability

The simplest problem involving instability of a shear flow is the case of a bounded homogeneous flow. This type of flow was first analyzed by Rayleigh (1945), and is sometimes called the *Rayleigh instability* problem. We consider a two-dimensional constant density flow with vertical boundaries at heights z_1 and z_2 such that $w_1 = 0$ at these heights. One can imagine a flow in an enclosed tank; however, z_1 and z_2 can be infinitely large if the flow perturbations, *i.e.*, the streamline displacements, vanish somewhere. This means that w_1 decreases exponentially above some height. The Taylor–Goldstein equation (2.42) takes the form

$$\frac{\partial^2 w_1}{\partial z^2} + \left[\frac{u_0'}{(c - u_0)} - k^2\right] w_1 = 0, \tag{5.60}$$

where primes denote differentiation with respect to z and the perturbations are due to waves. We assume c is complex. We now multiply (5.60) by w_1^*, the complex

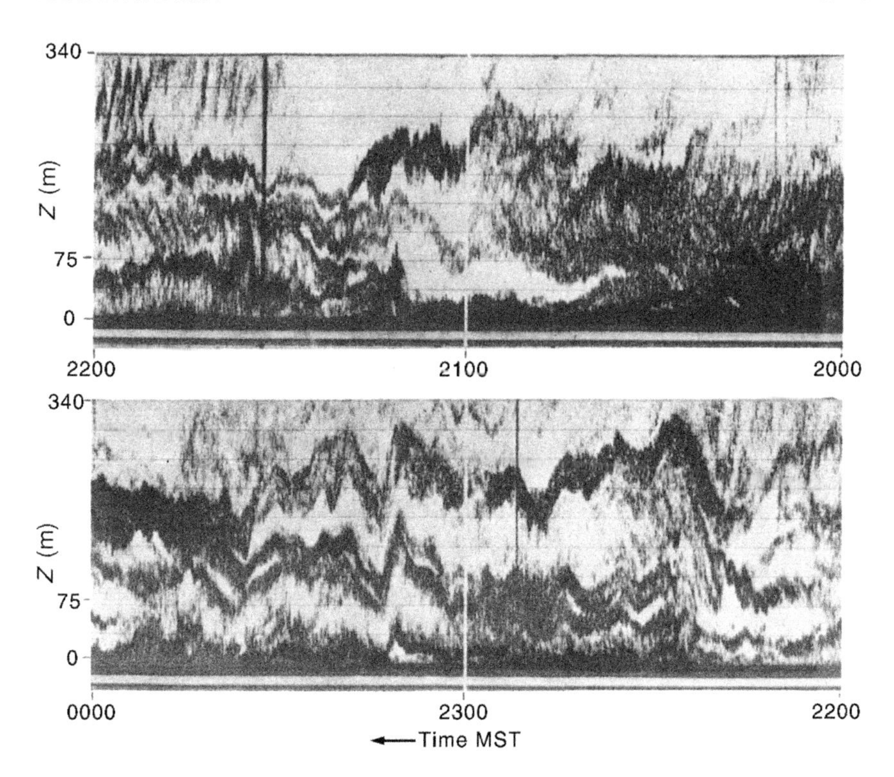

FIGURE 5.7 Kelvin–Helmholtz waves observed near Boulder, Colorado with an acoustic sounder. (Taken from Zamora (1983).)

conjugate of w_1, and integrate from z_1 to z_2 to get

$$-\int_{z_1}^{z_2}\left|\frac{\partial w_1}{\partial z}\right|^2 \mathrm{d}z - \int_{z_1}^{z_2} k^2|w_1|^2\mathrm{d}z = \int_{z_1}^{z_2}\frac{u_0'}{c_{\mathrm{R}}+ic_{\mathrm{I}}-u_0}|w_1|^2\,\mathrm{d}z, \quad (5.61)$$

where we have integrated the first term by parts, *i.e.*,

$$\int_{z_1}^{z_2} w_1^*\frac{\partial^2 w_1}{\partial z^2}\mathrm{d}z = -\int_{z_1}^{z_2}\left|\frac{\partial w_1}{\partial z}\right|^2 \mathrm{d}z. \quad (5.62)$$

The imaginary part of (5.61) is

$$c_{\mathrm{I}}\int_{z_1}^{z_2}\frac{u_0'}{|c-u_0|^2}|w_1|^2\mathrm{d}z = 0. \quad (5.63)$$

For instability, $c_{\mathrm{I}} \neq 0$, and if (5.63) is to hold, and if $|w_1| \neq 0$, then $u_0' = 0$ somewhere between z_1 and z_2. If u_0' is continuous, then for instability there must be at least one inflection point between the two vertical boundaries. Inflection points are often associated with velocity jets. At an inflection point, the velocity shear has a maximum, and we can expect a high probability of shear instability. However, the presence of a stable stratification, which is absent in the above analysis, can suppress the destabilizing effect of the wind shear. This is why instabilities are not always found near inflection points in stratified flows.

5.3.3.2 Instability of Stratified Shear Flows

The introduction of a stable stratification into the shear flow problem results in an immediate complication; however, the complication is more algebraic than physical. The problem was solved in detail by Miles (1961) who developed 10 theorems relating to stratified shear flow. Miles' Theorem 10 states that sufficient conditions for stability of a heterogeneous shear flow are the mean wind speed is nonzero and the Richardson number is greater than 0.25. Howard (1961) analyzed this problem from a more simple perspective. Here, we shall follow Howard's analysis, but the interested reader should look at Miles paper for details and insights not contained in Howard (1961). We assume a two-dimensional frictionless, irrotational flow, and make the Boussinesq approximation. Let ζ_1 be the first-order vertical displacement of a fluid particle from its equilibrium position. This displacement is given by

$$\zeta_1(x, z, t) = F(z)e^{ik(x-ct)}, \tag{5.64}$$

where k is real, and phase speed c can be a complex number. The boundary conditions are that $F(z) = 0$ at $z = z_1$ and $z = z_2$ which can recede to $\pm\infty$. Then noting that

$$\frac{D\zeta_1}{Dt} = \frac{\partial\zeta_1}{\partial t} + u_0\frac{\partial\zeta_1}{\partial x} = w_1, \tag{5.65}$$

and using (5.65) in (2.6)–(2.8) we obtain

$$u_1 = [(c - u_0)\zeta_1]', \tag{5.66}$$

$$w_1 = -ik(c - u_0)\zeta_1, \tag{5.67}$$

$$p_1 = \rho_0(c - u_0)^2\zeta_1' + p_0. \tag{5.68}$$

Equations (5.66)–(5.68) can be combined along with the hydrostatic equation, and solved for ζ to give

$$[\rho_0(c - u_0)^2 F']' + \rho_0[N^2 - k^2(c - u_0)^2]F = 0. \tag{5.69}$$

If we define a new variable as $G = W^{1/2}F$ where $W = u_0 - c$, then (5.69) can be written as

$$(\rho_0 WG')' - \left[\frac{1}{2}(\rho_0 u_0')' + k^2\rho_0 W + \rho_0 w^{-1}\left(\frac{1}{4}u_0'^2 - N^2\right)\right]G = 0. \tag{5.70}$$

If we now multiply (5.70) by G^* and integrate from z_1 to z_2 just as we did in the previous section, we get

$$\int \rho_0 W[|G'|^2+k^2|G|^2]+\int \frac{1}{2}(\rho_0 u_0')'|G|^2+\int \rho_0\left[\frac{1}{4}{u_0'}^2 - N^2\right] W^* \left|\frac{G}{W}\right|^2 = 0, \tag{5.71}$$

where it is understood that the limits on the integrals are still z_1 and z_2, and we have dropped dz in the integrals. From (5.64) it is clear that for an instability to occur, the imaginary part of c must be positive, *i.e.*, $c_{\rm I} > 0$. The imaginary part of (5.71) is

$$\int \rho_0[|G'|^2 + k^2|G|^2] + \int \rho_0 \left[N^2 - \frac{1}{4}{u_0'}^2\right]\left|\frac{G}{W}\right|^2 = 0, \tag{5.72}$$

where we have divided (5.72) by W. The only way (5.72) can be true is if the term $N^2 - \frac{1}{4}{u_0'}^2$ changes sign somewhere between z_1 and z_2. Thus a necessary but not sufficient condition for instability is that

$$\frac{N^2}{{u_0'}^2} = Ri < \frac{1}{4}. \tag{5.73}$$

We have formally proved the conjecture in Section 6.1, *i.e.*, for flow instability a necessary condition is that $Ri < \frac{1}{4}$.

When an instability develops in a stratified flow, a gravity wave is excited, but what is the phase speed of this wave? Howard (1961) elegantly answered this question. The method is straightforward. We multiply (5.69) by F^* and integrate over height to get

$$\int \rho_0(u_0 - c)^2[|F'|^2 + k^2|F|^2] - \int \rho_0 N^2|F|^2 = 0. \tag{5.74}$$

Now separate (5.74) into real and imaginary parts to get

$$\int [(u_0 - c_{\rm R})^2 - c_{\rm I}^2][|F'|^2 + k^2|F|^2] - \int N^2|F|^2 = 0 \tag{5.75}$$

and

$$2ic_{\rm I}\int (u_0 - c_{\rm R})[|F'|^2 + k^2|F|^2] = 0. \tag{5.76}$$

If (5.76) is to hold, then $[u_0(z) - c_{\rm R}]$ must change sign at least once between the limits z_1 and z_2. This means that $c_{\rm R}$ must lie between the minimum and maximum values of u_0. Thus, the wave speed cannot be arbitrary. When $u_0 = c$, the terms $\frac{1}{c-u_0}$ in the Taylor–Goldstein equation become singular points, *i.e.*, they are not defined. The requirement that $[u_0(z) - c_{\rm R}]$ must change sign between z_1 and z_2 means that there must exist a level where $u_0 - c_{\rm R} = 0$. Such a level is called a *critical level*. Now let $Q = |F'|^2+k^2|F|^2$, and assuming $c_{\rm I} > 0$, (5.75) and (5.76) become

$$\int (u_0^2 - 2u_0c_{\rm R} + c_{\rm R}^2 - c_{\rm I}^2)Q = \int N^2|F|^2 \tag{5.77}$$

and

$$\int u_0 Q = c_R \int Q. \tag{5.78}$$

Now if we multiply (5.78) by $2c_R$, and use this result in (5.77) the resulting equations become

$$\int u_0^2 Q = (c_R^2 + c_I^2) \int Q + \int N^2 |F|^2 \tag{5.79}$$

and

$$\int u_0 Q = c_R \int Q, \tag{5.80}$$

respectively. Now suppose that between z_1 and z_2, u_0 lies between $u_{\min}$ and $u_{\max}$. Then using (5.79) and (5.80) and some algebra we have

$$\begin{aligned}
0 &\leqslant \int (u_0 - u_{\min})(u_{\max} - u_0) \\
&= \int u_0^2 Q - (u_{\min} + u_{\max}) \int u_0 Q + u_{\min} u_{\max} \int Q \\
&= \left[c_R^2 + c_I^2 - (u_{\min} + u_{\max}) c_R + u_{\min} u_{\max}\right] \int Q + \int N^2 |F|^2 \\
&= \left\{ \left[c_R - \frac{1}{2}(u_{\min} + u_{\max})\right]^2 + c_I^2 - \left[\frac{1}{2}(u_{\min} - u_{\max})\right]^2 \right\} \\
&\quad \times \int Q + \int N^2 |F|^2.
\end{aligned} \tag{5.81}$$

Because $Q > 0$, the only way (5.81) can be true is if

$$\left[c_R - \frac{1}{2}(u_{\min} + u_{\max})\right]^2 + c_I^2 \leqslant \left[\frac{1}{2}(u_{\min} - u_{\max})\right]^2. \tag{5.82}$$

Equation (5.82) says that the complex phase speed of any unstable mode must lie inside a semi-circle of radius $\frac{1}{2}(u_{\max} - u_{\min})$ centered at the point $c_I = 0$, $c_R = \frac{1}{2}(u_{\min} + u_{\max})$. This famous result is known as *Howard's semi-circle theorem* as illustrated in Fig. 5.8 illustrates the result. The semi-circle theorem tells us that if a gravity wave is excited by wind shear, then there must be a critical level for that wave since $u_{\min} < c < u_{\max}$. If, for example, one observes a wave with a speed of, say, 30 m s^{-1}, then we can conjecture that the wave was generated at that height where the horizontal wind component is in the direction of horizontal wave vector and that $Ri < 0.25$ at that point.

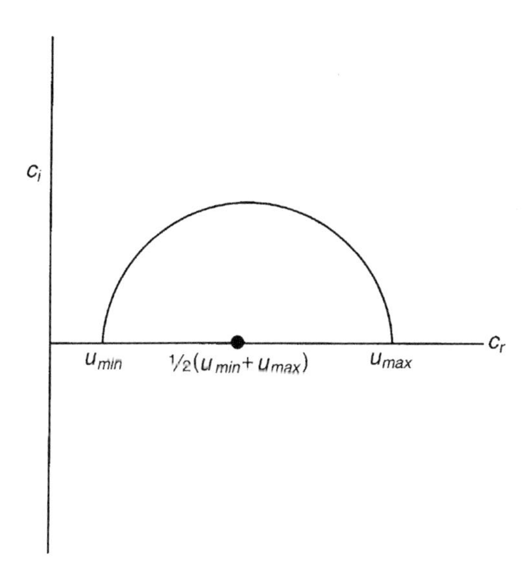

FIGURE 5.8 Schematic illustration of Howard's semi-circle theorem. If a gravity wave is generated by wind shear, then the phase speed of the wave must lie between some minimum and maximum wind speed.

5.4 THE CRITICAL LEVEL

We have noted that when $c = u_0$ the buoyancy and curvature terms in the Taylor–Goldstein equation $\rightarrow \infty$. In this section, we examine the behavior of a gravity wave as it approaches a critical level. The analysis follows that given by Booker and Bretherton (1967). If we ignore the terms involving the scale height, H_s, (2.43) is

$$\frac{d^2\hat{w}}{dz^2} + \left[\frac{N^2}{(c - u_0)^2} + \frac{u_0'}{(c - u_0)} - k^2\right]\hat{w} = 0. \tag{5.83}$$

There is a singularity (a second order pole) in (5.83) where $c - u_0 = 0$, and we must in some way account for this if we are to solve the equation over a reasonable depth of the atmosphere. As we shall see, critical levels are quite common in the atmosphere.

Let the wave approach the critical level from above as illustrated in Fig. 5.9. At a distance ζ above z_c we expand the background wind speed to second order[1] to get

$$u_0(\zeta + z_c) = c + \left.\frac{du_0}{dz}\right|_{z_c} \zeta + + \frac{1}{2}\left.\frac{d^2u_0}{dz^2}\right|_{z_c} \zeta^2 + \cdots . \tag{5.84}$$

[1] In some cases it might be necessary to expand to third or even higher orders.

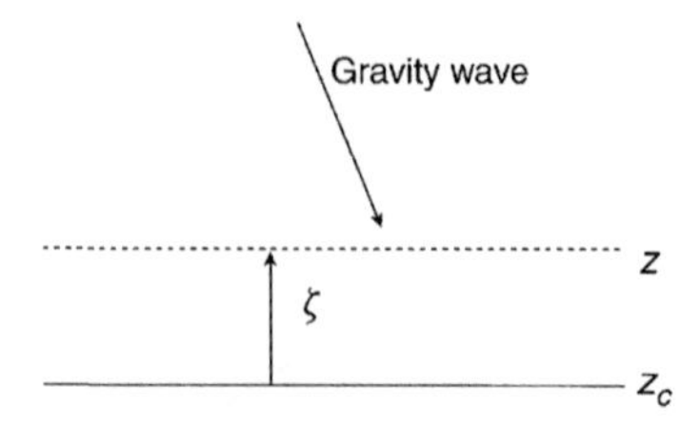

FIGURE 5.9 Schematic of a gravity wave approaching a critical level from above.

It is easily shown that

$$\frac{1}{c - u_0(\zeta + z_c)} \approx -\frac{1 - \frac{1}{2}\frac{a_2}{a_1}\zeta}{a_1\zeta}, \tag{5.85}$$

where

$$a_1 = \left.\frac{du_0}{dz}\right|_{z_c} \tag{5.86}$$

and

$$a_2 = \left.\frac{d^2u_0}{dz^2}\right|_{z_c}. \tag{5.87}$$

Using (5.85)–(5.87) in (5.83) leads to

$$\frac{dw_1}{d\zeta^2} + \left[\frac{R_c}{\zeta^2} - \frac{\alpha}{\zeta} + \gamma\right] w_1 = 0, \tag{5.88}$$

where

$$R_c = \frac{N^2}{a_1^2} \tag{5.89}$$

is the Richardson number at $z = z_c$,

$$\alpha = \frac{a_2}{a_1}(R_c + 1), \tag{5.90}$$

and

$$\gamma = \frac{1}{2}\frac{a_2^2}{a_1^2}\left(\frac{R_c}{2} + 1\right) - k^2. \tag{5.91}$$

We see in (5.88) that a singularity $\zeta \rightarrow 0$, remains in the Taylor–Goldstein equation.

We now introduce the Frobenius expansion (see, for example, Bender and Orszag, 1999),

$$w_1(k, \zeta + z_c) = \sum_n C_n \zeta^{n+\lambda}, \tag{5.92}$$

which will be used to provide a series solution over the critical level. Using (5.92) in (5.88) and ordering terms according to powers of ζ gives

$$[\lambda(\lambda-1)+R_c]C_0\zeta^{\lambda-2}+\{[\lambda(\lambda+1)+R_c]C_1-\alpha C_0\}\zeta^{\lambda-1}$$
$$+\{[(\lambda+1)(\lambda+2)+R_c]C_2-\alpha C_1+\gamma C_0\}\zeta^{\lambda}+\cdots=0. \quad (5.93)$$

For a non-trivial solution to (5.93), each coefficient of ζ must be zero. This leads to the recursion relations

$$C_1=\frac{\alpha}{\lambda(\lambda+1)+R_c}C_0 \quad (5.94)$$

and

$$C_2=\frac{\alpha^2[\lambda(\lambda+1)]^{-1}-\gamma}{(\lambda+1)(\lambda+2)+R_c}C_0. \quad (5.95)$$

The indicial equation is obtained from the first term in (5.93)

$$\lambda(\lambda-1)+R_c=0 \quad (5.96)$$

with solution

$$\lambda=\frac{1}{2}\pm i\mu, \quad (5.97)$$

where

$$\mu=[R_c-0.25]^{1/2}. \quad (5.98)$$

From (5.97) we see that the powers in the Frobenius expansion are complex numbers. We now expand (5.92) to second order to get the equations

$$w_1(k,z_c+\zeta)=C_0^+\zeta^{\lambda}\psi_1(\zeta)+C_0^-\zeta^{\lambda^*}\psi_1^*(\zeta), \quad (5.99)$$

and

$$\frac{dw_1}{dz}(k,z_c+\zeta)=C_0^+\zeta^{\lambda}\psi_2(\zeta)+C_0^-\zeta^{\lambda^*}\psi_2^*(\zeta), \quad (5.100)$$

where

$$\psi_1(\zeta)=1+\frac{C_1}{C_0}\zeta+\frac{C_2}{C_0}\zeta^2. \quad (5.101)$$

and

$$\psi_2(\zeta)=\lambda\zeta^{-1}+\frac{C_1}{C_0}(\lambda+1)+\frac{C_2}{C_0}(\lambda+2)\zeta. \quad (5.102)$$

In the above equations, C_0^+ refers to values using $\lambda=1/2+i\mu$, and C_0^- refers to values using $\lambda=1/2-i\mu$. Below the critical level where $\zeta<0$ we have

$$w_1(k,z_c-\zeta)=C_0^+(-\zeta)^{\lambda}\psi_1(-\zeta)+C_0^-(-\zeta)^{\lambda^*}\psi_1^*(-\zeta), \quad (5.103)$$
$$\frac{dw_1}{dz}(k,z_c-\zeta)=-C_0^+(-\zeta)^{\lambda}\psi_2(-\zeta)-C_0^-(-\zeta)^{\lambda^*}\psi_2^*(-\zeta). \quad (5.104)$$

We now have almost all we need to carry the wave solutions across the critical level. What remains is how to describe $(-\zeta)^{\lambda}$ and $(-\zeta)^{\lambda^*}$ in (5.103) and (5.104). This is explained in the Appendix where it is shown that the solutions are function of wind shear. If $a_1 > 0$ (positive wind shear) then

$$-\zeta = \zeta\, e^{-i\pi} \tag{5.105}$$

and

$$(-\zeta)^{\lambda} = \zeta^{\lambda}\, e^{-i\pi\lambda}. \tag{5.106}$$

Using the positive branch of (5.97) for λ in (5.106) gives

$$(-\zeta)^{\lambda} = \zeta^{\lambda}\, e^{-i\pi(1/2+i\mu)} = \zeta^{\lambda}\, e^{-i\pi/2}\, e^{\pi\mu} = -i\zeta^{\lambda}\, e^{\pi\mu}. \tag{5.107}$$

If $a_1 < 0$ (negative wind shear), then in a similar way we can show that

$$(-\zeta)^{\lambda} = i\zeta^{\lambda}\, e^{-\pi\mu}. \tag{5.108}$$

It is necessary to know what terms in (A.38) correspond to upward and downward moving waves. This can be unambiguously determined by calculating the wave stress. Considering ζ real and $\mu > 0$, then (A.38) takes the form

$$w_1 = \psi_1\zeta^{1/2}\zeta^{i\mu} + \psi_2\zeta^{1/2}\zeta^{-i\mu}. \tag{5.109}$$

The factor $\zeta^{i\mu}$ can be expressed as $e^{i\mu\ln(z-z_c)}$. Using this and the continuity equation (2.16) gives

$$w_1 = \zeta^{1/2}\left[\psi_1\, e^{i\mu\zeta} + \psi_2\, e^{-i\mu\zeta}\right], \tag{5.110}$$

$$u_1 = \zeta^{-1/2}\left[\left(i\frac{1}{2} - \mu\right)\frac{\psi_1}{k}e^{i\mu\ln\zeta} + \left(i\frac{1}{2} + \mu\right)\frac{\psi_2}{k}e^{-i\mu\ln\zeta}\right]. \tag{5.111}$$

Using (5.110) and (5.111) in (3.79) the wave stress is

$$\tau = \frac{0.5\mu}{k^2}(\psi_1^2 - \psi_2^2). \tag{5.112}$$

We see that the ψ_1-term represents an upward moving wave (positive wave stress), and the ψ_2-term represents a downward moving wave.

The final result is a function of the sign of the wind shear. Consider the case when the wind shear at the critical level is positive, *i.e.*, $a_1 > 0$, then above the critical level we can write

$$w_1^+ = |z - z_c|^{1/2}\left(\psi_1\, e^{i\mu\ln|z-z_c|} + \psi_2\, e^{-i\mu\ln|z-z_c|}\right), \tag{5.113}$$

and below the critical level

$$w_1^- = -i|z - z_c|^{1/2}\left(\psi_1\, e^{\pi\mu}\, e^{i\mu\ln|z-z_c|} + \psi_2\, e^{-\pi\mu}\, e^{-i\mu\ln|z-z_c|}\right). \tag{5.114}$$

In (5.113) and (5.114) we can consider $\mu\ln|z - z_c|$ as the vertical wavenumber, m, near the critical level. Thus, $\psi_1\, e^{im}$ represents the upward moving wave, and $\psi_2\, e^{-im_c}$ represents the downward moving wave. Comparing (5.114) with (5.113) we see that the amplitude of the upward moving wave decreases by an amount $e^{\pi\mu}$ as is passes through the critical level. However, as Booker and Bretherton (1967) point out, the wave stress is a truer measure of wave amplitude. Then, above the critical level, and using the continuity equation (2.8) we see that

$$u_1 \propto \frac{dw_1}{dhz} \propto \zeta^{\lambda-1}, \tag{5.115}$$

so that the wave stress $\tau^+ \propto \zeta^{2\lambda-1}$. Similarly, below the critical level the wave stress is $\tau^- \propto e^{2\pi\mu}\zeta^{2\lambda-1}$. Thus in passing through the critical wave level, the wave stress is reduced an amount $e^{-2\pi\sqrt{R_c-0.25}}$. This is a large reduction. For example, if $R_c = 0.5$, the reduction is 0.04; if $R_c = 1$, the reduction is 0.004; and if $R_c = 10$, the reduction is 3×10^{-9}. Thus, for almost all cases the wave is essentially dissipated on passing through a critical level. This is an extremely important result. We have seen in Chapter 3 that terrain features can generate gravity waves which in turn vertically transport horizontal momentum. However, unless the wave is dissipated, the momentum transport is constant. Critical levels provide this dissipation. For terrain-generated gravity waves, critical levels exist where $u_0 = 0$.

Let us now take a more physical rather than mathematical look at what happens when a wave approaches a critical level; however, we still remain within the linear theory. Imagine now an upward propagating wave approaching z_c from below. Then, using (2.8), (5.114) and (5.115) and taking real parts we have,

$$w_1 \propto |z - z_c|^{1/2} \sin(\mu \ln |z - z_c|), \tag{5.116}$$

$$u_1 \propto |z - z_c|^{-1/2} \sin(\mu \ln |z - z_c|). \tag{5.117}$$

As we have seen, the term $\sin(\mu \ln |z - z_c|)$ behaves as if the local wavenumber[2] is

$$m = \frac{\mu}{|z - z_c|}. \tag{5.118}$$

Then, as illustrated in Fig. 5.10, we see that as the wave approaches the critical level $w_1 \to 0$, $u_1 \to \infty$, and the vertical wavelength $\lambda_z \to 0$ resulting in increasingly rapid oscillations. However, in reality the vertical shear in the horizontal wind speed perturbation, which varies as $|z - z_c|^{-3/2}$, will become large and exceed the shear of the background wind. When this happens, the Richardson number near the critical level will become small indicating the production of turbulence. Thus, the wave field breaks down before reaching z_c. Using (5.118) in (2.77) we see that the vertical group velocity below the critical level ($z < z_c$) is

$$w_g = \frac{k\mu}{N^2} \frac{(c - u_0)^3}{|z - z_c|}. \tag{5.119}$$

As the wave approaches the critical level, $(u_0 - c)^3 \to 0$ more rapidly than $(z_c - z) \to 0$ so that $w_g \to 0$. In effect, the wave packet never reaches the critical

[2] As $z \to z_c$, $\ln |z - z_c| \to \infty$, but this is that same as $1/|z - z_c|$.

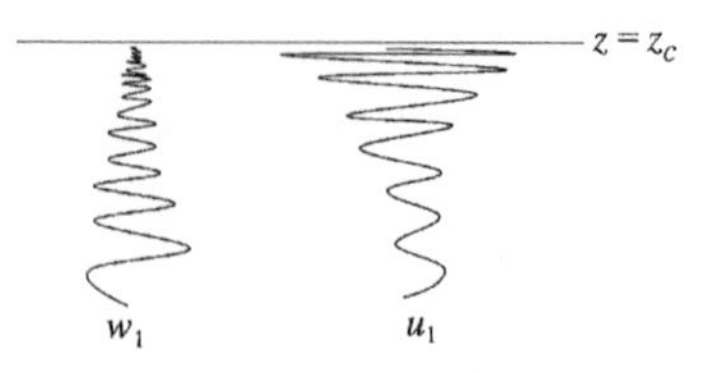

FIGURE 5.10 Gravity wave approaching a critical level from below showing behavior of u_1 and w_1.

level. Figure 5.11 illustrates the changes in a wave packet as it approaches a critical level from below. Because the vertical wavelength $(z - z_c)$ decreases, the vertical wavenumber increases as the wave packet approaches z_c. This causes the wave fronts to draw closer as $(z - z_c) \to 0$ as sketched in Fig. 5.11. We see also that the wave fronts become increasingly horizontally oriented, and the vertical group velocity becomes less. However, as pointed out by Booker and Bretherton (1967), these infinities at the critical level disappear if $c_I \neq 0$. These singularities are characteristic of standing waves, *i.e.*, a mountain waves, that have persisted for a long time. The waves generated by a mountain have the same intrinsic frequency, and they combine to form a standing disturbance of finite magnitude. Transient waves each with first-order (infinitesimal) amplitudes and dispersive wavenumbers do not "build-up," and singularities in the total integrated disturbance may never appear. These waves might pass through a critical level only lightly modified.

We can determine the changes in the characteristics of a transient wave when it passes through a critical level by noting the changes in the wavenumbers. Consider first the case where $a_1 > 0$. Then above the critical level (5.113) can be written as

$$w_1 = \psi_1 (z - z_c)^{\frac{1}{2}+i\mu} + \psi_2 (z - z_c)^{\frac{1}{2}-i\mu}. \tag{5.120}$$

Using (5.118) in (2.71) the phase equations are

$$kx + \frac{\mu}{(z - z_c)} z = c_i t \quad \text{for the } \psi_1 \text{ solution} \tag{5.121}$$

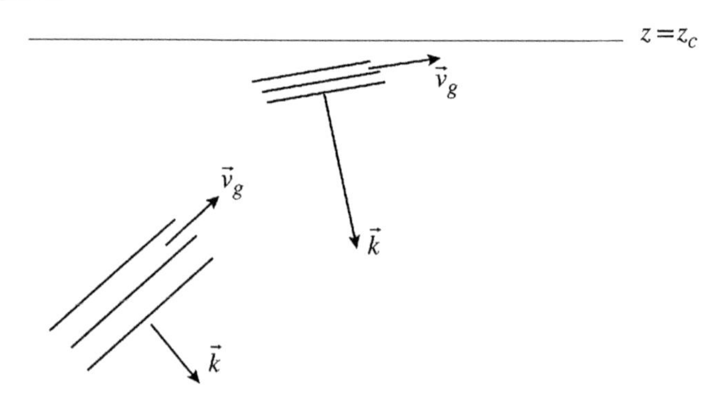

FIGURE 5.11 Behavior of a wave packet as it approaches a critical level.

and

$$kx - \frac{\mu}{(z - z_c)} z = c_i t \quad \text{for the } \psi_2 \text{ solution,} \tag{5.122}$$

where $c_i = c - u_0$ is the intrinsic phase speed, and we have taken phase angle ϕ equal to zero. Below z_c, *i.e.*, $z < z_z$, c_i is positive since $c > u_0$, and $\frac{\mu}{|z-z_c|}$ is negative for both the ψ_1 and ψ_2 solutions. Then, k is positive for both solutions, but from (5.121) and (5.122) we see that the vertical phase velocity is negative for the ψ_1 solution, *i.e.*, $m < 0$, and positive for the ψ_2 solution, *i.e.*, $m > 0$. Above z_c, c_i, and k are both negative for both solutions; however, m is still negative for the ψ_1 solution and positive for the ψ_2 solution. Thus, we see that the ψ_1 solution corresponds to a downward moving wave, and the ψ_2 solution corresponds to an upward moving wave.

If $a_1 < 0$, (5.113) takes the form

$$w_1 = \psi_1 (z - z_c)^{\frac{1}{2} - i\mu} + \psi_2 (z - z_c)^{\frac{1}{2} + i\mu}. \tag{5.123}$$

Applying the same arguments as above on (5.121) and (5.122) we see that now the ψ_1 solution corresponds to an upward moving wave, and the ψ_2 solution corresponds to a downward moving wave. These results are summarized in Fig. 5.12 which illustrates the orientation of the wave vector on either side of the critical level for positive and negative wind shears.

The above results were developed for a frictionless adiabatic atmospheric flow. If we included viscosity and heat conduction then as shown by Hazel (1967) and Fritts and Geller (1976) the equation for the vertical velocity perturbation is

$$\left[ik(u_0 - c) - K\left(\frac{\partial^2}{\partial z^2} - k^2\right)\right]\left\{\frac{\partial^2}{\partial z^2}\left[-ik(u_0 - c) + \nu\left(\frac{\partial^2}{\partial z^2} - 2k^2\right)\right]\right.$$
$$\left. + ik[u_0' + k^2(u_0 - c)] + \nu k^4 \right\} w + k^2 N^2 w = 0, \tag{5.124}$$

where ν is the kinematic viscosity and K is the thermal conductivity. Equation (5.124) is a sixth-order differential equation, which must be solved numerically. We see that the singularity at the critical level has been removed. Now, instead of $w \to 0$ and $u \to \infty$ as the critical level is approached the they vary continuously. However, the reductions in wave stress predicted by the linear model still occur. Baines (1995) points out that if a gravity wave approaches a critical level slowly, the

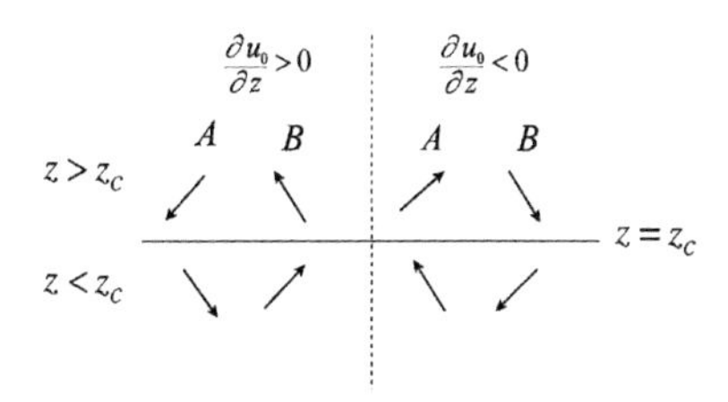

FIGURE 5.12 Wave changes on either side of a critical level.

viscous dissipation in the critical layer may be large enough to prevent explosive growth and overturning. In this case, the wave never reaches the critical level.

5.5 NEUTRAL, STABLE, AND UNSTABLE MODES

In a stably-stratified atmosphere with velocity shear u_0', there can be reflections of vertically propagating waves. As we have seen in Chapter 4, these reflections can lead to wave ducting, and the solutions to the Taylor–Goldstein equation are horizontally propagating modes. The stability of these modes in the atmosphere, oceans, and laboratories is very important. In many cases, unstable or growing modes can generate turbulence and affect vertical transports of heat, momentum, and other constituents. These instabilities are observed at all heights of the atmosphere and depths of the oceans. In all cases, velocity shear and stratification act to transfer flow kinetic energy into wave kinetic energy. Thus, in order to explain these observations we must understand the physical processes leading to flow instability. Studies of inviscid stratified shear flows have been made by authors too numerous to cite; however, some examples of papers and texts are Jones (1968), Hazel (1972), Turner (1973), Chimonas (1974, 2002), Gossard and Hooke (1975), Fua, Einaudi, and Lalas (1976), Lalas and Einaudi (1976), Mastrantonio, Einaudi, and Fua (1976), Fua and Einaudi (1984), Chimonas and Grant (1984), Sutherland, Caulfield, and Peltier (1994), Drazin (2002), Newsom and Banta (2003), and Xu (2007).

We have seen that if the phase speed of a wave is complex, *i.e.*, $c_I \neq 0$, wave amplitude can grow exponentially with time. Such a condition is unstable. Linear stability analysis seeks to find those eigenvalues (c, k) that are solutions to the Taylor–Goldstein equation for a given background velocity and temperature profile and boundary conditions. In particular, we seek the locus of points in the (c, k) plane that separate stable and unstable modes. Only under very restricted conditions, for example Kelvin–Helmholtz instability, can an analytical solution be found; in all other cases computers are required. The vertical boundary conditions require $w_1(z \to \pm\infty) = 0$ or that at some distance above and below the shear layer the radiation condition is satisfied. As we have seen, the only waves that can satisfy these boundary conditions are normal modes. In Chapter 4, we studied two simple types of modes; the pure temperature duct and the pure wind duct. In the pure temperature duct, the background winds were calm, and a discontinuity in stratification was present. In the pure wind duct, the background stratification was constant and the background wind profile contained a jet. For both cases, we found that wave solutions were modes with higher harmonics. The amplitudes of these modes remained constant with time because we did not consider critical levels or Richardson numbers less than 0.25. In the general case, critical levels can exist and there will be cases where $0 \leqslant Ri \leqslant 0.25$ at a critical level. However, the linear theory does easily lend itself to real atmospheric conditions (see, for example, Chimonas and Grant, 1984); therefor idealized profiles of wind and stratification are studied. For example Jones (1968), Hazel (1972), and Gossard and Hooke (1975), use multiple layers each with uniform stratification and wind shear. Fig. 5.13 illustrates the three wind speed profiles used by Jones. The stratification was uniform. Chimonas (1974), Lalas and Einaudi (1976), and Chimonas and Grant (1984) used hyperbolic-tangent wind profiles and exponential density profiles. The computations are straightforward in concept but complicated in detail. If $c_{\mathrm{I}} < 0$, the modes, if any, are unstable. Although there may be a critical level where $c_{\mathrm{R}} - u_0 = 0$, a singularity is not present because of the finite value of c_{I}.

Generally, a shooting method is used to find the modes. These calculations are almost always done in a non-dimensional space. Thus, for example, $y = z/h$ where z is the height above the lower boundary; h is the depth of the shear layer; $\alpha = hk$ where k is the horizontal wavenumber; $J = N^2h^2/V^2$ where V is a velocity scale and J is a scale Ri such that

$$Ri = J\left(\frac{\mathrm{d}u_0}{\mathrm{d}y}\right)^{-2}. \tag{5.125}$$

J is the Richardson number at the inflection point. At the bottom boundary, values of J and α are chosen and held fixed, and initial guesses for c_{R} and c_{I} are used to evaluate the Taylor–Goldstein equation up to the top boundary. There, the solution is tested for convergence with the top boundary condition. Usually, the first guess is wrong. The process is repeated with new values of c_{R} and c_{I} until convergence at the top boundary is achieved. The process is then repeated with different values of J and α to obtain the dispersion curve for that mode.

A slightly more complicated method was used by Lalas and Einaudi (1976) to find the singular neutral modes, *i.e.*, when $c_{\mathrm{I}} = 0$. For a given α, values of c_{R} and J are chosen. The integrations start at the bottom and top of the model, and each proceed within a small distance above and below the critical level. The Frobenius expansion is used to match the solutions at the critical level. If a matching does not occur, the process is repeated with new values of c_{R} and J until a matching is achieved. Usually J is the main stability parameter, and once c_{R} is found the Richardson number at the critical level, R_{c}, is calculated using $R_{\mathrm{c}} = J/(1 - c_{\mathrm{R}}^2)^2$. The variables in these models include $u_0(z)$, $N(z)$, k_x, c_{R}, and c_{I}. It would be impractical to study all possible combinations of these parameters, and so some are fixed and others are varied. Thus, the vertical wavenumbers above and below these scale heights are constants. As an illustration of the stability analysis, we shall draw on Lalas and Einaudi (1976). They used a hyperbolic-tangent wind profile,

$$u_0(z) = V\tanh\left(\frac{z}{h}\right), \tag{5.126}$$

and an exponential density profile

$$\rho_0(z) = \rho_{\mathrm{g}}\exp\left(-\frac{z}{h}\right), \tag{5.127}$$

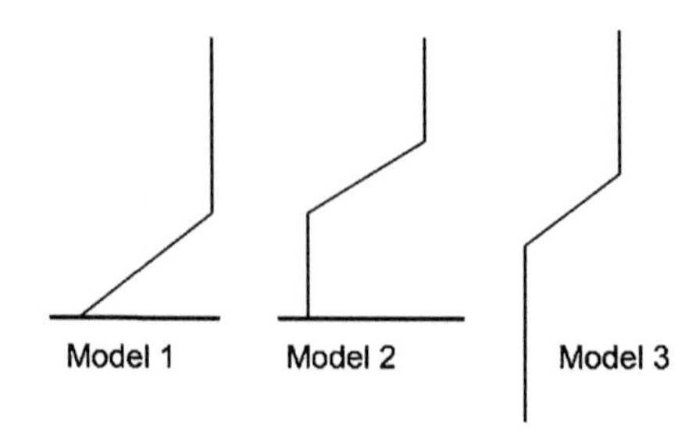

FIGURE 5.13 Model wind speed profiles studies by Jones (1968).

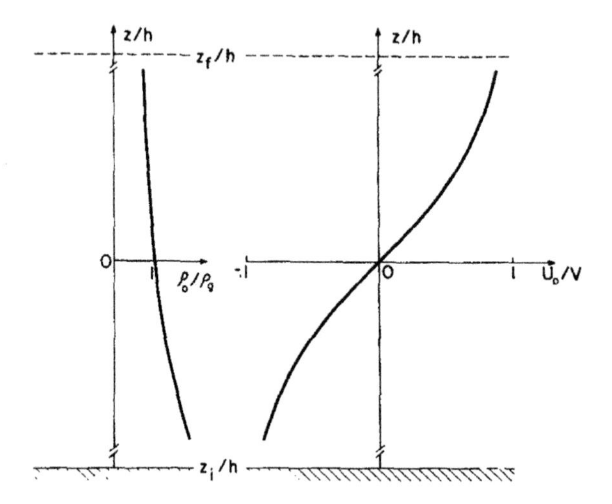

FIGURE 5.14 Dimensionless wind speed and density profiles used by Lalas and Einaudi (1976).

where h is the scale height for the change of density. These profiles, taken from Lalas and Einaudi (1976)) are sketched in Fig. 5.14 in terms of dimensionless height, z/h, dimensionless speed, u/V, and dimensionless density, ρ_0/ρ_g, where ρ_g is the density at the inflection point. The origin of the (x, z) coordinate system is located vertically at the inflection point of (5.126) where $u_0(0) = 0$. The inflection point lies some distance z_i above the lower boundary. For $|z/h| \geqslant 5$ the wind speed is essentially constant, and from (5.127) N is constant. By adjusting the values of h and z_i the velocity profiles can approximate the three Jones (1968) models shown in Fig. 5.13.

Because the forgoing stability analysis is complicated and possibly not intuitive, let us step back and review what we have seen. We are confronted with an inviscid stably-stratified flow, with a velocity shear layer of thickness $2h$. Below the bottom of the layer, there may be either a solid boundary or a semi-infinitely deep layer with uniform wind and stratification. The shear layer is caped by another semi-infinitely deep layer with constant wind and stratification. We know that the Richardson

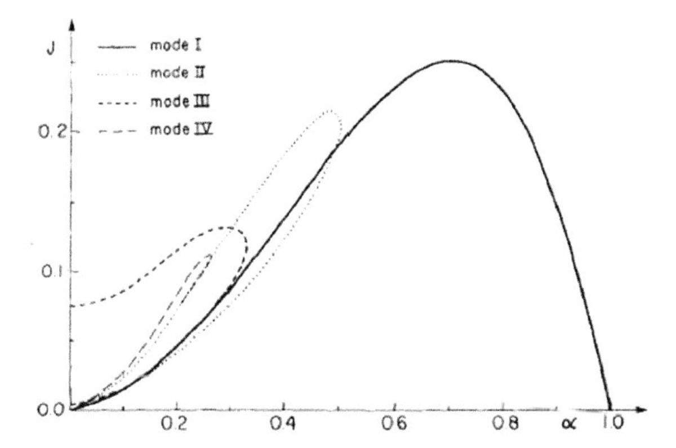

FIGURE 5.15 Stability boundaries, in the normalized wavenumber, minimum Richardson number (α, J) plane for modes I–IV. Note that the lower branches of modes II and III overlap in part with mode I and the lower branches of mode IV with the upper branch of mode II. (Taken from Lalas and Einaudi (1976).)

number will be between 0 and 0.25 throughout the flow and there might be a critical level somewhere in the shear zone. At solid boundaries we require $w_1 = 0$. If the boundaries are not solid, then we require either $w_1(z \to \pm\infty) \to 0$ or only outgoing radiation of energy. Because of these conditions, solutions to the Taylor–Goldstein equation will be either fundamental modes plus higher harmonics, or no solutions at all. We want to know the horizontal wavenumbers of the modes and whether or not they are stable, neutral, or unstable. If a mode is stable then $c_{\mathrm{I}} > 0$ for that mode. If the mode is unstable, then $c_{\mathrm{I}} < 0$ and the amplitude of the mode grows in time as $e^{c_I kt}$. On a two-dimensional graph that plots Ri against k all the points for the unstable modes will lie within a contour line that begins and ends on the vertical and horizontal axes or begins and ends on the same axis. All the points that are outside of this boundary line are for either stable modes or no modes at all. The boundary line is the locus of points for the singular neutral mode, *i.e.*, a mode with $c_{\mathrm{I}} = 0$. Once c_{I} is known for a mode, its growth rate is easily calculated.

Figure 5.15 from Lalas and Einaudi (1976) shows the neutral boundaries for the first four harmonics for the profiles shown in Fig. 5.14. All points (J, α) within a neutral boundary are unstable for that harmonic. Figure 5.16a plots J, c_{R}, and R_{c}, the Richardson number at the inflection point, as functions of α for the first mode. Figure 5.16b plots the unstable growth rate αc_{I} and c_{R} against α for various values of stability, J, for the first mode. It is clear that as stability increases, *i.e.*, larger J, the spectrum of unstable waves decreases, as it should.

5.6 WAVE-MODULATED RICHARDSON NUMBER

In the above analyses of shear instability, the Richardson number was defined in terms of constant background quantities. From (2.45) we see that the amplitudes of

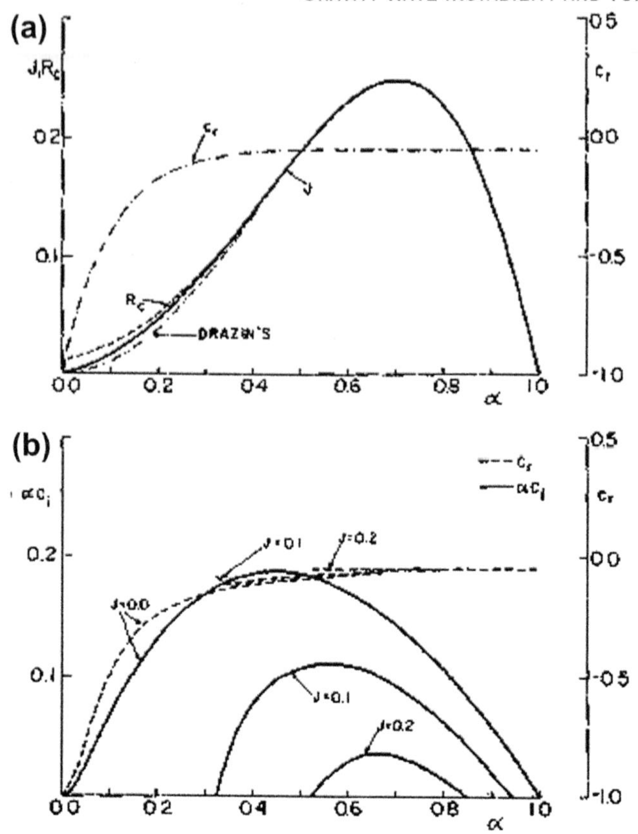

FIGURE 5.16 Mode I stability characteristics. (a) J (solid line), R_c (dashed line: non-dimensional Richardson number at the critical level, see Fig. 5.10) and corresponding c_r (dot-dashed line) as a function of $\alpha = kh$ for the neutral waves. (b) Normalized αc_i (solid lines) and phase speeds c_r (dashed lines) as a function of α for different values of J for the unstable waves. (Taken from Lalas and Einaudi (1976).)

upward moving gravity waves grow exponentially with height due to the decrease in atmospheric density. Hines (1960, 1963) proposed that gravity waves propagating into the upper atmosphere can generate turbulence there. Hodges (1967) used this fact to demonstrate how gravity waves can modulate local Richardson numbers with the possibility of generating turbulence. Consider now the local Richardson number defined by the total temperature and velocity, *i.e.*,

$$Ri = \frac{g}{\theta} \frac{\frac{\partial \theta}{\partial z}}{\left(\frac{\partial u}{\partial z}\right)^2} = \frac{g}{\theta_0} \frac{\frac{\partial}{\partial z}[\theta_0 + \theta_1]}{\left[\frac{\partial}{\partial z}(u_0 + u_1)\right]^2}. \tag{5.128}$$

If we assume a wave-perturbation vertical velocity of the form $w_1 = A\cos\phi$ where $\phi = kx + mz - \omega t$, then using (2.21) and (2.22) along with (1.66) we have

$$u_1 = -\frac{Am}{k}\cos\phi \tag{5.129}$$

and

$$\theta_1 = \frac{A}{\omega - u_0 k}\frac{\partial\theta_0}{\partial z}\sin\phi. \tag{5.130}$$

Using (5.129) and (5.130) in (5.128) gives

$$Ri = \frac{g}{\theta_0}\frac{\frac{\partial}{\partial z}\left(\theta_0 + \frac{A}{\omega - u_0 k}\frac{\partial\theta_0}{\partial z}\sin\phi\right)}{\left[\frac{\partial}{\partial z}\left(u_0 - \frac{Am}{k}\cos\phi\right)\right]^2}. \tag{5.131}$$

Clearly, for certain values of the constants in (5.131) we can expect that $Ri < 0.25$ over some range of ϕ. To illustrate this consider a constant flow over a corrugated surface. Then using (3.31) and (3.32) in (5.128) and assuming that $\partial^2\theta_0/\partial z^2 \ll 1$ gives

$$Ri = \frac{1 - m_s H\sin\phi}{\left[\frac{u_0 m_s}{N}(m_s H\cos\phi)\right]^2}, \tag{5.132}$$

where $\phi = k_s x + m_s z$; k_s is the wavenumber for the surface corrugation, and H is the corrugation amplitude. If the stratification is strong and the wavelength of the corrugation large, then

$$m_s^2 = \frac{N^2}{u_0^2} - k_s^2 \approx \frac{N^2}{u_0^2}, \tag{5.133}$$

so that (5.132) takes the simple form

$$Ri = \frac{1 - m_s H\sin\phi}{(m_s H\cos\phi)^2}, \tag{5.134}$$

which is identical to that developed by Hodges (1967). If $m_s H \geqslant 1$, the flow will be convectively unstable over some range of ϕ. The quantity $m_s H = NH/u_0 = F^{-1}$ where F is similar to the *internal Froude number* (Turner, 1973).

The ordinary Froude number is defined as the ratio of a characteristic speed to the speed of long waves on a free surface (5.57), *i.e.*, $V/\sqrt{g\lambda}$ where V is a velocity scale and λ is a wavelength. The Froude number can also be defined as the square root of the inertia force to the gravity force. The internal Froude number, F_i is defined as

$$F_i = \frac{V}{\sqrt{\frac{\rho_2 - \rho_1}{\rho_2 + \rho_1}\lambda}}, \tag{5.135}$$

where ρ_1 and ρ_2 are the densities in adjoining fluid layers. In the case of a continuous stratification, it is common to define V/NH as the Froude number.

Thus, we can write (5.134) as

$$Ri = \frac{F^2 - F \sin \phi}{\cos^2 \phi}. \tag{5.136}$$

We see that convective instability over the surface corrugation occurs when $F < 1$. Rottman and Smith (1989) showed experimentally that wave breaking occurs over a two-dimensional ridge in a linearly stratified flow for $F = 0.8$, 0.9, and 1.0 for steep ($H/L = 0.56$), intermediate ($H/L = 0.34$), and gentle ($H/L = 0.16$) ridges respectively. Here L is the half-width of the ridge and H/L is the aspect ratio of the ridge. Figure 5.17 shows plots of (5.134) as functions of ϕ for $F = 0.9$ and 1.1. Ri is always greater than 0.25 for $F = 0.9$, but for $F = 1.1$ the ranges of ϕ for dynamic and convective instability are evident. Figure 5.18 shows the regions of convective instability ($Ri \leqslant 0$) over a corrugated surface calculated using the linear wave model described in the Appendix. The calculation used $u_0 = 1$ m s^{-1}, $H = 60$ m, $\lambda_s = 500$ m, and $N = 0.022$ s^{-1}, and for this case, $F = 0.8$. The unstable regions tilt upwind with an angle $\tan^{-1}(m_s/k_s)$. For this case, the half-width of corrugation is 250 m so that the aspect ratio is 0.24. This is gentle topography, yet gravity waves can generate regions of *clear air turbulence*. Figure 5.19 plots the convectively unstable regions over a two-dimensional Gaussian ridge with $H = 50$ m, $b = 1000$ m, $u_0 = 1$ m s^{-1}, and $N = 0.022$ s^{-1}. The unstable patches remain directly above the ridge. The Froude number for this

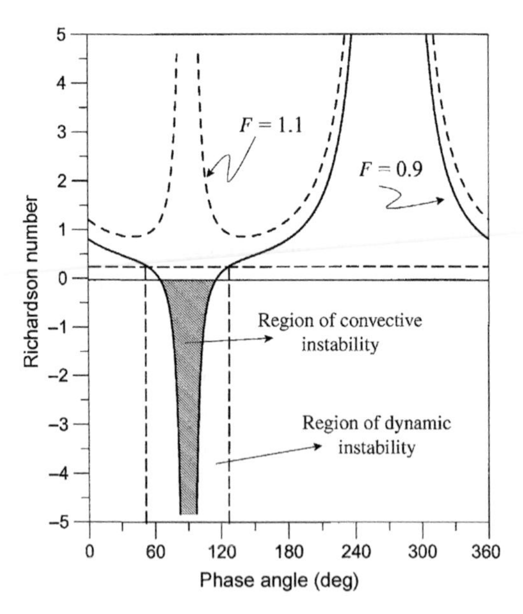

FIGURE 5.17 Richard number as a function of phase angle for Froude numbers of 0.9 and 1.1. Region of dynamic instability extends below $Ri = 0.25$ and between the dashed lines. Region of convective instability is shaded.

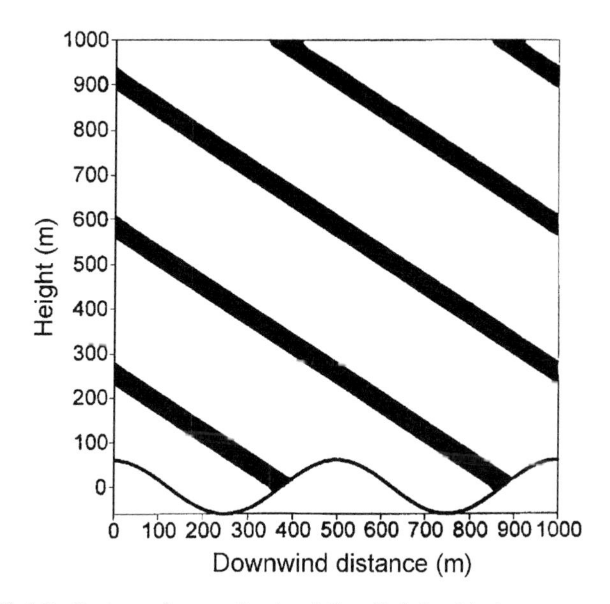

FIGURE 5.18 Regions of convective instability (dark bands) above a corrugated surface. $u_0 = 1$ m s^{-1}, $H = 60$ m, $N = 0.022$ s^{-1}, and $\lambda_s = 500$ m; Froude number $F = 0.8$.

case is 0.9. The half-width of the ridge is $0.83b$ so the aspect ratio for this case is 0.06. This is a very gentle topography, but even so turbulence can be generated by the ridge.

Chimonas (1972) used the concept of wave-modulated Richardson number in a quasi-linear theory of wave-generated breakdowns of a stratified flow. He included terms in the linearized equations (2.6)–(2.9) to account for the fluxes of heat and momentum due to wave-generated turbulence. These flux terms are switched on whenever the Richardson number drops below 0.25. When this happens, the turbulence feeds momentum and energy from the background flow into the wave causing the wave amplitude to increase. This in turn increases the intensity and spatial extent of the regions of turbulence. The result is a rapid rise of turbulence and wave amplitude; however, the linear theory cannot follow this process to a conclusion because the wave amplitudes become too large.

5.7 WAVE–TURBULENCE COUPLING

In a series of articles including Einaudi and Finnigan (1981, 1993), Finnigan and Einaudi (1981), and Fua *et al.* (1982), Finnigan, Einaudi, and Fua (1984), and Finnigan (1988), Finnigan and Einaudi examined the interactions of waves and turbulence. The mechanism by which waves and turbulence interact and modify

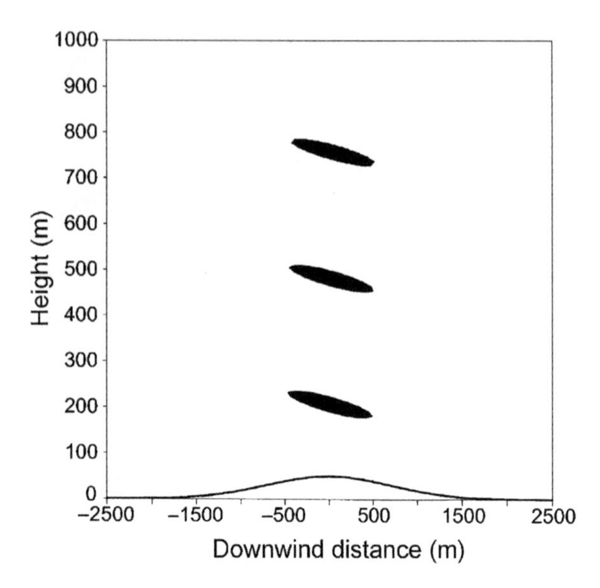

FIGURE 5.19 Regions of convective instability above a Gaussian-shaped ridge. $u_0 = 1\ m\ s_{-}1$, $H = 50$ m, $b = 1000$ m, $N = 0.022\ s_{-}1$; Froude number $F = 0.9$.

the mean flow is commonly referred to as *wave–turbulence coupling*. Fua *et al.* (1982) lists the four basic ideas underlying the process:

1. The occurrence of turbulence can be related to the local (wave-modified) Richardson number.
2. Turbulence occurs with a mean and periodic component.
3. Turbulence extracts energy from the wave limiting its growth, or feeds energy into the wave.
4. Turbulence modifies the mean fields.

The analysis of wave–turbulence coupling is based on a triple decomposition of the flow variables based on the work of Reynolds and Hussain (1972). The decomposition separates the atmospheric variables into mean, $\bar{a}$, turbulence, a', and wave, $\tilde{a}$, components, *i.e.*,

$$a(x_i, t) = \bar{a}(x_i) + a'(x_i, t) + \tilde{a}(x_i, t), \tag{5.137}$$

where x_i represents the spatial coordinates. The triple decomposition puts requirements on the wave field. It must be linear, monochromatic, and have constant frequency and amplitude. The problem involves writing equations for turbulence and wave variables, and examining the terms that connect these equations. The time average or background is defined in the usual way, *i.e.*,

$$\bar{a}(x_i) = \lim_{t\to\infty} \left\{ \frac{1}{2t} \int_{-t}^{+t} a(x_i, t')\mathrm{d}t' \right\}. \tag{5.138}$$

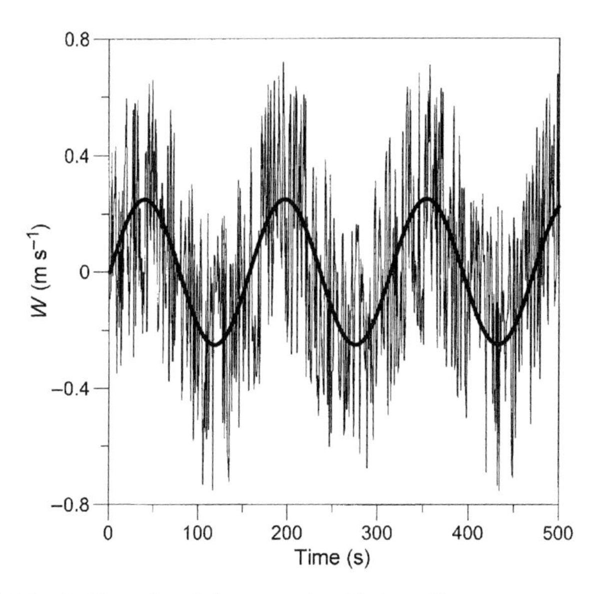

FIGURE 5.20 An illustration of phase averaging. The heavy line represents an underlying wave signal which would be revealed by phase averaging.

If the wave has period τ, then the *phase average* is defined as

$$\langle a(x_i,t)\rangle = \lim_{N\to\infty}\left\{\frac{1}{N}\sum_{n=1}^{N} a(x_i,t+n\tau)\right\}, \tag{5.139}$$

where now N is the number of wave cycles, not the Brunt–Väisälä frequency. Figure 5.20 illustrates the phase averaging process. A turbulence time series with zero mean and standard deviation 0.6 m s^{-1} is added to a sine wave with period 157 s and amplitude 0.25 m s^{-1}. The value of $\langle a\rangle$ at time t is the average of N data points, with each point separated by time period τ. Because the turbulence is random, it is removed by the ensemble averaging. The illustration in Fig. 5.20 is conceptual; in order to reveal the actual wave signal from the raw data, averaging over many cycles would be required. After phase averaging, the *wave average* component and the turbulent component of the raw signal are calculated using

$$\tilde{a}(x_i,t) = \langle a(x_i,t)\rangle - \bar{a}(x_i), \tag{5.140}$$

$$a'(x_i,t) = a(x_i,t) - \langle a(x_i,t)\rangle\,. \tag{5.141}$$

Using the triple decomposition defined by (5.137) in (2.1)–(2.4) along with (1.66) gives equations for the wave field. However, several averaging identities are required, specifically for variables a and b:

$$\text{(i) } \langle a'\rangle = 0,\quad \text{(ii) } \bar{\tilde{a}} = 0,\quad \text{(iii) } \overline{a'} = 0,$$
$$\text{(iv) } \overline{\bar{a}b} = \bar{a}\bar{b},\ \text{(v) } \langle \tilde{a}b\rangle = \tilde{a}\langle b\rangle,\ \text{(vi) } \langle \bar{a}b\rangle = \bar{a}\langle b\rangle,$$
$$\text{(vii) } \langle \bar{a}\rangle = \bar{a},\ \text{(viii) } \overline{\langle a\rangle} = \bar{a},\quad \text{(ix) } \overline{\tilde{a}b'} = \overline{\langle \tilde{a}b'\rangle} = 0.$$

Using these identities, and assuming that the density is constant over the small vertical scales considered, and that the flow is inviscid, the dynamical equations for the wave and turbulence fields are (Reynolds and Hussain, 1972)

$$\frac{\partial \tilde{u}_i}{\partial t}+u_{0,j}\frac{\partial \tilde{u}_i}{\partial x_j}+\tilde{u}_j\frac{\partial u_{0,i}}{\partial x_j}=-\frac{\partial \tilde{p}}{\partial x_i}-\frac{\partial \tilde{r}_{ij}}{\partial x_j}-\frac{\partial}{\partial x_j}(\tilde{u}_i\tilde{u}_j-\overline{\tilde{u}_i\tilde{u}_j})+\frac{g}{T_0}\tilde{\theta}\delta_{i3}, \quad (5.142)$$

and

$$\frac{\partial u'_i}{\partial t}+u_{0,j}\frac{\partial u'_i}{\partial x_j}+u'_j\frac{\partial u_{0,i}}{\partial x_j}=-\frac{\partial p'}{\partial x_i}+\frac{\partial \tilde{r}_{ij}}{\partial x_j}+\frac{g}{T_0}\theta'\delta_{i3}, \quad (5.143)$$

where

$$\tilde{r}_{ij}=\langle u'_i u'_j\rangle-\overline{u'_i u'_j}, \quad (5.144)$$

and p is the kinematic pressure. Similar equations can be written for the temperature wave and turbulence fields. The term $(\tilde{u}_i\tilde{u}_j-\overline{\tilde{u}_i\tilde{u}_j})$ in (5.142) represents the fluctuation part of the wave stress. Note that except for $\tilde{r}_{ij}$ all the terms in the equation for the organized wave (5.142) involve wave terms, and all the terms in the turbulence equation (5.143) involve turbulence terms. The term $\tilde{r}_{ij}$ represents the oscillating part of the Reynolds stress due to the presence of the wave, *i.e.*, the wave-modified turbulence. The term appears with opposite sign in the two equation. Thus, if it is a source term in the wave equation (5.142), then it is a sink term in the turbulence equation (5.143). We see that the wave-modified Reynolds stress is a mechanism for transferring energy between the turbulence and the wave. The term $(\tilde{u}_i\tilde{u}_j-\overline{\tilde{u}_i\tilde{u}_j})$ in the wave equation does not have a counterpart in the turbulence equation. This term represents wave-like fluctuations in the wave stress; however, similar organized motions do not appear in the turbulence field. Recall it is the integral of the wave stress, *i.e.*, (5.9) that is constant with height. In (5.142) we see that wave amplitude changes according to the variations of the wave stress from its average value.

The equations for the wave and turbulence kinetic energies are

$$\frac{D\overline{\tilde{q}^2}}{Dt}=\frac{D}{Dt}\left(\frac{\overline{\tilde{u}_i\tilde{u}_i}}{2}\right)=-\frac{\partial}{\partial x_j}\left\{\overline{\tilde{u}_j\left(\tilde{p}+\frac{\tilde{u}_i\tilde{u}_i}{2}\right)}\right\}-\overline{\tilde{u}_i\tilde{u}_j}\frac{\partial u_{0,i}}{\partial x_j}$$
$$+\overline{\tilde{r}_{ij}\frac{\partial \tilde{u}_i}{\partial x_j}}-\frac{\partial}{\partial x_j}(\overline{\tilde{u}_i\tilde{r}_{ij}})+\frac{g}{\theta_0}\overline{\tilde{\theta}\tilde{u}_i}\delta_{i3}+\text{viscous terms} \quad (5.145)$$

and

$$\frac{D\overline{q'^2}}{Dt}=\frac{D}{Dt}\left(\frac{\overline{u'_i u'_i}}{2}\right)=-\frac{\partial}{\partial x_j}\left\{\overline{u'_j\left(p'+\frac{u'_i u'_i}{2}\right)}\right\}-\overline{u'_i u'_j}\frac{\partial u_{0,i}}{\partial x_j}$$
$$-\overline{\tilde{r}_{ij}\frac{\partial \tilde{u}_i}{\partial x_j}}-\overline{\tilde{u}_j\frac{\partial}{\partial x_j}\left(\frac{\tilde{r}_{ii}}{2}\right)}+\frac{g}{\theta_0}\overline{\theta' u'_i}\delta_{i3}+\text{viscous terms}. \quad (5.146)$$

Equations (5.145) and (5.146) are similar to conventional turbulence budgets; however, the term $\overline{\tilde{r}_{ij}\frac{\partial \tilde{u}_i}{\partial x_j}}$ is the direct result of the triple decomposition. This is a

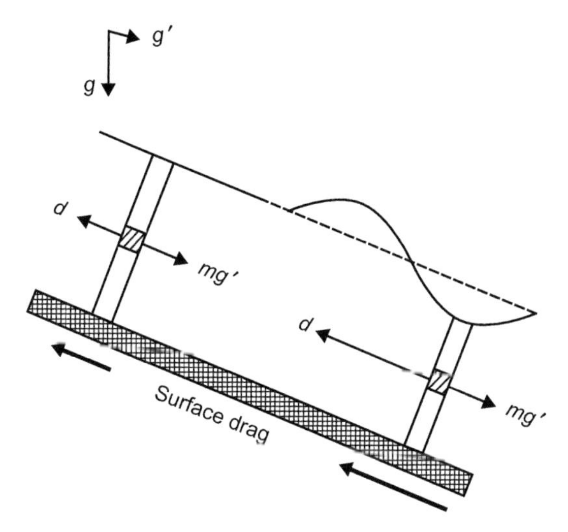

FIGURE 5.21 "Drag waves" on a downslope flow. The instabilities develop because variations the surface drag results in a readjustment of the downslope mass flow. (Taken from Chimonas (1993)).

production term and represents the rate of work of wave-modified Reynolds stress against the wave rate of strain. It is this term that is responsible for the transfer of energy between the wave and turbulence fields. Because $\tilde{r}_{ij}$ and $\frac{\partial \tilde{u}_i}{\partial x_j}$ represent linear wave fluctuations, their average product is zero if the two terms have a phase difference of $\pi/2$. Then, there is little energy transfer between the wave and turbulence fields. Finnigan (1988) reports the observations of four episodes observed at the Boulder Atmospheric Observatory with substantial gravity waves. Phase differences of about $\pi/4$ between $\tilde{r}_{ij}$ and $\frac{\partial \tilde{u}_i}{\partial x_j}$ were observed to occur with significant transfer of energy from the wave to the turbulence.

In the wind tunnel experiments on wave turbulence–turbulence coupling reported on by Hussain and Reynolds (1972), the turbulence component was very week, and so a large number of wave cycles (typically 10^5) was required in the averaging process. In the atmosphere, the conditions required for phase averaging, *i.e.*, a linear, monochromatic wave with constant frequency and amplitude will hardly ever be observed. Thus, wave–turbulence coupling is perhaps more useful as a theoretical rather than an observational tool.

5.8 JEFFERYS' ROLL-WAVE INSTABILITY MECHANISM

Chimonas (1993, 1994) calls attention to a wave instability mechanism proposed by Jefferys (1925) to explain *progressive waves* on the surface of water flowing in a channel. As Jefferys (1925) describes "... in certain Alpine conduits, with plane bottoms and rectangular sections (water), does not flow in a steady

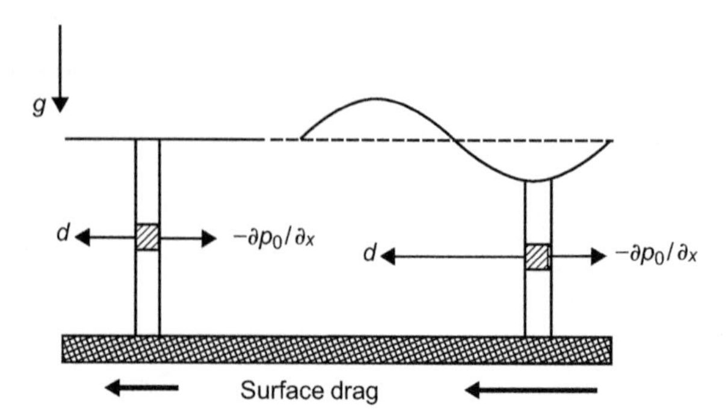

FIGURE 5.22 "Drag waves" in the ageostrophic boundary-layer flow. The instability mechanism is similar to that illustrated in Fig. 5.20 except now the gravity force is replaced by the pressure gradient force. (Taken from Chimonas (1993).)

stream but in a series of waves." Chimonas (1993) draws a parallel between progressive waves in conduits and the wave systems that appear on water-covered slopes on a road during a heavy rainfall. He notes that the waves "... seem to be in a continuous state of breaking, like rollers approaching a beach and somehow archive a lasting state of overturning." These progressive waves are instabilities that develop on the turbulent sheet of water as the water flows downslope. The downslope acceleration of flow is balanced by the aerodynamic drag at the ground surface as illustrated in Fig. 5.21. The instabilities develop because variations in the surface drag result in a readjustment of the downslope mass flow. An increase in drag causes an upslope build-up of water. This flow convergence can lead to wave excitation. Once initiated, the wave produces periodic variations in surface drag, and a resonance develops. Chimonas (1993) applied this mechanism in the atmosphere where the turbulent water is replaced by a turbulent planetary boundary layer (PBL), and the downslope acceleration is replaced by the atmospheric mean pressure gradient. Figure 5.22 illustrates this case. The atmospheric waves illustrated in Fig. 5.22 are different than the *roll waves* studied by, for example, Brown (1980), Mason and Sykes (1982), and Etling and Raasch (1987). These roll waves are associated with a critical level; however, the Jefferys waves have phase speeds considerably greater than the flow speeds. An assumption of Jefferys mechanism is a fully turbulent flow so that variations in surface drag are felt immediately through the depth of the water layer. A similar assumption is required for the atmospheric case.

We now assume a steady, two-dimensional, hydrostatic, irrotational, atmospheric background flow above a level ground surface. Figure 5.23 illustrates the flow geometry. A three-layer model provides the possibility for a ducting region with little change in the model physics. A turbulent surface layer of depth h and mean depth h_0 is in contact with the ground surface. In the two-layer model, a capping inversion separates the surface layer from the free atmosphere which has constant wind speed and stratification. In the case of three layers, the middle layer

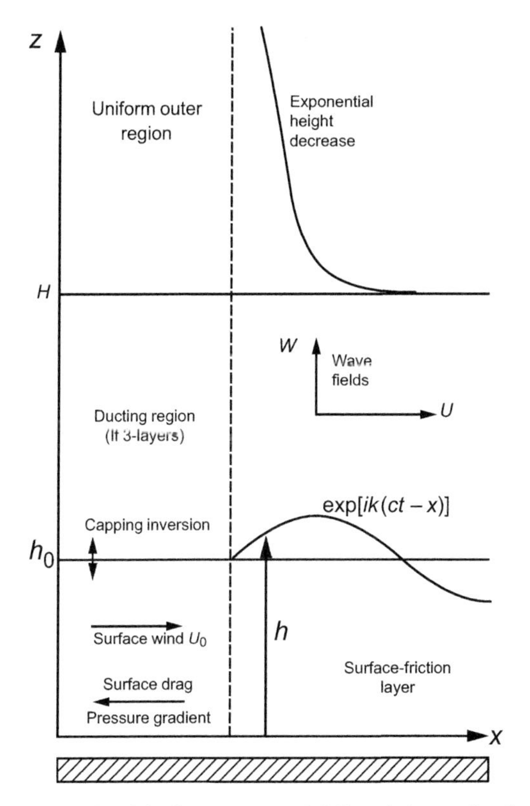

FIGURE 5.23 Schematic of the flow geometry. A fully turbulent surface has a mean depth h_0. In the two-layer case, the surface layer is capped by a semi-infinite layer with constant wind and stable stratification. The optional middle layer allows for the ducting of gravity waves. (Taken from Chimonas, 1993).

has intermediate wind and stratification between the surface layer and the semi-infinite layer. The middle layer allows for the ducting of gravity waves, but the physics is the same as for the two layer case. The balance of forces in the surface layer is

$$\frac{\partial p_0}{\partial x} + \rho_0 d_0 = 0, \tag{5.147}$$

where d_0 is the force per unit mass due to the Reynolds stress. The turbulence in the surface layer is the result of surface friction so that

$$\int_0^{h_0} \rho_0 d_0 \, \mathrm{d}z = 0.5 \rho_0 C_\mathrm{D} U^2, \tag{5.148}$$

where C_D is the surface drag coefficient and U is a velocity scale for the surface layer. In the free atmosphere, the Taylor–Goldstein equation (2.42) applies, and at any layer of discontinuity in wind speed or stratification, *i.e.*, at $z = h$ we impose the kinematic and dynamic boundary conditions. The equation of continuity for the surface layer is (Jefferys, 1925)

$$\frac{\partial h}{\partial t} = -\frac{\partial uh}{\partial x}. \tag{5.149}$$

We assume that in the surface layer the horizontal velocity is vertically uniform, and the vertical acceleration is balanced by the hydrostatic force so that

$$p(z) = p(h_0) + \rho_0 g(h_0 - z). \tag{5.150}$$

We integrate the horizontal momentum equation with height, and linearize in terms of a mean background and wave perturbation, for example u_0 and u_1, to get

$$ikh_0(c - u_0)u_1 = -h_1\frac{\partial p_0}{\partial x} + ikh_0 p_1 - \rho_0 C_D u_0 u_1, \tag{5.151}$$

where scale velocity U in (5.148) has been replaced by u_0, and pressures p_0 and p_1 are evaluated at $z = h_0$. Combining (5.149) with (5.151) gives the relation between $p_1(h_0)$ and h_1, *i.e.*,

$$[kh_0(c - u_0)^2 + i0.5u_0^2 C_D - iu_0(c - u_0)C_D]\frac{h_1}{h_0} = \frac{1}{\rho_0}kh_0 p_0. \tag{5.152}$$

The most simple model has two layers. In the upper layer, $N = 0$ and $u_0 = u_\infty =$ constant. Such a model approximates a shallow nocturnal boundary layer beneath a residual convective layer (Stull, 1988) or a density current undercutting well-mixed air. The solutions in the upper layer are

$$w_1(z) = A\,\mathrm{e}^{-kz}\,\mathrm{e}^{-ik(x-ct)}, \tag{5.153}$$

and

$$p_1 = i(c - u_\infty)\rho_0 w_1, \tag{5.154}$$

where $u_\infty\hat{x}$ is the constant undisturbed velocity in the upper layer. The vertical displacement of the interface is

$$\zeta_1 = -i\frac{w_i}{k(c - u_\infty)}. \tag{5.155}$$

At the interface between the two layers we apply the dynamic and kinematic boundary conditions, and with (5.152)–(5.155) we get the dispersion relation

$$(c - u_0)^2 + i\frac{C_D u_0}{kh_0}\left(\frac{3}{2}u_0 - c\right) + (c - u_0)^2 kh_0(1 - f) - g'h_0 = 0, \tag{5.156}$$

where f is the fractional decrease in density between the two layers, and $g' = fg$ is the reduced gravity.

The stability analysis of the interface displacements proceeds as before. We divide the dispersion relation into real and imaginary parts, and look for conditions where $c_I < 0$. The imaginary part of (5.156) with the assumption that $kh_0 < 1$ (*i.e.*, the horizontal wavelength is large compared to the fluid depth) is

$$c_I = -C_D u_0 \frac{\left(\frac{3}{2}u_0 - c_R\right)}{2kh_0(c_R - u_0)}. \tag{5.157}$$

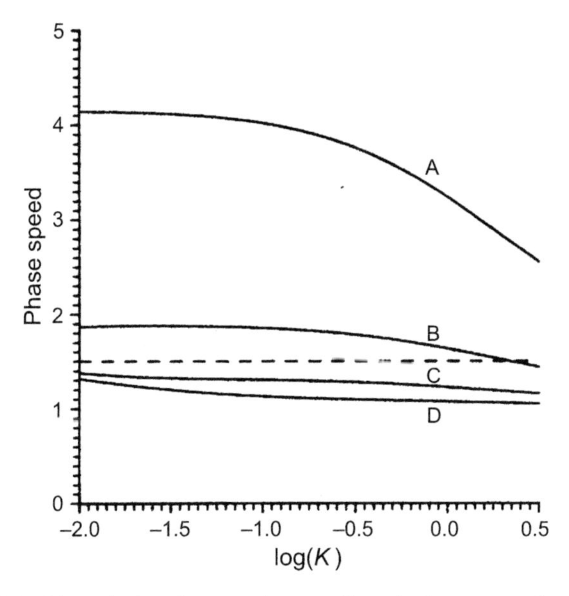

FIGURE 5.24 Dimensionless phase speeds versus dimensionless wavenumber for the two-layer drag-wave model with CD = 0.005. The dashed line indicates the threshold wave speed of $1.5u_0$ for wave growth. (A) $g' = 10$, the waves are similar to shallow-waver surface waves. (B) $g' = 0.8$, the phase speed crosses the threshold value of 1.5. (C) $g' = 0.1$, surface drag begins to have an effect. (D) $g' = 0.01$, the wave is dominated by the surface drag. (Taken from Chimonas (1993).)

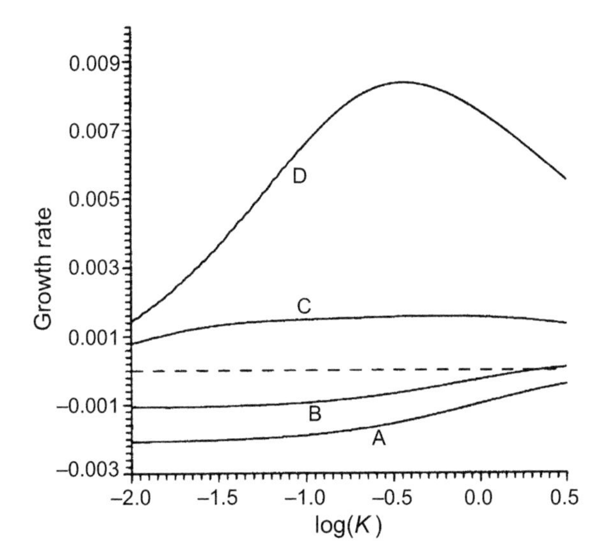

FIGURE 5.25 Growth rates versus dimensionless wavenumber. Surface friction causes instabilities in waves traveling slower than $1.5u_0$, but dampens waves traveling faster. (Taken from Chimonas (1993).)

The phase speeds for the instabilities are bounded by the *neutral modes* where $c_I = 0$. Thus, instabilities occur within the region $u_0 \leqslant c \leqslant 1.5u_0$. Figures 5.24 and 5.25 plot the normalized phase speed c/u_0 and growth rates respectively as functions of normalized wavenumber $K = kh_0$ for various values of the dimensionless group $g'h_0/u_0^2$. For these results, the wind speed in the upper layer is set equal to the mean speed of the surface layer. The dashed line in Fig. 5.24 marks the value $c = 1.5u_0$, and it is seen from Fig. 5.25 that this is indeed a boundary between stable and unstable modes. It is seen that the Jefferys instabilities favor high surface-layer speeds and week capping inversions, *i.e.*, small f. Chimonas (1994) expands on the Jefferys mechanism, and presents some example of observed wave events that resemble to a high degree of this mechanism.

PROBLEMS

1. In an isothermal atmosphere with scale height 8 km, a mid-latitude air parcel is moved adiabatically from the top of the tropopause to the bottom of the mesosphere. The average wind shear is 2.5×10^{-3} s^{-1} and $N = 0.02$ s^{-1}. How much work per unit mass is required?
2. Use (5.26)–(5.28) to get (5.29).
3. Derive an expression for the fastest growing wavenumber in a Kelvin–Helmholtz billow cloud.
4. Above the PBL, where is there likely to be an inflection point instability for a mountain wave?
5. What is the greatest horizontal phase speed a ducted wave can have in the troposphere?
6. What is the internal Froude number for a gravity wave with cut-off horizontal wavelength of 2 km and an amplitude of 200 m?
7. What is the horizontal speed of a surface wave with wavelength 10 km?
8. A gravity wave approaches a critical level where $R_c = 1.0$ and $N = 0.02$ s^{-1}. At what height below the critical level is the wave likely to become convectively unstable?
9. As a wave packet approaches a critical level, z_c, the intrinsic frequency falls to zero, the vertical wavenumber $m \to \infty$, and

$$\frac{\partial \omega}{\partial m} = -N \frac{|k|m}{|k^2 + m^2|^{\frac{3}{2}}}.$$

Show that when $z - z_c$ is small and N and k are constants, the vertical speed of the packet is

$$\frac{dz}{dt} \propto (z - z_c)^2.$$

10. In the wave–turbulence theory of Einaudi and Finnigan, show that when the wave-modulated Reynolds stress and the wave rate of train are out of phase by 90° the transfer of energy from the wave to the turbulence is zero.

6

WAVE STRESS

6.1 INTRODUCTION

The greatest impact gravity waves have on the atmosphere is the vertical transport of mean-flow horizontal momentum, *i.e.*, *wave stress*. Wave stress affects all scales of atmospheric motions, ranging from the stable PBL to the general circulation. The upward propagating of waves generated by flow over orography is the major source of wave stress in midlatitudes. In the tropics, nonorographic gravity waves are the major sources of wave stress. When gravity waves break they can generate turbulence where turbulence is not expected. In the stable PBL, wave stress can be as great as the friction stress at the ground surface. Examples of the importance of wave stress in the PBL can be found, for example, in Chimonas and Nappo (1989), Grisogono (1994), and Steeneveld, Nappo, and Holtslag (2009). In the troposphere and lower stratosphere wave stress is the source of clear air turbulence (CAT) (see for example, Dunkerton, 1984; Keller, 1994; Chun and Baik, 2002). In the upper atmosphere (60–110 km), momentum transport plays an integral role in the general circulation and the global climate (see, for example, McLandres, 1998; Meek and Manson, 1989; Chun *et al.*, 2001; Fritts and Alexander, 2003). In the thermosphere and ionosphere, momentum transport leads to high winds and turbulence which are manifested by the distortion of meteor trails, and by traveling ionospheric disturbances which disrupt radio communications (Justus and Woodrum, 1973; Hines, 1974; Gossard and Hooke, 1975). The principal source of wave stress is orography; however, nonorographic wave stress generated by, for example, large convective systems, diabatic heating, Kelvin–Helmholtz instability, and interactions with transient eddies, also contribute to the wave stress (see, for example, Büshler, McIntyre, and Scinocca, 1999; Chun *et al.*, 2001; Horinouchi, 2003; McLandress and Scinocca, 2005; Hertzog, 2008; Hardiman *et al.*, 2010). Mountain waves strongly affect the atmospheric momentum balance. Sawyer (1959) was one of the first to recognize the necessity of accounting for the effects of mountain-generated gravity waves in numerical forecast models. Bretherton (1969) computed the wave stress over the hilly terrain in north Wales, and showed that for a 19 ms^{-1} wind speed the wave stress can reach 0.4 Pa of which 0.3 Pa probably acted on the atmosphere above 20 km. To scale the effect of this stress on the atmosphere, let us assume that this stress is dissipated over a 5 km height range. Then, the atmospheric flow would be slowed at a rate of 5 ms^{-1}/day. Lilly (1972) calculated the lee-wave stress over the Front Range of the Colorado Rockies and reported values of between 0.5 and 1 Pa in the troposphere averaged over a horizontal distance of 100–200 km. Blumen and McGregor (1976) calculated a wave drag coefficient of 0.21×10^{-2} over a two-dimensional ridge and 0.11×10^{-2} over a three-dimensional hill. They estimated a wave stress on the order of 1 Pa over a series of ridges situated 25 km apart.

6.2 MATHEMATICAL DERIVATION

We begin by considering a steady, frictionless, non-rotating, and horizontally homogeneous background flow over an isolated two-dimensional ridge of arbitrary cross section height $h(x)$. Under the Boussinesq approximation, the linearized horizontal momentum equation is

$$u_0 \frac{\partial u_1}{\partial x} + w_1 \frac{\mathrm{d}u_0}{\mathrm{d}z} + \frac{1}{\rho_0}\frac{\partial p_1}{\partial x} = 0. \tag{6.1}$$

Because the terrain obstacle is isolated, $h(x) \to 0$ as $x \to \pm\infty$. Next we multiply (6.1) by $h(x)$ and integrate over x to get

$$\int_{-\infty}^{+\infty} u_0 h \frac{\partial u_1}{\partial x}\,\mathrm{d}x + \int_{-\infty}^{+\infty} w_1 h \frac{\mathrm{d}u_0}{\mathrm{d}z}\,\mathrm{d}x + \int_{-\infty}^{+\infty} \frac{h}{\rho_0}\frac{\partial p_1}{\partial x}\,\mathrm{d}x = 0. \tag{6.2}$$

Integrate the first integral by parts to get

$$I_1 = \int_{-\infty}^{+\infty} u_0 h \frac{\partial u_1}{\partial x}\,\mathrm{d}x = -\int_{-\infty}^{+\infty} u_1 u_0 \frac{\partial h}{\partial x}\,\mathrm{d}x. \tag{6.3}$$

Using the linear boundary condition (3.17), (6.3) becomes

$$I_1 = -\int_{-\infty}^{+\infty} u_1 w_1\,\mathrm{d}x, \tag{6.4}$$

where w_1 is the vertical perturbation velocity created by the obstacle. Again using (3.17) in the second integral in (6.2) and noting that $h(x) \to 0$ as $x \to \pm\infty$ gives

$$I_2 = \frac{\mathrm{d}u_0}{\mathrm{d}z}\int_{-\infty}^{+\infty} u_0 h \frac{\mathrm{d}h}{\mathrm{d}x}\,\mathrm{d}x = \frac{1}{4}\frac{\mathrm{d}u_0^2}{\mathrm{d}z}\int_{-\infty}^{+\infty}\frac{\mathrm{d}h^2}{\mathrm{d}x}\,\mathrm{d}x = 0. \tag{6.5}$$

Integrating the third integral in (6.2) by parts gives:

$$I_3 = \int_{-\infty}^{+\infty}\frac{1}{\rho_0} h \frac{\mathrm{d}p_1}{\mathrm{d}x}\,\mathrm{d}x = -\int_{-\infty}^{+\infty}\frac{p_1}{\rho_0}\frac{\mathrm{d}h}{\mathrm{d}x}\,\mathrm{d}x. \tag{6.6}$$

Adding $I_1 + I_2 + I_3$ gives

$$-\int_{-\infty}^{+\infty}\rho_0 u_1 w_1\,\mathrm{d}x = \int_{-\infty}^{+\infty} p_1 \frac{\mathrm{d}h}{\mathrm{d}x}\,\mathrm{d}x. \tag{6.7}$$

The right-hand side of (6.7) is the *form drag* per unit length of ridge exerted on the ridge by the flow, and the left-hand side is the drag per unit length of ridge exerted on the flow by the ridge, but this is simply a statement of Newton's third law. This response to the drag on the obstacle is transported upward by the gravity waves launched at the ground surface. If the mountain height and the pressure distribution are symmetric in x so that $h(x) = h(-x)$ and $p_1(x) = p_1(-x)$, then the form drag will be zero, and propagating gravity waves will not be launched; instead, the waves will be evanescent. In this case, the flow uniformly follows the terrain. If, however, $h(x)$ and $p_1(x)$ are asymmetric, then the waves can be launched. If we divide (6.7) by some horizontal length scale, ℓ, then the *wave stress* over the length ℓ is

$$\tau(z) = -\frac{1}{\ell}\int_{-\ell/2}^{\ell/2}\rho_0 u_1 w_1\,\mathrm{d}x = -\rho_0\overline{u_1 w_1}. \tag{6.8}$$

A useful formula for evaluating terms such as (6.8) is

$$\overline{w_1 u_1} = \frac{1}{2}\Re(w_1 u_1^*). \tag{6.9}$$

However, for this stress to be physically meaningful ℓ must be defined in a meaningful way. Over a corrugated surface, we can take ℓ to be the wavelength of the corrugation. Over an isolated mountain, we can take ℓ as some scale of the mountain width. Recall that "stress" represents a flux of momentum across some surface. It is sometimes stated that wave stress is a vertical flux of wave momentum, but as discussed by McIntyre (1981) such a statement is not accurate. As we have seen in Section 2.5.2.2, gravity waves poses pseudo-momentum. From (3.29) and (3.30) we see that the vertical and horizontal velocity perturbations are of opposite sign so that *cross correlation* $\overline{u_1 w_1}$ is negative. This term may be interpreted as either the downward propagation of mean-flow positive momentum or the upward propagation of mean-flow negative momentum. Now consider a steady surface-layer flow with $u_0(z)$ increasing with height above the ground

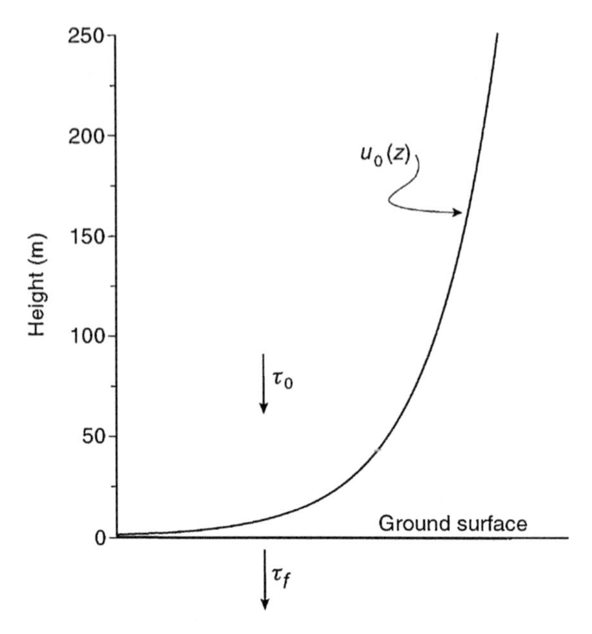

FIGURE 6.1 In the surface layer, a constant downward flux of momentum due to turbulence, τ_0, is required to balance the loss of momentum at the ground surface due to friction, τ_f.

surface. The theoretical wind profile is (Garratt, 1992)

$$u_0(z) = \frac{1}{k}\frac{\tau_0}{\rho}\ln(z - z_0), \tag{6.10}$$

where k is the Von Karman constant, z_0 is the aerodynamic surface roughness height (the effective height where $u_0 = 0$), and $\tau_0 = -\rho\overline{u'w'}$ is the surface stress due to turbulence. Then, as illustrated in Fig. 6.1, a constant downward flux of momentum, τ_0, is required to maintain the velocity profile and balance the momentum lost to friction at the ground surface, τ_f.

6.3 VARIATION OF WAVE STRESS WITH HEIGHT

In this section, we present the famous theory of Eliassen and Palm (1960) for the variation of wave stress with height. We begin with the energy equation (2.91), *i.e.*,

$$\frac{\mathrm{D}E}{\mathrm{D}t} = -\frac{\partial}{\partial x}(u_1 p_1) - \frac{\partial}{\partial z}(w_1 p_1) - \rho_0 u_1 w_1 \frac{\mathrm{d}u_0}{\mathrm{d}z}. \tag{6.11}$$

When E is independent of time, as in the case of mountain waves, (6.11) becomes

$$\frac{\partial}{\partial x}(E + u_1 p_1) + \frac{\partial}{\partial z}(w_1 p_1) = -\rho_0 u_1 w_1 \frac{\mathrm{d}u_0}{\mathrm{d}z}. \tag{6.12}$$

The right-hand side of (6.12) represents the product of the wave stress $(-\rho_0 u_1 w_1)$ times the rate of strain $(\mathrm{d}u_0/\mathrm{d}z)$. Now integrate (6.12) as

$$\int_{-\infty}^{+\infty} \frac{\partial}{\partial x}(u_0 E + u_1 p_1)\,\mathrm{d}x + \frac{\partial}{\partial z}\int_{-\infty}^{+\infty} w_1 p_1\,\mathrm{d}x = -\frac{\mathrm{d}u_0}{\mathrm{d}z}\int_{-\infty}^{+\infty} \rho_0 u_1 w_1\,\mathrm{d}x. \tag{6.13}$$

Because we require the perturbation terms to vanish as $x \to \pm\infty$ the first integral in (6.13) is zero, and we then have

$$\frac{\partial}{\partial z}\int_{-\infty}^{+\infty} w_1 p_1\,\mathrm{d}x = -\frac{\mathrm{d}u_0}{\mathrm{d}z}\int_{-\infty}^{+\infty} \rho_0 u_1 w_1\,\mathrm{d}x. \tag{6.14}$$

Now consider the linearized first-order momentum equation, *i.e.*, (6.1), which can be written as

$$\frac{\partial}{\partial x}(\rho_0 u_0 u_1 + p_1) + \rho_0 w_1 \frac{\mathrm{d}u_0}{\mathrm{d}z} = 0. \tag{6.15}$$

Now multiply (6.15) by $(\rho_0 u_0 u_1 + p_1)$ to get

$$\frac{1}{2}\frac{\partial}{\partial x}(\rho_0 u_0 u_1 + p_1)^2 + \rho_0 w_1 \frac{\mathrm{d}u_0}{\mathrm{d}z}(\rho_0 u_0 u_1 + p_1) = 0. \tag{6.16}$$

Now integrate (6.16) to get

$$\frac{1}{2}\int_{-\infty}^{+\infty} \frac{\partial}{\partial x}(\rho_0 u_0 u_1 + p_1)^2\,\mathrm{d}x + \rho_0 \frac{\mathrm{d}u_0}{\mathrm{d}z}\int_{-\infty}^{+\infty} \rho_0 u_0 u_1 w_1\,\mathrm{d}x + \rho_0 \frac{\mathrm{d}u_0}{\mathrm{d}z}\int_{-\infty}^{+\infty} w_1 p_1\,\mathrm{d}x = 0. \tag{6.17}$$

The first integral in (6.17) is a perfect differential and so evaluates to zero, leaving

$$u_0 \int_{-\infty}^{+\infty} \rho_0 u_1 w_1\,\mathrm{d}x = -\int_{-\infty}^{+\infty} w_1 p_1\,\mathrm{d}x. \tag{6.18}$$

Making use of the definition of an average as in (6.8), we can write (6.14) and (6.18), respectively as

$$\frac{\partial}{\partial z}(\overline{w_1 p_1}) = -\rho_0 \overline{u_1 w_1}\frac{\mathrm{d}u_0}{\mathrm{d}z} \tag{6.19}$$

$$\overline{w_1 p_1} = -\rho_0 u_0 \overline{u_1 w_1}. \tag{6.20}$$

Substituting (6.20) into (6.19) gives

$$u_0 \frac{\partial}{\partial z}(\rho_0 \overline{u_1 w_1}) = 0. \tag{6.21}$$

Thus, if $u_0 \neq 0$, the wave stress $-\rho_0 \overline{u_1 w_1}$ is constant unless the wave is dissipating. Dissipation occurs during wave breaking or near a critical level.

6.4 MOUNTAIN WAVE STRESS

The vertical transport of mean-flow horizontal momentum has been studied since the middle of the 20th Century. When mountain waves break or encounter a critical level, the momentum they transport is transferred to the mean flow in the form of a drag. This wave drag balances the form drag generated by the horizontal pressure differences across the mountain. Wave stress is an essential component of the global momentum balance of the atmosphere and the general circulation. However, wave stress is also important on the meso- and microscales. Mountain wave stress is special because it can be unambiguously calculated and measured. This feature allows testing of theoretical and numerical models. The equations to be developed in the following sections are limited to the lower atmosphere; however extensions to hydrostatic gravity waves and consideration of compressibility $1/H_s$ in the middle and upper atmosphere are straightforward. These equations are also applicable to the WKB method.

6.4.1 WAVE STRESS OVER A SURFACE CORRUGATION

The simplest wave stress problem is uniform flow with constant stratification over a surface corrugation. From Section 3.2, we have

$$w_1(x,z) = -u_0 H k_s \sin(k_s x + m_s z),$$
$$u_1(x,z) = u_0 H m_s \sin(k_s x + m_s z),$$

where k_s and H are the corrugation wavenumber and amplitude respectively, and

$$m_s = \left(\frac{N^2}{u_0^2} - k_s^2 \right).$$

Using the above equations in (6.8) gives

$$\tau = \begin{cases} 0.5\rho_0 (u_0 H)^2 k_s \left[\frac{N^2}{u_0^2} - k_s^2 \right]^{1/2} & \text{if } \frac{N}{u_0} > k_s, \\ 0 & \text{if } \frac{N}{u_0} \leqslant k_s. \end{cases} \quad (6.22)$$

Figure 6.2 (from Chimonas and Nappo (1989)) shows a plot of dimensionless wave stress $\frac{\tau}{\rho_0 (NH)^2}$ over a corrugated surface as a function of non-dimensional background wind speed $u_0 k_s/N$. The maximum wave stress occurs when $u_0 k_s/N = 1/\sqrt{2}$. For low u_0, (6.22) takes the linear form

$$\tau \approx 0.5\rho_0 H^2 k_s N u_0, \quad (6.23)$$

which is also plotted in Fig. 6.2. We can compare the magnitude of the wave stress with the *friction stress* at the ground surface if we let the friction stress, τ_f, be given by, for example, Gill (1982)

$$\tau_f = 0.5\rho_0 C_D u_0^2, \quad (6.24)$$

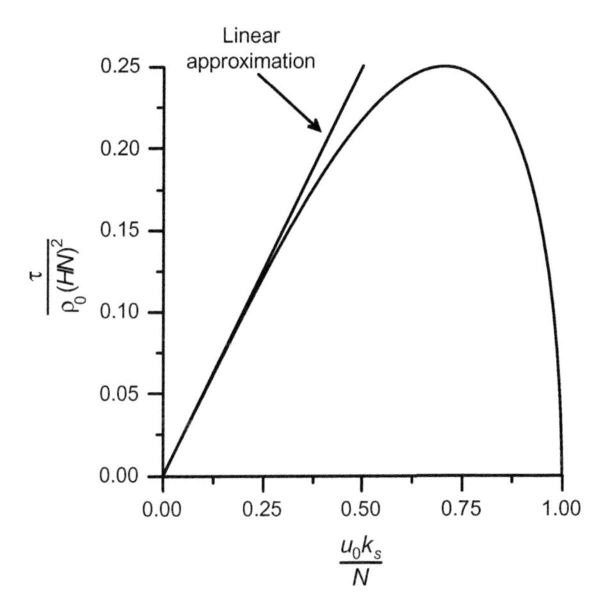

FIGURE 6.2 Non-dimensional wave stress as a function of non-dimensional wind speed over a surface corrugation (taken from Chimonas and Nappo (1989)).

where C_D is the *drag coefficient*. If we let τ_w be the wave stress, then

$$\frac{\tau_w}{\tau_f} = \frac{0.5\rho_0 k_s (u_0 H)^2 \left[\frac{N^2}{u_0^2} - k_s^2\right]^{1/2}}{0.5\rho_0 C_D u_0^2}. \tag{6.25}$$

For $C_D = 0.005$, which corresponds to lawn grass up to 1 cm high (Sutton, 1953) and $H/\lambda_s = 0.01$,

$$\frac{\tau_w}{\tau_f} \approx \left[\left(\frac{N}{u_0 k_s}\right)^2 - 1\right]^{1/2}. \tag{6.26}$$

If $u_0 < \frac{N}{\sqrt{2}k_s}$, then for this particular case $\frac{\tau_w}{\tau_f} > 1$, *i.e.*, the wave stress is greater than the surface friction stress. If for example, $N = 0.03\ \mathrm{s}^{-1}$, $\lambda_s = 1000$ m, and $H = 10$ m, then for wind speeds less than 3 ms^{-1} the wave stress will be greater than the friction stress. This is a significant effect in the nighttime planetary boundary layer, especially under conditions of weak winds and strong stratification.

6.4.1.1 Wave Stress Over an Isolated Two-Dimensional Mountain

We now consider the wave stress over an isolated two-dimensional mountain.[1] By isolated, we mean that the terrain is flat upwind and downwind of the mountain. The drag per unit length of mountain D/l is

$$\frac{D}{l} = -\int_{-\infty}^{+\infty} \rho_0 u_1 w_1 \,\mathrm{d}x. \tag{6.27}$$

Using the inverse Fourier transform, we write

$$w_1(x, z) = \frac{1}{2\pi}\int_{-\infty}^{+\infty} \hat{w}(k, z)\mathrm{e}^{ikx}\,\mathrm{d}k \tag{6.28}$$

and

$$u_1(x, z) = \frac{1}{2\pi}\int_{-\infty}^{+\infty} \hat{u}(k, z)\mathrm{e}^{ikx}\,\mathrm{d}k. \tag{6.29}$$

Using the continuity equation (2.24) in (6.29) we get

$$u_1(x, z) = \frac{1}{2\pi}\int_{-\infty}^{+\infty} \frac{i}{k}\frac{\mathrm{d}\hat{w}}{\mathrm{d}z}\mathrm{e}^{ikx}\,\mathrm{d}k. \tag{6.30}$$

Then using (6.28) and (6.30) in (6.27) and using the sifting function (3.20) gives

$$\begin{aligned}
\frac{D}{l} &= -\frac{\rho_0}{4\pi^2}\int_{-\infty}^{+\infty}\mathrm{d}x\int_{-\infty}^{+\infty}\hat{w}(k, z)\mathrm{e}^{ikx}\,\mathrm{d}k\int_{-\infty}^{+\infty}\frac{i}{k'}\frac{\mathrm{d}\hat{w}(k', z)}{\mathrm{d}z}\mathrm{e}^{ik'x}\,\mathrm{d}k' \\
&= -i\frac{\rho_0}{4\pi^2}\int_{-\infty}^{+\infty}\mathrm{d}k\int_{-\infty}^{+\infty}\mathrm{d}k'\,\hat{w}(k, z)\frac{\mathrm{d}\hat{w}(k', z)}{\mathrm{d}z}\frac{1}{k'}\int_{-\infty}^{+\infty}\mathrm{e}^{i(k+k')x}\,\mathrm{d}x \\
&= -i\frac{\rho_0}{2\pi}\int_{-\infty}^{+\infty}\mathrm{d}k\int_{-\infty}^{+\infty}\mathrm{d}k'\,\hat{w}(k, z)\frac{\mathrm{d}\hat{w}(k', z)}{\mathrm{d}z}\frac{1}{k'}\delta(k + k').
\end{aligned} \tag{6.31}$$

Integration of (6.31) with respect to $\mathrm{d}k'$ gives

$$\frac{D}{l} = i\frac{\rho_0}{2\pi}\int_{-\infty}^{+\infty}\hat{w}(k, z)\frac{\mathrm{d}\hat{w}(-k, z)}{\mathrm{d}z}\frac{1}{k}\,\mathrm{d}k. \tag{6.32}$$

The reality condition requires

$$\frac{\mathrm{d}\hat{w}(-k, z)}{\mathrm{d}z} = \frac{\mathrm{d}\hat{w}^*(k, z)}{\mathrm{d}z}, \tag{6.33}$$

and then we can write

$$\frac{D}{l} = \frac{\rho_0}{2\pi}\Im\int_{-\infty}^{+\infty}\frac{1}{k}\hat{w}(k, z)\frac{\mathrm{d}\hat{w}^*(k, z)}{\mathrm{d}z}\,\mathrm{d}k, \tag{6.34}$$

where $\Im$ is the imaginary part of the integral. An alternate form of (6.34) used by Bretherton (1969) is

$$\frac{D}{l} = \frac{\rho_0}{2\pi}\int_{-\infty}^{+\infty}\frac{1}{2ki}\left(\hat{w}\frac{\mathrm{d}\hat{w}^*}{\mathrm{d}z} - \hat{w}^*\frac{\mathrm{d}\hat{w}}{\mathrm{d}z}\right)\mathrm{d}k. \tag{6.35}$$

[1] The same mathematics also applies to two-dimensional valleys.

If the isolated mountain has a Gaussian-shaped cross section and if the background variables are constant, then we can use (3.40) and (3.42) in (6.34) to get

$$\frac{D}{l} = \rho_0(u_0H)^2 \int_0^{k_c} \left[b^2 \mathrm{e}^{-(kb)^2/2}\right] k \left[\frac{N^2}{u_0^2} - k^2\right]^{1/2} \mathrm{d}k, \qquad (6.36)$$

where $k_c = N/u_0$ is a cutoff wavenumber. Mountain waves with $k > k_c$ will be evanescent. Figure 6.3 shows a plot of dimensionless wave drag per unit length of mountain $(D/l)/[\rho_0 b(HN)^2]$ as a function of dimensionless wind speed $u_0/(Nb)$ for the case of flow over a Gaussian-shaped mountain. The curve is similar to that shown in Fig. 6.2; however, now the drag approaches zero asymptotically for large values of $u_0/(Nb)$. Numerical evaluation of (6.36) gives $0.34\rho_0 b(HN)^2$ for the maximum wave drag, which occurs at a wind speed of $0.54Nb$.

If the wind speed is small and the mountain is wide such that $Nb/u_0 \gg 1$, then (6.36) reduces to

$$D/\ell = \rho_0 u_0 H^2 b^2 N \int_0^{k_c} k\mathrm{e}^{-k^2b^2/2}\, \mathrm{d}k. \qquad (6.37)$$

The integration is elementary giving

$$D/\ell = \rho_0 u_0 H^2 N \left[1 - \mathrm{e}^{-k_c^2 b^2/2}\right]. \qquad (6.38)$$

Then because $k_c b = Nb/u_0 \gg 1$, (6.38) further reduces to

$$D/\ell = \rho_0 u_0 H^2 N. \qquad (6.39)$$

which represents D/l for hydrostatic gravity waves. We see that for hydrostatic mountain waves the wave drag is linear in u_0 and independent of mountain width.

When the wind speed is high and the mountain is narrow such that $Nb/u_0 \ll 1$ only the very long wavelengths contribute to the integral in (6.36). In this case, the exponential term approaches unity and

$$D/\ell = \rho_0(u_0Hb)^2 \frac{N}{u_0} \int_0^{k_c} k\, \mathrm{d}k \qquad (6.40)$$

and

$$D/\ell = 0.5\rho_0(HNb)^2 \frac{N}{u_0}, \qquad (6.41)$$

where we have used $k_c = N/u_0$. Thus, for high winds, the wave drag decreases as u_0^{-1} as seen in Fig. 6.3. This behavior is the most significant difference between wave drag over a corrugated surface and over an isolated mountain.

If we divide both sides of (6.36) by $2b$, the result will be a wave stress, *i.e.*,

$$\tau_w = 0.5\rho_0(u_0H)^2 \int_0^{k_c} \left[kb\mathrm{e}^{-(kb)^2/2}\right] \left[\frac{N^2}{u_0^2} - k^2\right]^{1/2} \mathrm{d}k. \qquad (6.42)$$

In Chapter 3, we examined the term $kb\mathrm{e}^{-(kb)^2/2}$ as a function of kb. This result is graphed in Fig. 3.12. Using that result, the maximum contribution to τ_w occurs

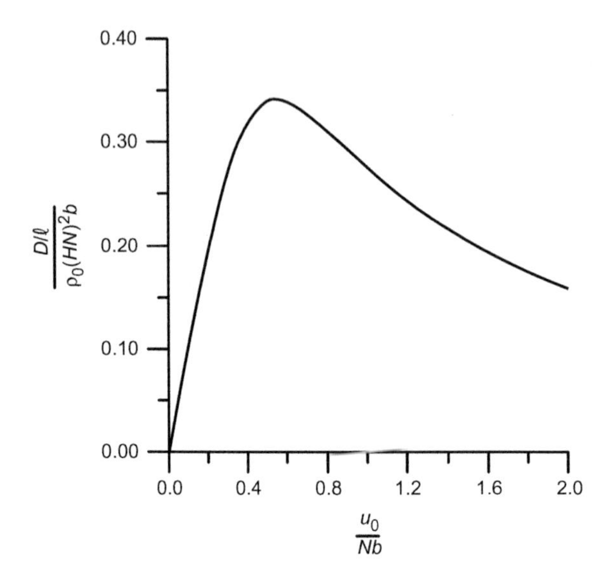

FIGURE 6.3 Same as Fig. 6.2 but for flow over a Gaussian ridge.

when $kb = \sqrt{2}$. For scaling purposes, it is sufficient to estimate the wave stress maximum over a mountain by setting $k = 1/b$ in (6.42). Then integration gives

$$\tau_w \approx 0.5\rho_0(u_0 H)^2 \frac{1}{b}\left[\frac{N^2}{u_0^2} - \left(\frac{1}{b}\right)^2\right]^{1/2}, \tag{6.43}$$

which looks very much like the stress over a surface corrugation (6.22) if we replace k_s by $1/b$. We see that a surface corrugation and a two-dimensional mountain with approximately equal height and length scales will generate approximately equal values of wave stress. This presents a computational simplification since we can replace the integration over wavenumbers by a single term.

6.4.2 WAVE STRESS OVER THREE-DIMENSIONAL OBJECTS

In reality, isolated mountains or corrugated surfaces are seldom seen. Rather we see orography similar to the contrasting types ranging from Alpine peaks to the rolling hills of the Appalachian Mountains shown in Figs. 6.4 and 6.5, respectively. Each horizontal component of the surface wind that passes over an obstacle launches gravity waves. In complex terrains, mountains and hills can be orientated in many directions, and so we expect many gravity waves each generated by a component of the mean surface wind. The directions of these components are confined to $\chi(0) \pm \pi/2$, where $\chi(0)$ is the azimuth of the surface wind vector. Bretherton (1969) and Shutts (1995) developed equations for the

FIGURE 6.4 In the Alps. Photo by Ben Smith.

FIGURE 6.5 Complex terrain in the Appellation foothills. Photo by Emily Pass, mlewallpapers.com

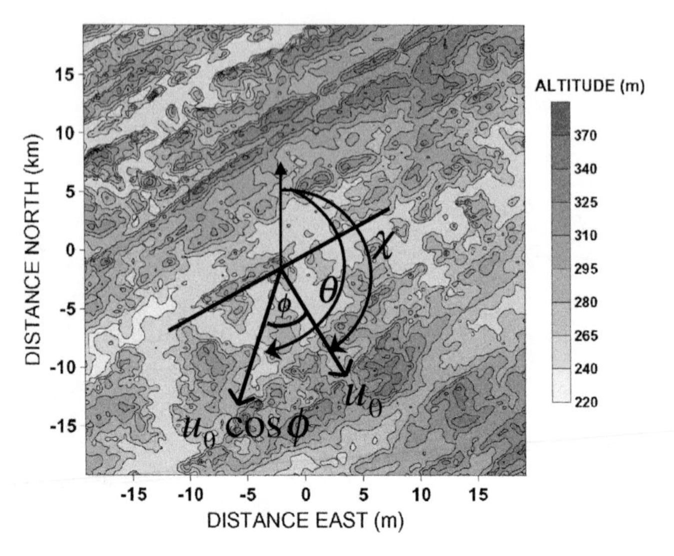

FIGURE 6.6 Surface wind component in direction $\theta - \chi$.

horizontal-average wave stress of the form

$$\bar{\tau}_x = -\rho \frac{8\pi^2}{XY} U_0^2 \int_0^{\infty} \int_{-\pi/2}^{\pi/2} |\hat{h}|^2 F(K, \phi) \cos^2(\phi - \chi_0) \cos\phi K^2 \, \mathrm{d}k\mathrm{d}\phi, \quad (6.44)$$

where XY is the area of the domain, U_0 is the magnitude of the background surface wind speed, χ is the azimuth of the surface wind, and F is a response function

$$F(K, \phi) = \frac{1}{2i|\hat{w}(K, \phi, 0)|^2} \left(\frac{\partial \hat{w}}{\partial z} \hat{w}* - \frac{\partial \hat{w}*}{\partial z} \hat{w} \right). \quad (6.45)$$

We describe here a method while lacking the mathematical elegance of Bretherton and Shutts is simple and easy to comprehend. Rather than first developing stress equations in cartesian coordinates (x, y, z) and then transforming into cylindrical-polar coordinates (r, θ, z), we go directly to the latter reference system. Let the background surface wind have magnitude $U_0(z)$ and direction $\chi(z)$. Note that the background wind is free to change direction with height. The component of the surface wind speed in some direction θ from north is

$$u_\theta(0) = U_0(0) \cos[\theta - \chi(0)]. \quad (6.46)$$

We let $\phi = \theta - \chi$ and require that $-\pi/2 < \phi < \pi/2$ as illustrated in Fig. 6.6.[2] The wave stress must be evaluated for all values of ϕ. For a given direction θ, the

[2] The contours for Fig. 6.6 represent terrain elevation of the central portion Tennessee River Valley. Knoxville, Tennessee, USA lies about in the center of the terrain map.

Taylor–Goldstein equation is

$$\frac{d^2\hat{w}(\kappa,\theta,z)}{dz^2}+\left[\frac{N^2}{u_\theta^2}-\frac{u_\theta''}{u_\theta}-\kappa^2\right]\hat{w}(\kappa,\theta,z)=0, \tag{6.47}$$

and the bottom boundary condition is

$$\hat{w}(\kappa,\theta,0)=i\kappa u_\theta\hat{h}(\kappa,\theta), \tag{6.48}$$

where $\hat{h}$ is the two-dimensional Fourier transform of the terrain height in polar coordinates, *i.e.*, (κ,θ). The calculations are simplified by introducing a shape function Φ, *i.e.*,

$$\hat{w}(\kappa,\theta,z)=\hat{w}(\kappa,\theta,0)\Phi(\kappa,\theta,z) \tag{6.49}$$

with the bottom boundary condition $\Phi(\kappa,\theta,0)=1$. Using (6.49) in (6.47) gives

$$\frac{d^2\Phi}{dz^2}+\left[L_s^2-\kappa^2\right]\Phi=0, \tag{6.50}$$

where we have introduced the Scorer parameter,

$$L_s^2=\frac{N^2}{u_\theta^2}-\frac{u_\theta''}{u_\theta}. \tag{6.51}$$

The area-averaged wave stresses are now

$$\overline{\tau}_x(z)=\frac{2\pi^2 i}{XY}\int_0^\infty\int_{-\pi/2}^{\pi/2}\rho_0|\hat{w}(\kappa,\theta,0)|^2\sin\theta\left(\frac{\partial\Phi}{\partial z}\Phi^*-\frac{\partial\Phi^*}{\partial z}\Phi\right)d\kappa d\theta, \tag{6.52}$$

$$\overline{\tau}_y(z)=\frac{2\pi^2 i}{XY}\int_0^\infty\int_{-\pi/2}^{\pi/2}\rho_0|\hat{w}(\kappa,\theta,0)|^2\cos\theta\left(\frac{\partial\Phi}{\partial z}\Phi^*-\frac{\partial\Phi^*}{\partial z}\Phi\right)d\kappa d\theta. \tag{6.53}$$

Although the preceding is straightforward, the actual calculations are complicated. A technique developed by Nappo and Svensson (2008) divides the semicircle $\pm\pi/2+\theta$ into a number of wedges or sectors as illustrated in Fig. 6.7. They used 10° sectors. The component of the surface wind in the middle of each sector, represented by the broken lines in Fig. 6.7, defines $u_0(\theta)$. Equation (6.49) is then written as

$$\hat{w}(\kappa_n,\theta_i,z)=\hat{w}(\kappa_n,\theta_n,0)\Phi(\kappa_n,\theta_n,z), \tag{6.54}$$

where subscript n represents the nth sector. The bottom boundary condition (6.48) must be calculated for each sector, and this requires the Fourier transform of the terrain height $\hat{h}(k_n,\theta_n)$ within each sector. This is done by first calculating the two-dimensional Fourier transform of the topography within giving $\hat{h}(k,l)$. A polar coordinate system consisting of equally spaced annular rings is superimposed on the (k,l) wavenumber plane. A schematic of this procedure is given in Fig. 6.8. The broken horizontal and vertical lines represent values of $\hat{h}(k,l)$. Sector S_i is

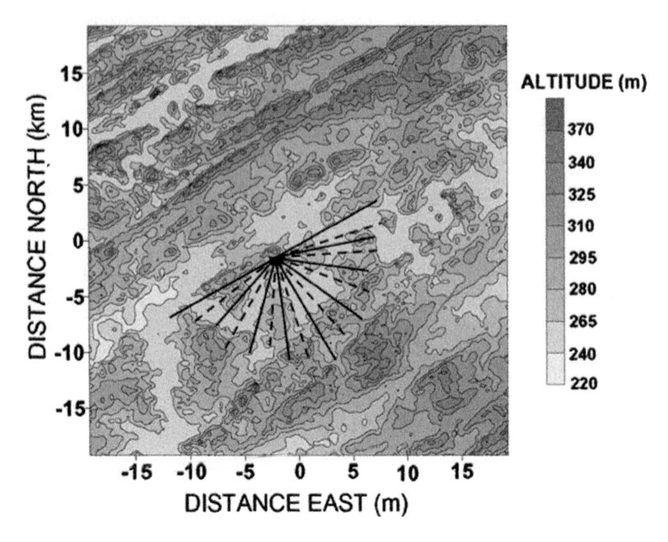

FIGURE 6.7 Solid lines mark the sectors within which the wave stress is calculated. The dashed lines in the center of each sector represent the direction of the sector average horizontal wavenumbers and surface wind speed.

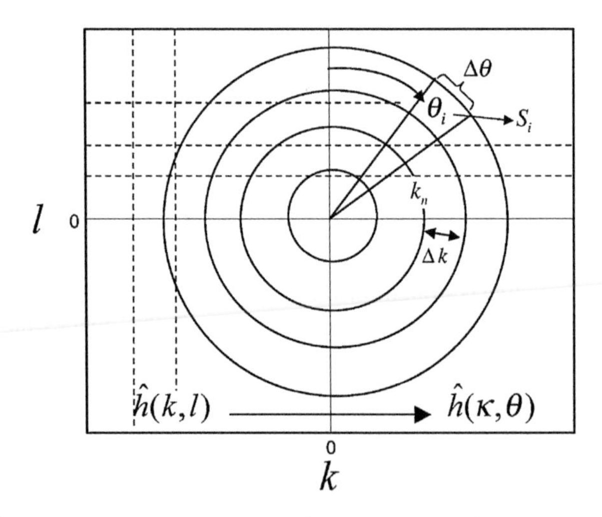

FIGURE 6.8 Illustration of the conversion of Cartesian wavenumber space to rectangular polar coordinate space.

defined by the area $k_i(k_i + \Delta k)\Delta\theta$ The numerical average of the values of $\hat{h}(k, l)$ that lie within sector S_i is taken to be $\hat{h}(\kappa_i, \theta_i)$.

Bretherton (1969), Hines (1988), and Shutts (1995) recognized that a turning of the horizontal wind with height, *i.e.*, wind-direction shear, can result in critical

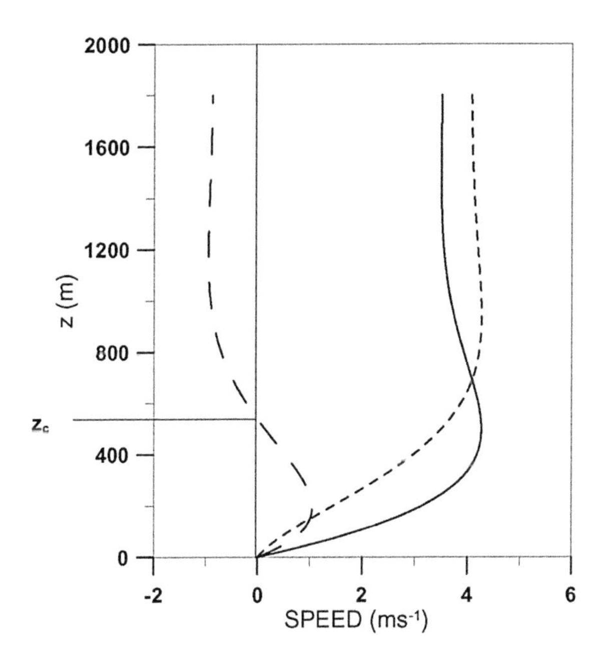

FIGURE 6.9 Wind speed profiles in an Ekman spiral. Wind profile in the direction of the surface wind (solid line); wind profile in the direction of 125° (dashed line); wind profile in the direction of 350° (broken line).

levels for mountain waves, *i.e.*, $u_0 = 0$ for some horizontal projection of the mean-wind profile. To illustrate, consider the Ekman spiral defined by (4.74) and (4.75) and illustrated in Fig. 4.21. The surface wind vector is directed toward 45° from north, and so the wave stress must be calculated for wind projections from 315° to 135°. The wind speed profile in the direction of the surface wind is shown in Fig. 6.9. Also plotted in Fig. 6.9 are the profiles projected in the directions of 350° and 125° from north. Because the Ekman spiral turns to the right in the northern hemisphere we do not see critical levels until after 135°. But to the left of the spiral, critical levels exist as shown in Fig. 6.9.

As an example of the effects of wind-direction shear, let us consider the wave stress created by a three-dimensional horizontally symmetric Gaussian hill as illustrated in Fig. 6.10. The hill has a maximum height of 300 m and a scale width of 100 m. The Ekman wind spiral shown in Fig. 4.21 was used and the thermal stratification was constant. Figure Fig. 6.11 shows the profile of the horizontally averaged wind stress. Because of the horizontal symmetry of the hill, one would expect the wave stress to be constant with height, which is indeed the case for constant direction winds. However, from Fig. 6.11, we see that the stress decreases continually but not continuously with height. These decreases are due to wave absorption at critical levels, and the discontinuities are attributed to the partitioning of the winds and terrain into 10° sectors. The mean wind changes

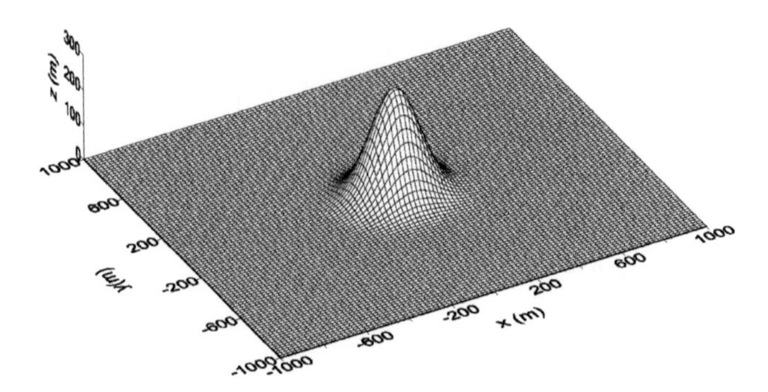

FIGURE 6.10 Three-dimensional Gaussian hill. Maximum height is 300 m and scale width is 100 m.

direction smoothly with height as seen in Fig. 6.11, but the direction of the wave stress shifts discontinuously in a clockwise direction. An important effect due to wind-direction shear is that the stress does not act in opposition to the wind direction. In our example, the surface wind is directed toward 54° but the wave stress at the ground surface is directed toward 214° which is about 20° south of the "wind-apposition case." This difference in the wind-stress directions will act in time to force the winds to the north toward lower atmospheric pressure. But this is exactly what the surface stresses do.

6.5 SECONDARY EFFECTS OF WAVE DRAG

6.5.1 DIRECTION FORCING

The main effect of terrain-induced wave drag on the atmosphere is to reduce flow speeds. However, because the wave drag acts locally and in a given direction secondary flow effects can be created. As discussed by Chimonas and Nappo (1989) one such effect created by the *directional selectivity* of wave drag is to turn flow over complex terrain in a direction parallel to the surface contours. This process is illustrated in Fig. 6.12. The turbulent drag always acts against the flow, *i.e.*, opposite to the direction of the mean wind; however, the wave drag acts against the component of the mean wind normal to the terrain contours. This directional selectivity results in a tendency for the flow to follow the terrain contours rather than go over the terrain, an effect which is enhanced by a stable stratification. If this drag continues, the flow will become parallel to the mountains, and the wave drag will vanish. However, this process might take some time to be realized. We should also note that in some numerical models the wave stress is taken to be in opposition to the mean wind. But this is not correct. For a two-dimensional mountain, the wave stress is in opposition to the component of the mean wind perpendicular to the mountain.

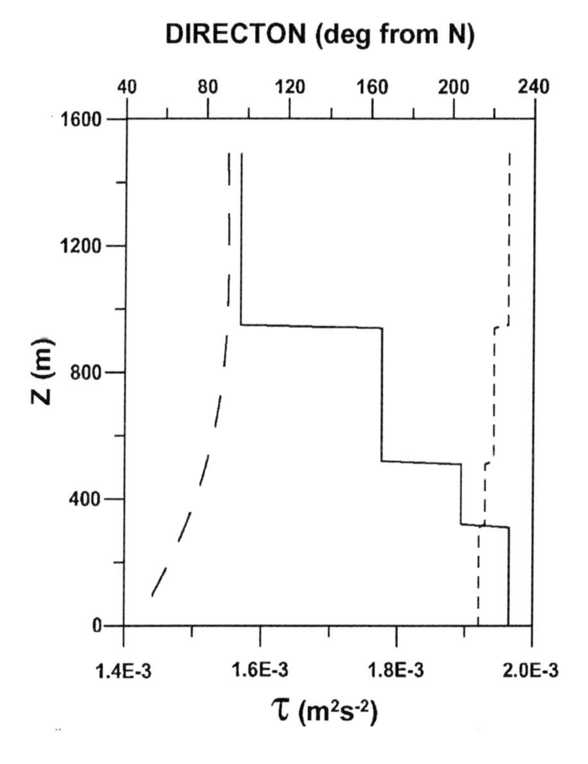

FIGURE 6.11 Vertical profile of average horizontal gravity wave stress (solid line); profile of wave stress direction (dashed line); profile of mean wind direction (broken line).

6.5.2 LEE WAVE DRAG OVER A TWO-DIMENSIONAL MOUNTAIN

Smith (1976, 1978) gives a detailed analysis of lee wave drag downwind of a two-dimensional ridge of moderate height wand width. The following argument is based on Smith's 1976 analysis. Using (3.38) in (3.44) we can write

$$w_1(x, z) = \Re \int_{\infty}^{\infty} iku_0\hat{h}(k_n)\frac{\hat{w}(k_n, z)}{\hat{w}(k, 0)}e^{ikx}\,dk, \tag{6.55}$$

where $\hat{h}$ is the Fourier transform of the mountain height profile. At some point on a two-dimensional mountain, $w_1(x, z)$ will have a maximum value and $u_1(x, z) = 0$. Then the drag per unit length is

$$D/\ell = \int_0^h \Delta p(z)\mathrm{d}z, \tag{6.56}$$

where $\Delta p(z)$ is the atmospheric pressure differences on either side of the mountain at height z, and h is the height of the mountain. Bernoulli's equation along this

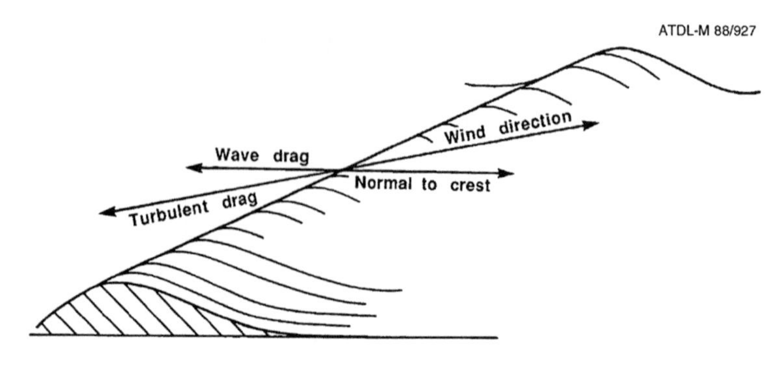

FIGURE 6.12 The directional preference introduced by wave drag. Conventional turbulent drag acts along the wind direction, while the wave drag is always normal to the surface contours.

streamline is

$$\frac{p_0}{\rho} + gz + \frac{1}{2}[(u_0 - u_1)^2 + w_1^2)] = \text{constant}. \tag{6.57}$$

Using (6.57) in (6.56) gives[3]

$$D/\ell = \int_0^\infty \frac{1}{2}\rho w_1^2\, \mathrm{d}z. \tag{6.58}$$

Using (4.76) for w_1 gives

$$D/\ell = \frac{2\rho\pi^2 u_0^2 |\hat{h}(k_n)|^2}{\overline{Z}_n^3}, \tag{6.59}$$

where

$$\overline{Z}_n = \left[\frac{\int_0^\infty \hat{w}^2(k_n, z)\mathrm{d}z}{(\partial w/\partial z)|_{0^2}}\right]^{\frac{1}{2}}. \tag{6.60}$$

Expanding on Smith (1976), we identify (6.59) as a production term for wave stress, and $\overline{Z}_n$ as a measure of the distance from the ground surface to the height of maximum wave energy. Note that the lee wave drag is a function of a single horizontal wavenumber k_n which is the wavenumber of the lee wave.

6.5.3 MOMENTUM FLUX DUE TO MOUNTAIN LEE WAVES

During the Mesoscale Alphine Program, Smith *et al.* (2002) observed a case when lee waves off Mount Blanc were expected none were observed. At other times the lee waves extended only a short way from the mountain. These were interesting observations since the conditions for lee waves were satisfied. Smith

[3] Note that Smith (1976) does not justify the relation $\Delta p = 1/2\rho w_1^2$; however, Broad (2002) describes this result in detail.

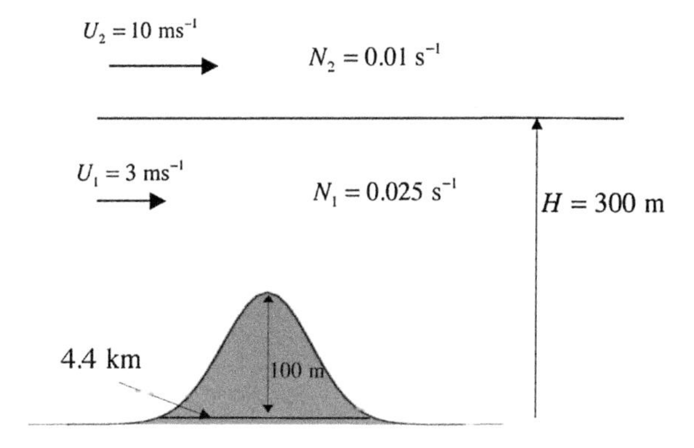

FIGURE 6.13 Conditions for a lee wave stress illustration and calculation.

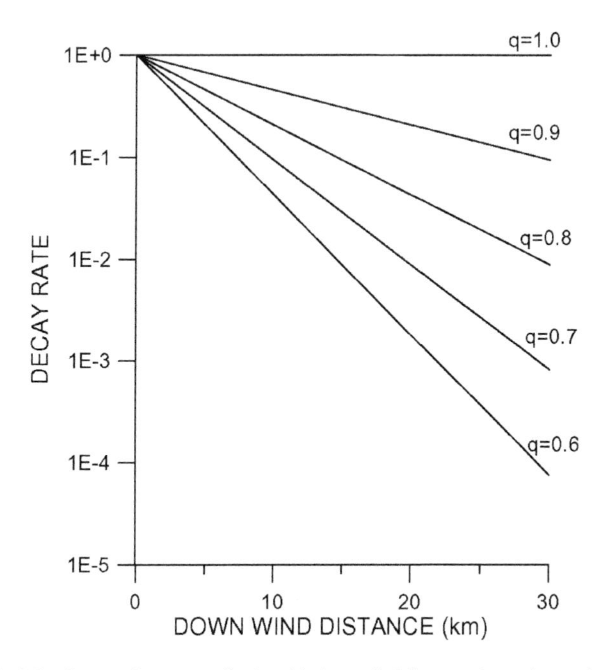

FIGURE 6.14 Decay of wave amplitude with downwind distance for various values of reflection coefficient, q.

concluded that momentum was being extracted from the lee wave in the boundary layer. Smith, Jiang, and Doyle (2006) proposed a theory for lee wave absorption based on a two-layer model, *i.e.*, a thin homogeneous boundary layer beneath a deep free atmosphere. Jiang, Doyle, and Smith (2006) extended this work with a numerical simulation and theoretical formulation. The main conclusion was that a viscous boundary layer can absorb momentum from a lee wave through a reflection

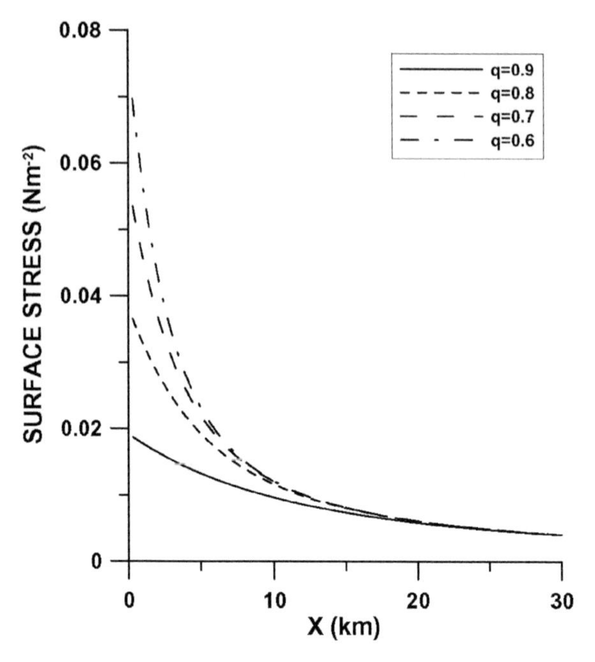

FIGURE 6.15 Decay of lee-wave stress with downwind distance for various values of q.

coefficient, $0 < q < 1$ that is a function of wind speeds and vertical wavenumbers in each layer, *i.e.*, in the bottom layer

$$\hat{w}_1 = A\mathrm{e}^{im_1 z} + B\mathrm{e}^{-im_1 z}, \tag{6.61}$$

and $q = -A/B$. The reflection coefficient is given by

$$q\cos(2m_1 H) = \frac{s^2 - m_1^2/m_2^2}{s^2 + m_1^2/m_2^2}, \tag{6.62}$$

where $s = U_1^2/U_2^2$ and H is the depth of the boundary layer. Equation (6.62) is similar to the reflection coefficient r defined by (4.9), except q represents a loss of momentum since there is no wave transmission into the ground. Thus the lee wave looses amplitude at each trough in the boundary layer. To illustrate this decay consider the problem illustrated in Fig. 6.13. A two-dimensional Gaussian-shaped ridge has a maximum height of 100 m and a base width of 4.4 km. The ridge sits in a boundary layer 300 m deep with $U_1 = 3$ ms^{-1} and $N_1 = 0.25$ s^{-1}. The free atmosphere above has $U_2 = 10$ ms^{-1} and $N_2 = 0.01$ s^{-1}. A linear mountain wave program calculated the spectrum of mountain waves. The lee wave is taken to be the most energetic mountain wave; in this case the wavelength of the lee wave is 4.4 km. Figure 6.14 shows the rate of decay of lee-wave amplitude for various values of reflection coefficient, q, as functions of downwind distance. Figure 6.15

shows the decay of wave stress as functions of downwind distance for various values of q.

PROBLEMS

1. Prove that $\rho_0 \overline{u_1 w_1} = \frac{\rho_0}{2ik}\left(\frac{\partial \hat{w}^*}{\partial z}\hat{w} - \frac{\partial \hat{w}}{\partial z}\hat{w}^*\right)$, where $w_1 = \Re\left[\hat{w}(k,z)e^{ikx}\right]$ and $\hat{w}^*$ denotes the complex conjugate.
2. Show that $k\frac{\overline{E}}{\omega - u_0 k}\frac{\partial \omega}{\partial m}$ is equal to the wave Reynolds stress.
3. Calculate the wave drag per unit length of a mountain with maximum height of 200 m if the atmospheric pressure difference between the upwind and downwind sides of the mountain is 50 Pa.
4. Show that $\overline{w_1 u_1} = \frac{1}{2}\Re(w_1 u_1{}^*)$.
5. Give an approximate value of the wave stress over a Gaussian mountain with $N = 0.02\ \mathrm{s}^{-1}$, $u_0 = 10\ \mathrm{ms}^{-1}$, H = 1000 m, and $b = 2000$ m.
6. If the Scorer parameter is $8\times10^{-5}\ \mathrm{m}^{-1}$ what is the maximum mountain-wave wavelength that a Gaussian mountain can generate if $H = 1000$ m and $b = 2$ km?
7. A constant wind of 5 ms^{-1} is blowing over a number of parallel ridges. Each ridge is 100 m high and the valleys between the ridges are about 500 m wide. The ridges are orientated N–S, and the wind is blowing at 50° from north. What is the wave stress if $N = 0.15\ \mathrm{s}^{-1}$?
8. What is the maximum wind speed for gravity wave generation for the case in Problem 7?
9. By how much is the wave stress reduced if the wave crosses a critical level where the Richardson number is 1.5?
10. At some time, t_0 a constant west wind of 20 ms^{-1} blows perpendicular over a long mountain range. The wind is in geostrophic balance and the wave stress is 20 Pa. Estimate how long it will take for the wind direction to change from 270 to 260?
11. Calculate q for the case illustrated in Fig. 6.13.
12. A wave with amplitude $w_1 = 3\ \mathrm{ms}^{-1}$, wavelength $\lambda_x = 5$ km, and period 6 min moves in an atmosphere with $N = 0.15\ \mathrm{ms}^{-1}$, $u_0 = 5\ \mathrm{ms}^{-1}$, and $\rho_0 = 1000$ hPa. Calculate the wave stress when the phase difference between w_1 and u_1 is (a) 0, (b) 30, (c) 60, and (d) 90.
13. Show that a ducted gravity has zero wave stress.
14. By how much is the wave stress reduced if a "downward" propagating gravity wave crosses a critical level?
15. Calculate the increase of $\overline{u_1 w_1}$ if a wave packet moves upward from 5 km to 50 km and the average temperature of the atmosphere is 280 K.

shows the decay of wave stress as functions of downwind distance for various values of q.

PROBLEMS

[illegible]

7

GRAVITY WAVES IN THE MIDDLE AND UPPER ATMOSPHERE

7.1 INTRODUCTION

It is now well established through observations and theory that gravity waves play an essential role in the global circulation of the atmosphere. Although the horizontal scales of the most active gravity waves are much smaller than the planetary and Rossby waves which dominate the general circulation, gravity waves transport energy and momentum from the troposphere and deposit it in the mesosphere and thermosphere. These transports have a large impact on the spatial and temporal characteristics of the middle and upper atmosphere. In almost all cases, the horizontal resolutions of general circulation models and global climate models are too course to resolve the scales of the important part of the gravity wave spectrum. Thus, their necessary contributions to global simulations are missed. As discussed by McLandres (1998), the modeler is left with two choices; either increase the horizontal resolution of the models or parameterize the effects of the gravity waves. To date, the highest model resolution is 20 km (Murakami *et al.*, 2012), but this is still greater than the wavelengths of important waves. Thus, for the forceable future modelers will be using parameterizations of the effects of gravity wave, and this requires a comprehensive understanding of the complex links between gravity wave fluxes of energy and momentum and the dynamics of the middle and upper atmosphere.

Until now we have dealt mostly with theoretical issues and simple examples. However, when we enter the middle and upper atmosphere (MUA) it becomes difficult to satisfy the requirements for linear theory. Background flows cannot be assumed constant and horizontally homogeneous. Compressibility is generally not an issue in the lower atmosphere; however, in the MUA decreasing air density becomes important, and terms involving the isothermal scale height, H_s, cannot be ignored. We have previously applied linear theory to cases involving single waves or wave packets described by a single dominant wave. However, in the MUA we are confronted with a spectrum of gravity waves with frequencies ranging from the Coriolis frequency f to the local Brunt–Väisälä frequency N. Disturbances have horizontal scales ranging from tens to thousands of kilometers and vertical scales ranging from a few to several tens of kilometers. Because of $e^{z/2H_s}$ growth, the large amplitude waves can break. Wave breaking almost always occurs in the upper and middle atmosphere. Yet, within the face of these complexities, the linear theory is robust and useful.

7.2 BACKGROUND

In the early days of long-distance radio transmission, variations and interruptions of over-the-horizon transmissions initiated the study of traveling ionospheric disturbances (TID). In his classic paper, Hines (1960) proposed that TIDs were caused by upward propagating gravity wave disturbances originating in the troposphere. Gossard (1962) calculated the gravity wave spectrum of vertical energy flow out of the troposphere. He found that the parts of the spectrum corresponding to propagating gravity waves have periods between about 10 min to 2 h, and can at times transport large amounts of energy through most of the MUA. His results were in agreement with data at D-layer heights (80–100 km) taken by Hargreaves (1961) using radio wave reflections in the ionosphere, and by Greenhow and Neufeld (1959) using meteor trails. These and the great number of papers published since agree that most of the energy driving the general circulation originates in the troposphere, and that the majority of this energy is transported upward by gravity waves. One of the major problems faced by global-scale and synoptic-scale modelers is the incorporation of gravity wave fluxes into their models. These models require relatively large horizontal grid cells; however, because these cells are generally larger than the wavelengths of the energy-carrying gravity waves, the wave fluxes cannot be directly calculated. They must be parameterized, which will be discussed in Chapter 8.

Gravity wave transport of energy throughout the atmospheric layers results in a coupling between these layers. This coupling is the essence of the general circulation. Vincent (2009) reviewed the coupling mechanisms and concluded:

1. The majority of the waves observed in the lower stratosphere couple energy and momentum upward into the middle and upper atmosphere.
2. Largest momentum fluxes are observed over regions of high topography, but these regions have the greatest wave variability.
3. On a zonally averaged basis, momentum fluxes over mountains and oceans are approximately equal.
4. Thermospheric gravity waves are selectively filtered by the kinematic dissipation. Only the high frequency, long vertical wavelength components penetrate to the highest altitudes.
5. There is a strong solar cycle effect in gravity wave propagation into the thermosphere. Gravity waves propagate to higher altitudes during high sunspot conditions than during solar minimum conditions.

An interesting example of atmosphere–earth coupling is given by Artu *et al.* (2005) who found a correlation between observed gravity waves in the ionosphere and a tsunami. This was the first observation of a coupled tsunami-gravity wave.

The study of synoptic and global scale motions is far beyond the scope of this book, and it is impossible to give a comprehensive understanding of gravity waves above the troposphere in a single chapter. Indeed, text books and extensive reviews are devoted to this study (see for example, Hines, 1974; Lilly and Gal-Chen, 1983; Fritts, 1984; Andrews, Holton, and Leovy, 1987; Randall, 2000; Fritts and Alexander, 2003; Holton, 2004; Brassure and Solomon, 2005; Mohankumar, 2008; Alexander, 2010; Warner, 2011 and references therein). In light of this, we present here only those details that pertain to gravity waves keeping in mind that these details are still under active investigation, and many questions remain unanswered.

Figure 7.1 illustrates the climatological vertical temperature structure of the atmosphere which is divided on the bases of dynamics and chemistry into three regions, the *lower*, *middle*, and *upper* atmosphere. The lower atmosphere contains the troposphere and tropopause; the middle atmosphere extends from the base of the *stratosphere* to the top of the *mesosphere*; the upper atmosphere includes the *thermosphere* and the *ionosphere*; it extends from the top of the mesopause to the exosphere and outer space. Each region experiences different dynamics and different responses to gravity waves. Figure 7.2 shows the profile of Brunt–Väisälä frequency from the ground surface to the thermosphere. We see that N_2 increases with height in the troposphere, mesosphere, and lower thermosphere; however, in the thermosphere N^2 decreases rapidly with height above about 115–120 km. In the stratosphere, N^2 decreases with height. The wave stability of the various layers

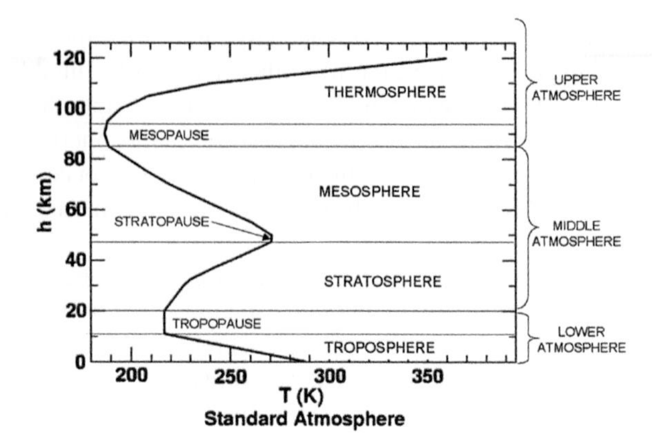

FIGURE 7.1 Midlatitude yearly average temperature profile in the middle and upper atmosphere. Also showing atmospheric zones based on temperature profile.

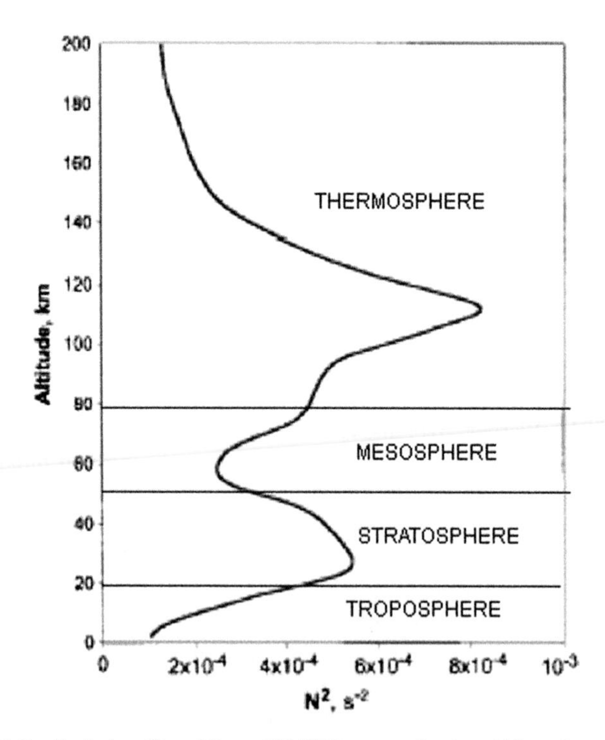

FIGURE 7.2 Vertical profiles of Brunt–Väisälä frequency for the middle and upper atmosphere.

can be simply scaled as suggested by Fritts (personal communication). We make the following assumption: horizontal wavelengths are much greater than vertical wavelengths so that $k \ll m$, and the vertical scale for changes in the background variables is greater than a vertical wavelength, *i.e.*, the WKB assumption holds. We now write the Richardson number in terms of the perturbations,

$$Ri = \frac{N^2}{\rho_0 \overline{u_1 w_1} \frac{\partial u_1}{\partial z}}. \tag{7.1}$$

From the polarization equation

$$u_1 = \frac{m}{k} w_1. \tag{7.2}$$

Multiplying (7.2) by w_1 gives

$$u_1 w_1 = \frac{m}{k} w_1^2. \tag{7.3}$$

The magnitude of the vertical derivative of 7.2 is

$$\frac{\partial u_1}{\partial z} \sim \frac{m^2}{k} w_1. \tag{7.4}$$

Using (7.3) and (7.4) in (7.1) gives

$$Ri \sim \frac{N^2}{\rho_0 \frac{m^3}{k^2} w_1^3}. \tag{7.5}$$

Under WKB, $m = Nk/\Omega$ where Ω is the intrinsic frequency. Then

$$Ri \sim \frac{N^2}{N^3} \frac{\Omega^3}{k w_1^3}. \tag{7.6}$$

We assume that Ω and k are constants or slowly varying with height. We also make the *ad hock* assumption that w_1 is slowly varying when no critical levels are present. We then get the result that

$$Ri \sim \frac{1}{N}. \tag{7.7}$$

Relation (7.7) indicates that in regions where N is decreasing with height Ri increases, *i.e.*, the upward propagating gravity wave becomes more stable, but in regions of increasing N, Ri decreases indicating less gravity wave stability and possible wave breaking. Thus, in the troposphere where N increases with height wave breaking is to be expected, which indeed is the case for the smaller (horizontal wavelength) waves. However, in the stratosphere wave breaking is suppressed by the increasing stability, and waves tend to propagate through it with little dissipation, but wave reflection is possible. In the mesosphere, wave stability decreases and because of the large horizontal wind shears, wave instability and transfers of pseudomomentum and drag occur. It is this deposition of momentum that drives the mesosphere winds and leads to seasonal wind direction reversals.

To first order, the atmospheric general circulation is driven by differential absorption of solar heating at the ground surface. In response, an upward

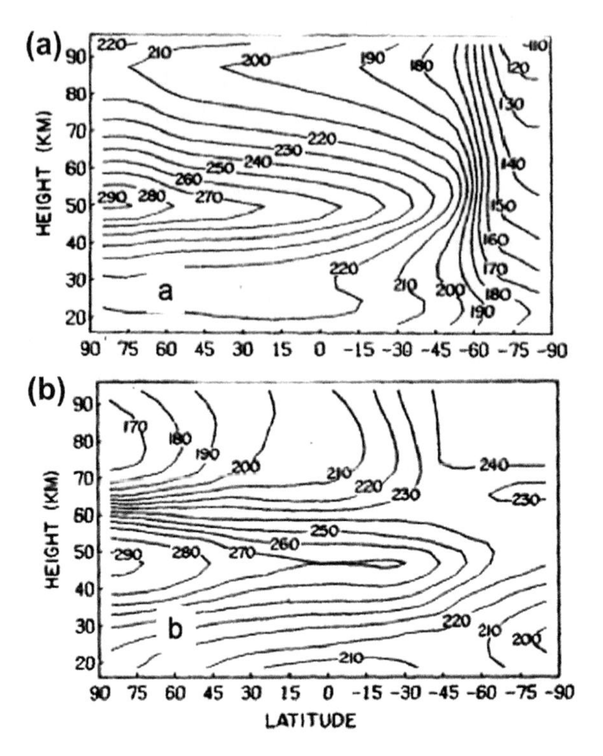

FIGURE 7.3 (a) The calculated MUA radiative equilibrium temperature distribution at solstice with winter hemisphere on the right. (b) The observed zonal mean MUA temperature distribution at solstice with the winter hemisphere on the right. Taken from Wehrbein and Levoy (1982).

heat flux lies above the warm equatorial regions and a downward heat flux is found above the cold polar regions. In the absence of eddy motions, the zonal mean temperatures in the middle atmosphere would be determined by a seasonally varying radiative equilibrium. Because of a balance between the Coriolis force and the meridional temperature gradient, the global circulation would be an averaged zonal wind with no meridional circulation. In such a situation, there would be no coupling between the middle and lower atmosphere. As an illustration consider Fig. 7.3 taken from Wehrbein and Levoy (1982). Figure 7.3a shows the calculated radiative equilibrium MUA temperature distribution at solstice with winter hemisphere on the right. Figure 7.3b shows the observed zonal mean temperature distribution in the MUA at solstice with the winter hemisphere on the right. The difference in temperatures between the radiative equilibrium winter hemisphere and the observed temperatures is quite large, the latter being much warmer. Mesosphere temperatures in the equilibrium summer hemisphere are much lower than the observed temperatures. Thus, we see that coupling exists between the lower and MUA such that energy and momentum are transported up from the lower atmosphere. The major transfers of momentum and energy into the middle atmosphere are through gravity waves in the equatorial regions. Additional transfers by gravity waves occur in midlatitudes. Fritts and Alexander (2003) describe several gravity wave sources in the troposphere including topography, convection, shear instability, geostrophic adjustment, and wave–wave interactions. Each of these generation mechanisms produce waves with individual characteristics. For example, hydrostatic topographic waves propagate upward almost directly above the terrain feature and have a narrow spectrum of horizontal wavelengths, but convection waves propagate vertically and horizontally with a wide spectrum of wavelengths. Globally, the troposphere is a 'soup' of nonlinear gravity waves with wide-ranging frequencies, wavelengths, and amplitudes such that individual waves and their sources are at best difficult to identify by observations.

7.3 INTERIA-GRAVITY WAVES IN THE MIDDLE ATMOSPHERE

Because of the wide spectrum of gravity waves launched from the lower atmosphere, modeling in terms of monochromatic waves is inappropriate. We cannot say a priori which waves will be reflected, absorbed at critical levels, or become dynamically unstable. Thus, we must consider a spectrum of gravity waves with intrinsic frequencies Ω such that $f < \Omega < N$. However, linear inviscid wave theory can be useful in isolating mechanisms and interpreting observations. Because we are dealing with large waves with low frequencies, the Coriolis force will have an effect. Thus we must solve for inertia-gravity waves. Following Holton (2004) the equations are:

$$\frac{\mathrm{d}u}{\mathrm{d}t} - fv + \frac{1}{\rho}\frac{\partial p}{\partial x} = X, \tag{7.8}$$

$$\frac{\mathrm{d}v}{\mathrm{d}t} + fu + \frac{1}{\rho}\frac{\partial p}{\partial y} = Y, \tag{7.9}$$

$$\frac{\mathrm{d}w}{\mathrm{d}t} + \frac{1}{\rho}\frac{\partial p}{\partial z} + g = 0, \tag{7.10}$$

$$\frac{1}{\rho} + \frac{\partial u}{\partial x} + \frac{\partial v}{\partial y} + \frac{\partial w}{\partial z} = 0, \tag{7.11}$$

$$\frac{\mathrm{d}\theta}{\mathrm{d}t} = Q, \tag{7.12}$$

$$\theta = \frac{p}{\rho R}\left(\frac{p_0}{p}\right)^{\kappa}, \tag{7.13}$$

where $\mathrm{d}/\mathrm{d}t = \partial/\partial t + u\partial/\partial x + v\partial/\partial y + w\partial/\partial w$, X, Y, and Q represent generalized forcings, θ is potential temperature (1.42) and the other symbols have their usual meanings. Linearizing (7.8)–(7.13) where the slowly varying background variables have subscript 0 and first-order perturbations have subscript 1 gives

$$\frac{\mathrm{D}u_1}{\mathrm{D}t} + w_1\frac{\partial u_0}{\partial z} - f v_1 + \frac{\partial}{\partial x}\left(\frac{p_1}{p_0}\right) = 0, \tag{7.14}$$

$$\frac{\mathrm{D}v_1}{\mathrm{D}t} + w_1\frac{\partial v_0}{\partial z} + f u_1 + \frac{\partial}{\partial y}\left(\frac{p_1}{p_0}\right) = 0, \tag{7.15}$$

$$\frac{\mathrm{D}w_1}{\mathrm{D}t} + \frac{\partial}{\partial z}\left(\frac{p_1}{p_0}\right) - \frac{1}{H_{\mathrm{s}}}\left(\frac{p_1}{p_0}\right) + g\frac{\rho_1}{\rho_0} = 0, \tag{7.16}$$

$$\frac{\mathrm{D}}{\mathrm{D}t}\left(\frac{\theta_1}{\theta_0}\right) + w_1\frac{N^2}{g} = 0, \tag{7.17}$$

$$\frac{\mathrm{D}}{\mathrm{D}t}\left(\frac{\rho_1}{\rho_0}\right) + \frac{\partial u_1}{\partial x} + \frac{\partial v_1}{\partial y} + \frac{\partial w_1}{\partial z} - \frac{w_1}{H_{\mathrm{s}}} = 0, \tag{7.18}$$

$$\frac{\theta_1}{\theta_0} = \frac{1}{c_s^2}\left(\frac{p_1}{p_0}\right) - \frac{\rho_1}{\rho_0}, \tag{7.19}$$

where

$$\frac{\mathrm{D}}{\mathrm{D}t} = \frac{\partial}{\partial t} + u_0\frac{\partial}{\partial x} + v_0\frac{\partial}{\partial y}, \tag{7.20}$$

and we have dropped the forcing terms and the background velocity shears. This is consistent with the WKB approximation. We now assume wave solutions of the form

$$A_1 = \widetilde{A}\exp\left[i(kx + ly + mz - \omega t) + \frac{z}{2H_z}\right]. \tag{7.21}$$

Using these wave solutions in (7.14) to (7.19) gives the polarization wave equations

$$-i\Omega\tilde{u} - f\tilde{v} + ik\tilde{p} = 0 \quad - i\Omega\tilde{v} + f\tilde{u} + il\tilde{p} = 0, \tag{7.22}$$

$$-i\Omega\tilde{w} + \left(im - \frac{1}{2H_{\mathrm{s}}}\right) = -g\tilde{\rho}, \tag{7.23}$$

$$-i\Omega\tilde{\theta} + \frac{N^2}{g}\tilde{w} = 0, \tag{7.24}$$

$$-i\Omega\tilde{\rho} + ik\tilde{u} + il\tilde{v} + \left(im - \frac{1}{2H_{\mathrm{s}}}\right) = 0, \tag{7.25}$$

$$\tilde{\theta} = \frac{\tilde{p}}{c_{\mathrm{s}}^2} - \tilde{\rho}. \tag{7.26}$$

Recall that Ω is the intrinsic frequency defined in (2.18). Solving (7.22) to (7.26) for $\tilde{w}$ and equating real and imaginary parts we get two equations. One equation is

$$\frac{g}{c_{\mathrm{s}}^2} = \frac{1}{H_{\mathrm{s}}} - \frac{N^2}{g}. \tag{7.27}$$

The second equation will be similar to (2.36), the Taylor–Goldstein equation, with dispersion relation

$$\Omega^2\left(K^2+m^2+\frac{1}{4H_s^2}-\frac{(\Omega^2-f^2)}{c_s^2}\right)=N^2K^2+f^2\left(m^2+\frac{1}{4H_s^2}\right), \tag{7.28}$$

where $K^2=k^2+l^2$. In the Boussinesq approximation, $c_s\to\infty$, (7.27) reduces to

$$\frac{1}{H_s}=\frac{N^2}{g}. \tag{7.29}$$

In Chapter 1, we defined an exponentially decreasing atmospheric density in (1.58). The vertical derivative of this density is

$$\frac{\partial\rho_0}{\partial z}=-\frac{\rho_0}{H_s}, \tag{7.30}$$

hence,

$$-\frac{1}{\rho_0}\frac{\partial\rho_0}{\partial z}=\frac{N^2}{g}=\frac{1}{H_s}. \tag{7.31}$$

Also, in the Boussinesq approximation the dispersion relation reduces to

$$\Omega^2=\frac{N^2K^2+f^2\left(m^2+\frac{1}{4H_s^2}\right)}{K^2+m^2+\frac{1}{4H_s^2}}. \tag{7.32}$$

From (7.32) the vertical wavenumber is

$$m^2=\frac{K^2(N^2-\Omega^2)}{(\Omega^2-f^2)}-\frac{1}{4H_s^2}. \tag{7.33}$$

If the wave is propagating, then m must be real, and so $f<\Omega<N$. In the notation of Fritts and Alexander (2003), the group velocities are

$$(c_{gx},c_{gy},c_{gz})=(u_0,v_0,0)+\frac{[k(N^2-\Omega^2),l(N^2-\Omega^2),m(\Omega^2-f^2)]}{\Omega\left(K^2+m^2+\frac{1}{4H_s^2}\right)}. \tag{7.34}$$

Using the polarization (7.22) to (7.26) we can develop the relations

$$\tilde{p}=\left(\frac{\Omega^2-f^2}{\Omega k+ifl}\right)\tilde{u}=\left(\frac{\Omega^2-f^2}{\Omega l-ifk}\right)\tilde{v}, \tag{7.35}$$

$$\tilde{w}=\frac{\left(m-\frac{i}{2H_s}\right)\Omega}{N^2-\Omega^2}\tilde{p}, \tag{7.36}$$

$$\tilde{u}=\left(\frac{i\Omega k-fl}{i\Omega l+fk}\right)\tilde{v}. \tag{7.37}$$

It is of interest to note the phase relation between $\tilde{u}$ and $\tilde{v}$ in (7.37). For a zonally propagating wave, the meridional velocity perturbation is

$$\tilde{v} = -i\frac{f}{\Omega}\tilde{u}, \tag{7.38}$$

and we see that the horizontal perturbation velocities are 90° out of phase. As the wave propagates upward the horizontal wave vector rotates cyclonical, in the northern hemisphere, resulting in a spiral *hodograph* (see, for example, Gill, 1982; Tsuda *et al.*, 1994; Sato, 1994; Guest *et al.*, 2000; Walterscheid, Schubert, and Brinkman, 2001). This spiraling behavior will be explained in Chapter 8 where we discuss observations in the MUA.

7.4 PLANETARY WAVES IN THE MIDDLE ATMOSPHERE

7.4.1 ROSSBY WAVE

As Holton (2004) points out, the predominant eddy motions in the midlatitude stratosphere are vertically propagating Rossby waves. However, Rossby waves can propagate only in the winter stratosphere. In this book, we cannot develop the theory from basic principles, and so we present here the essential features taken from Holton (2004). The geostrophic potential vorticity equation on a midlatitude β-plane in log-pressure coordinates, $z = -H_s \ln(p/p_s)$ is

$$\left(\frac{\partial}{\partial t} + \vec{V}_g \bullet \nabla\right) q = 0, \tag{7.39}$$

where

$$q = \nabla^2 \psi + f + \frac{f_0^2}{\rho_0 N^2}\frac{\partial}{\partial z}\left(\rho_0 \frac{\partial \psi}{\partial z}\right), \tag{7.40}$$

where $\psi = \Phi/f_0$ is the geostrophic stream function, f_0 is a constant midlatitude reference value of the Coriolis parameter, and Φ is the geopotential defined by $d\Phi = gdz$. Linearizing, we assume that the motion consists of small-amplitude disturbances superimposed on a constant zonal-mean flow. Then letting $\psi = -Uy + \psi'$ and $q = q_0 + q'$ in (7.40) we get

$$\left(\frac{\partial}{\partial t} + u_0\frac{\partial}{\partial x}\right) q' + \beta\frac{\partial \psi'}{\partial x} = 0, \tag{7.41}$$

where U is a constant zonal wind and

$$q' = \nabla^2 \psi' + \frac{f_0^2}{\rho_0 N^2}\frac{\partial}{\partial z}\left(\rho_0 \frac{\partial \psi'}{\partial z}\right). \tag{7.42}$$

We now take wave solutions of the form

$$\psi'(x, y, z, t) = \Psi(z) e^{i(kx+ly-kc_{Ix}t)+z/2H_s}, \tag{7.43}$$

where $c_{Ix} = U - c_x$ is the intrinsic zonal phase speed. Using (7.43) in (7.41) gives the vertically-propagating midlatitude Rossby wave,

$$\frac{d^2\Psi}{dt^2} + m^2\Psi = 0, \tag{7.44}$$

where

$$m^2 = \frac{N^2}{f_0^2}\left(\frac{\beta}{U} - K^2 - \frac{1}{4H_s^2}\right), \tag{7.45}$$

where we have dropped c_x because for Rossby waves it is generally small compared with U, and $K^2 = k^2 + l^2$. For propagating waves m must be real. This requires that $U > 0$, *i.e.*, only eastward moving waves are allowed. A *Rossby critical velocity*, U_c is found by setting $m = 0$ in (7.45), *i.e.*,

$$U_c = \frac{\beta}{K^2 + \alpha^2}, \tag{7.46}$$

where $\alpha^2 = f_0^2/(4N^2H_s^2)$. For the stratosphere, N^2 is typically 4×10^{-4} s^{-1} and the temperature is about 220 K. The scale height, H_s is about 6.4 km and the term $f/2NH_s$ corresponds to zonal wavenumber 2 in midlatitudes. At midlatitudes, $\beta \sim 1.6 \times 10^{-11}$, $K \sim 4.4 \times 10^{-7}$, and $\alpha \sim 4 \times 10^{-7}$. Using these values $U_c \sim 46$ ms^{-1}. Thus, midlatitude Rossby wave will not propagate into the MUA if the average zonal wind is greater than about 50 ms^{-1}. In the winter northern hemisphere Rossby waves can propagate upward to about 60–70 km, but they cannot move out of the summer troposphere.

7.4.2 TROPICAL ATMOSPHERE

The equatorial region is of great interest. In the tropics, upwelling of energy and momentum into the upper troposphere and stratosphere occurs. How this happens is a major scientific question since these details greatly impact global scale models. In the tropics, significant topography does not exist; therefore terrain-generated gravity waves need not be considered. The remaining mechanisms are convectively generated gravity waves and upward propagating planetary waves. Convective gravity waves have been studied by, for example, Karoly, Roff, and Reed, 1996; Alexander and Holton, 1997; Walterscheid, Schubert, and Brinkman, 2001; Nakamura *et al.*, 2003; Chun *et al.*, 2004; Chun *et al.*, 2011, and references therein. Fritts and Alexander (2003) give a comprehensive review of middle atmospheric gravity wave research since 1984. The planetary wave contributions to the upwelling have been studied by, for example, Matsuno, 1966; Holton and Lindzen, 1968; Andrews and McIntyre, 1976; Dunkerton, 1993; Alexander and Holton, 1997; Nakamura *et al.*, 2003, and references therein.

In the equatorial regions, $f \to 0$, and as seen in (7.32) in the absence of Coriolis forces very low-frequency atmospheric oscillations are possible. Now the equations for midlatitude inertia-gravity waves apply, but with $f = 0$. The linearized perturbations on an equatorial β-plane are given by Holton (2004). Using log-pressure coordinates, these equations are

$$\frac{\partial u_1}{\partial t} - \beta y v_1 = -\frac{\partial \Phi_1}{\partial x}, \tag{7.47}$$

$$\frac{\partial v_1}{\partial t} + \beta y u_1 = \frac{\partial \Phi_1}{\partial y}, \tag{7.48}$$

$$\frac{\partial u_1}{\partial x} + \frac{\partial v_1}{\partial y} + \rho_0^{-1}\frac{\partial(\rho_0 w_1)}{\partial z} = 0, \tag{7.49}$$

$$\frac{\partial^2 \Phi_1}{\partial t \partial z} + w_1 N^2 = 0. \tag{7.50}$$

Now substituting wave solutions of the form

$$q_1 = \hat{q}e^{i(kx+mz-\omega t)+z/2H_s} \tag{7.51}$$

gives the differential equations for the meridional structure

$$-i\omega\hat{u} - \beta y\hat{v} = -ik\hat{\Phi}, \tag{7.52}$$

$$-i\omega\hat{v} + \beta y\hat{u} = -\frac{\partial\hat{\Phi}}{\partial y}, \tag{7.53}$$

$$\left(-k\hat{u} + \frac{\partial\hat{v}}{\partial y}\right) + i\left(m + \frac{i}{2H_s}\right)\hat{w} = 0, \tag{7.54}$$

$$\omega\left(m - \frac{i}{2H_s}\hat{\Phi}\right) + \hat{w}N^2 = 0. \tag{7.55}$$

7.4.2.1 Vertically Propagating Rossby-Gravity Waves

Equations (7.52) to (7.55) can be combined to get

$$\frac{\partial^2\hat{v}}{\partial y^2} + \left[\left(\frac{\omega^2 m^2}{N^2} - k^2 - \frac{k}{\omega}\beta\right) - \frac{\beta^2 y^2 m^2}{N^2}\right]\hat{v} = 0. \tag{7.56}$$

The boundary conditions for (7.56) are $\hat{v} \to 0$ as $|y| \to \infty$. Thus, (7.56) is an eigenvalue problem with modal solutions. As shown by Matsuno (1966), these solutions will exist if

$$\frac{m}{N\beta}\left(-\frac{k}{\omega}\beta - k^2 + \frac{\omega^2 N^2}{m^2}\right) = 2n+1, \quad n = 0, 1, 2\ldots \tag{7.57}$$

where n is the modal index, *i.e.*, $n = 0$ is the gravest mode. For $n = 0$

$$|m| = \frac{N}{\omega^2}(\beta + \omega k). \tag{7.58}$$

When $\beta = 0$ we recover the dispersion relation for hydrostatic propagating gravity waves. For eastward propagating waves, $\omega > 0$, and for westward propagating waves, $\omega < 0$. Thus, eastward propagating Rossby waves have shorter vertical wavelengths than westward propagating Rossby waves. The $n = 0$ mode exists only if

$$c = \omega/k > -\beta/k^2. \tag{7.59}$$

Because $k = s/a$ where the *zonal wavenumber* s is the number of wavelengths around a latitude circle and a is the mean radius of the earth, (7.59) will hold only for frequencies

$$|\omega| < \frac{2\Omega_E}{s}, \tag{7.60}$$

where Ω_E is the earth's angular speed of rotation. For frequencies that do not satisfy (7.60), the waves will not decay away from the equator and the boundary conditions will not be satisfied. As always, for upward propagation $m < 0$.

Observations by Yanai and Maruyama (1966) show that for the Rossby-gravity wave, the periods are 4–5 days; the zonal wavenumber is 4; the vertical wavelength is 4–5 km, and the average phase speed relative to the ground is $-23\ \mathrm{ms}^{-1}$.

7.4.2.2 The Kelvin Wave

For the Kelvin wave, the motions are zonal, *i.e.*, $\hat{v} = 0$. The equations take the form

$$-i\omega\hat{u} = -ik\hat{\Phi}, \tag{7.61}$$

$$\beta y\hat{u} = \frac{\partial\hat{\Phi}}{\partial y}, \tag{7.62}$$

$$-\omega\left(m^2 + \frac{1}{4H_s^2}\right)\hat{\Phi} + \hat{u}kN^2 = 0. \tag{7.63}$$

Using (7.63) to solve for Φ and using this in (7.51) and (7.52) gives two solutions. The first solution is

$$\frac{1}{\hat{u}}\frac{\partial\hat{u}}{\partial y} = -\frac{\beta y}{c}, \tag{7.64}$$

and integration gives

$$\hat{u} = \hat{u}_0 \mathrm{e}^{-0.5\beta/cy^2}. \tag{7.65}$$

The horizontal wind perturbations are in the zonal direction and are Gaussian function of y where $\hat{u}_0$ is the perturbation speed at the equator. The second solution is

$$c^2\left(m^2 + \frac{1}{4H_s^2}\right) - N^2 = 0. \tag{7.66}$$

Observations show that for stratospheric Kelvin waves the vertical wavelength is between 6–10 km. Assuming a wavelength of 8 km and letting $H_s = 6$ km we see that $m^2 \sim 6 \times 10^{-7}$ m^{-2} and $1/4H_s^2 \sim 6 \times 10^{-9}$ m^{-2}, and so we can drop the scale-height term to get

$$m^2 = \frac{N^2}{c^2}, \tag{7.67}$$

but this is the same as (7.33) in the case of long horizontal waves. Observations by Wallace and Kousky (1968) show that for Kelvin waves the wave period is 15 days; the zonal wavenumber is 1–2; the vertical wavelength is 6–10 km, and the average phase speed is 25 ms^{-1}.

7.5 MIDLATITUDE WAVE SPECTRA

Gravity waves entering the middle atmosphere from below extend over a wide range of frequencies and wavelengths (see, for example, Dunkerton and Butchart, 1984; Dewan and Good, 1986; Fritts and VanZandt, 1987; Fritts *et al.*, 1988; Marks and Eckermann, 1995; Alexander and Dunkerton, 1999; Fritts and Alexander, 2003; Preusse and Ern, 2008) The inertia-gravity wave approaches a singularity as $\Omega^2 \to f^2$ and so the range of frequencies of internal waves is $f < \Omega < N$, where we are reminded that Ω is the intrinsic frequency, $\omega - u_0 k$. We see that the frequency range of middle atmospheric gravity waves is limited at the lower end, *i.e.*, $f = 2\Omega_E \sin\phi$, where ϕ is the latitude.

7.5.1 HIGH-FREQUENCY WAVES: $\Omega \gg f$

When $\Omega \gg f$ we can drop f from the equations in Section 7.3. From the basic definition, $\Omega = Kc_I = 2\pi/\lambda_h c_I$ where c_I is the intrinsic phase speed, and we see that high-frequency waves will have short horizontal wavelengths, λ_h. From Chapter 2, we saw that in the non-hydrostatic case short horizontal waves, *i.e.*, large K, will become evanescent or reflected more readily than longer waves. In the high frequency regime under the Boussinesq approximation (7.28) takes the form

$$\Omega^2 = \frac{N^2K^2}{\left(K^2 + m^2 + \frac{1}{4H_s^2}\right)}. \tag{7.68}$$

From (7.68) the vertical wavenumber is

$$m^2 = \frac{K^2N^2}{\Omega^2} - K^2 - \frac{1}{4H_s^2}. \tag{7.69}$$

Setting $m = 0$ we can solve (7.69) for the maximum intrinsic phase speed,

$$c_{I,\max} = \pm\frac{N}{K^2 + \frac{1}{4H_s^2}}. \tag{7.70}$$

Fritts and Alexander (2003) point out that only waves with horizontal wavelength > 10 km are important in the middle atmosphere. Using $\lambda_h = 10$ km, $H_s = 7$ km, and $N = 0.02\ s^{-1}$, $c_{I,max} \approx 33\ ms^{-1}$.

Neglecting f and for $m^2 \gg 1/(4H_s^2)$ the dispersion relation is

$$\Omega^2 = \frac{N^2 K^2}{K^2 + m^2} = N^2 \cos^2 \beta, \tag{7.71}$$

where β is the angle between the wave vector and the horizontal plane. Note that (7.71) is identical to (2.57) but for three dimensions. Under these same assumptions the group velocities and other wave characteristics are the same as in Chapter 2. Fritts and Alexander (2003) point out that for $H_s \sim 7$ km, the compressibility term becomes significant for waves with vertical wavelengths greater than about 30 km. They also point out that the compressibility effect is important in the study of airglow near the mesopause (Swenson, Alexander, and Haque, 2000). For long vertical wavelengths, the vertical scale of variations of background variables, especially N, will not be large in comparison to λ_z. Since this goes against the conditions for the WKB method, the applicability of the WKB become problematical. As the vertical wavelength becomes great $m \to 0$ which marks a turning point for the upward propagating wave, *i.e.*, wave reflection. Using (7.32) with $m = 0$ and ignoring rotation effects, the maximum horizontal intrinsic phase speed $|c_{I,\mathrm{max}}|$ at the reflection point is

$$|c_{I,\mathrm{max}}| = N\left(K^2 + \frac{1}{4H_s^2}\right)^{-1/2}, \tag{7.72}$$

which will be small at the turning point. Thus $|c_{I,\mathrm{max}}|$ will be small for small horizontal wavelengths. Shorter wavelength waves will be either reflected or ducted in the stratosphere. However, Ω cannot be so large or K be so small as to make m complex. As Fritts and Alexander (2003) point out, a turning point for the wave occurs where $m \to 0$ and this will occur where $\Omega \to KN/2H_s$. Thus, in the stratosphere where N decreases with height the high frequency long horizontal wavelength waves are likely to be reflected downward. As we have seen in Section 7.4.2, in the equatorial zones very low wave frequencies can exist. Then the vertical wavenumber (7.33) becomes

$$m \approx \frac{KN}{\Omega}. \tag{7.73}$$

7.5.2 MID-FREQUENCY WAVES: $f \gg \Omega \gg N$

In the middle frequency range, the effects of f and $1/2H_s$ will not be significant, although this may not be true in limiting cases. Following Fritts and Alexander (2003), we define the horizontal phase speed as c_h and the horizontal wavenumber as $k_h = \sqrt{k^2 + l^2}$. The dispersion then simplifies to

$$\Omega = N\left|\frac{k_h}{m}\right|, \tag{7.74}$$

and the vertical wavenumber becomes

$$|m| = \frac{N}{c_{I\mathrm{h}}}. \tag{7.75}$$

Following our convention in Chapter 3, if $k_\mathrm{h} > 0$ then $c_{I\mathrm{h}} = (c_h - u_0) > 0$ and for $k < 0$ $c_{I\mathrm{h}} < 0$. At a critical level $c_{I\mathrm{h}} = 0$. Using (7.74) and (7.75) it can be shown that the group velocities have the simple forms

$$u_\mathrm{g} = u_{0h} + \frac{\Omega}{k_\mathrm{h}} \tag{7.76}$$

and

$$w_\mathrm{g} = -\frac{\Omega}{m}. \tag{7.77}$$

The polarizations equations have the form

$$\frac{p_1}{w_1} = w_\mathrm{g} \tag{7.78}$$

and

$$\frac{w_1}{u_{1h}} = -\frac{k_\mathrm{h}}{m} = \frac{k_\mathrm{h} c_{I\mathrm{h}}}{N} = \frac{\Omega}{N}. \tag{7.79}$$

In the previous chapters, we almost always considered the horizontal winds constant leaving numerics to the more general condition. Now we note that as the wave moves *upshear, i.e.*, $\partial c_I/\partial z > 0$, group velocity, vertical wavelength, and intrinsic frequency increase. The revers holds in regions of downshear. Equations (7.79) show that as the wave moves upshear, the vertical wave speed perturbation becomes increasingly greater than the horizontal perturbation speed. We also see from (7.78) that the pressure perturbation becomes large for upshear propagation.

7.5.3 LOW-FREQUENCY RANGE: $\Omega \sim f$

For low-frequency waves where $\Omega^2 \ll N^2$, the vertical wavenumber (7.33) takes the form

$$m^2 = \frac{K^2N^2}{\Omega^2 - f^2} - \frac{1}{4h_\mathrm{s}^2}. \tag{7.80}$$

These waves are the inertia-gravity waves and are associated with large-scale topographic features and mesoscale to synoptic scale weather systems. The singularity where $\Omega \to f$ is important; however, as f decreases, as in the tropics, very low frequency gravity waves are possible. In the tropics where there is little significant topography low frequency gravity waves are generated by large convective systems. The compressibility term $1/2H_\mathrm{s}$ becomes important as m becomes small. We can find a critical frequency, Ω_c by following Marks and Eckermann (1995). We set $m = 0$ in (7.80) and solve for Ω_c to get

$$\Omega_\mathrm{c}^2 = \frac{K^2N^2}{\left(K^2 + \frac{1}{4H_\mathrm{s}^2}\right)}. \tag{7.81}$$

Now the range of allowed frequencies is $f < \Omega < \Omega_c \ll N$. Marks and Eckermann (1995) note that the compressibility term is small and is often dropped; however, this can lead to calculated frequencies lower than real frequencies. In this case, m can become large and the vertical wavelength λ_z can become small.

7.6 MODELING THE GRAVITY WAVE FLUXES IN THE MUA

As mentioned at the beginning of this chapter, global-scale atmospheric models require realistic parameterizations of the gravity wave fluxes into the MUA (McLandres, 1998). This requires a comprehensive understanding of gravity wave fluxes of energy and momentum. A vast literature on these subjects exists. Fritts and Alexander (2003) give a recent review of research into middle atmosphere gravity waves, and Fritts and Lund (2011) review recent research on gravity waves in the thermosphere and ionosphere. Here, we can summarize, only crudely, the fundamentals of this research. Basically, analyses proceed on two paths, hydrodynamic models and ray tracing.

7.6.1 HYDRODYNAMIC MODELS

In hydrodynamic models, differential Equations (7.8) to (7.12) are solved, and the body forces X, Y, and Q represent divergences of gravity fluxes. The temporal and spatial scales of these problems are arbitrary. Theoretical expressions or parameterizations of the forcing terms are proposed and used to match observations or develop further theories. Some of these parameterizations will be described in Chapter 8. Examples of these approaches are found, for example, in Andrews and McIntyre (1976), Lindzen (1981), Holton (1982), Tanaka (1996), Lott and Teitelbaum (1993), Alexander and Dunkerton (1999), Walterscheid, Schubert, and Brinkman (2001), Chen, Durran, and Hakim (2005), and Chun *et al.* (2011).

7.6.2 RAY TRACING

We have seen in Section 2.5.2 that when a wave packet moves along a path that is tangent to the group velocity vector, the wave action, $\rho_0 E/\Omega$ is constant. This forms the basis for *ray tracing*. Marks and Eckermann (1995) give an excellent description of the method based on Lighthill (1978). We assume a wave packet with mid-frequency waves (Section 7.5.2) moving within a slowly varying flow without gradients in the wind field. Then with the Boussinesq approximation

$$\Omega^2 = \frac{N^2K^2 + f^2(m^2+\alpha^2)}{K^2+m^2+\alpha^2}, \tag{7.82}$$

where $\alpha = 1/4H_s^2$. The vertical wavenumber is given by

$$m^2 = \frac{(\Omega_c^2 - \Omega^2)(K^2+\alpha^2)}{\Omega^2 - f^2}, \tag{7.83}$$

where Ω_c is the critical or cutoff frequency (7.81). The dispersion relation is $\omega = \omega(k_i, x_i)$, $i = 1, 2, 3$, and we assume that the background flow and the wave parameters do not vary with time so that ω is constant along the ray path. The equations for the ray path are then

$$\frac{dx}{dt} = \frac{\partial \omega}{\partial k} = u_g, \tag{7.84}$$

$$\frac{dy}{dt} = \frac{\partial \omega}{\partial l} = v_g, \tag{7.85}$$

$$\frac{dz}{dt} = \frac{\partial \omega}{\partial m} = w_g. \tag{7.86}$$

The refraction equations for the ray path are

$$\frac{dk}{dt} = -\frac{\partial \omega}{\partial x}, \tag{7.87}$$

$$\frac{dl}{dt} = -\frac{\partial \omega}{\partial y}, \tag{7.88}$$

$$\frac{dm}{dt} = -\frac{\partial \omega}{\partial z}. \tag{7.89}$$

These equations are in their simplest form. In application, there can be several rays each representing different origins or initial conditions. For each ray, initial longitudes and latitudes, elevations, horizontal wavenumbers (k_0, l_0), and initial ground-based frequency ω_0 are specified. Using these values, (7.83) is evaluated for m_0. The sign of m_0 is taken opposite to Ω so that $w_g > 0$, *i.e.*, upward propagating energy and momentum. The amplitude of the wave packet is calculated by the wave-action equation

$$\frac{\partial A}{\partial t} + \nabla \bullet (\vec{v}_g A) = -\frac{2A}{\tau}, \tag{7.90}$$

where $A = \overline{E}/\Omega$ is the wave action and τ is a dampening time scale. The initial value of A is the rms horizontal velocity perturbations at the starting point of the ray, *i.e.*, $A_0 = (\overline{u_1^2} + \overline{v_1^2})^{1/2}$. The position and amplitude of the wave packet along the ray path are calculated forward in time and space noting that the background wind speeds can slowly vary in time and location. Examples of ray tracing calculations can be found in Dunkerton and Butchart (1984), Dunkerton (1984), Miyhara (1985), Schoeberl (1985), Hines (1988), Marks and Eckermann (1995), and Preusse and Ern (2008).

PROBLEMS

1. What is the frequency range for hydrostatic inertia-gravity waves in the middle and upper atmosphere?

2. An upward propagating inertia-gravity wave in the upper stratosphere at 50° north has a period of 1 h and a zonal wave vector. Calculate the major and minor axis of the hodograph formed by the velocity perturbations. What is the direction of rotation of the perturbation velocity vector.
3. What is a β-plane, and explain its purpose.
4. Explain the observation that inertia-gravity waves with long horizontal wavelengths pass through the stratosphere but those with short horizontal wavelengths do not.
5. Account for the fact that internal hydrostatic gravity waves can propagate to greater altitudes in the tropics than in midlatitudes.
6. What is the vertical wavelength at the point of wave reflection?
7. What is the wave property that allows ray tracing?
8. In the absence of wave breaking, what two wave quantities are conserved.
9. In going from (7.82) to (7.83) what assumption has been made?
10. What distinguishes mid-frequency inertia-gravity waves from those wave with high or low frequencies?

8

WAVE STRESS PARAMETERIZATION

8.1 INTRODUCTION

In Chapter 6, we illustrated the importance of gravity wave stress on all scales of atmospheric motions. Numerical models of the atmosphere are designed to simulate atmospheric motions. While in general all vertically propagating waves transport stress, orographically-generated waves are the only types with unambiguous amplitudes. However, important as mountain waves may be, these effects will not be simulated unless the orography generating the waves can be resolved by the model. Generally this is not the case. Since the mid 20th century, our ability to numerically model the atmosphere has advanced to the point where it is limited mostly by computer capabilities. Models of the general circulation, mesoscale air quality models, and atmospheric chemistry models are *computationally bound*, *i.e.*, they spend most of their time solving many equations in addition to the dynamical equations. In order to make model execution times shorter, the number of points where these equations are solved must not be great. Accordingly, the spatial resolution of the model must be limited. The smallest horizontal feature that can be resolved by a numerical model has a scale size of about $4\Delta x$, where Δx is the horizontal length of a grid cell (see, for example, Pielke, 1984; Grasso, 2000). Rontu (2007) points out that a model with $\Delta x < 1$ km will be able to resolve terrain features with horizontal scales between 2 and 4 km. Operational weather forecast models even with grid size as small as 4 km will not resolve obstacles less than about 15 km in scale. Skamarock (2004) suggests that implicit and explicit terrain smoothing within a model makes the effective resolution even courser, maybe $(7\text{–}8)\Delta x$. Global-scale models have horizontal resolutions on the order 10 degrees of latitude and longitude, and accordingly only the largest terrain features, for example the large mountain ranges, will be resolved. Cumulus convective cells with horizontal scales of a few tens of kilometers will not be seen by these model grid sizes; however, their dynamical effects can be significant especially in the tropics (see, for example, Chun, Song, and Baik, 1999). While research-grade models can achieve high horizontal resolutions, operational models are still limited, and will continue to be until the next generation of super computers. Thus, for most numerical applications in regions of mountainous terrain or active cumulus convection there will be unresolved wave effects, and these *subgrid-scale* effects must be parameterized. In this chapter we shall examine the basics of these wave stress parameterizations. A detailed discussion of orographic effects in numerical weather prediction models is given by Rontu (2007).

8.1.1 WAVE BREAKING, WAVE SATURATION, AND EDDY DIFFUSIVITY

From (2.45), the general form of a two-dimensional upward propagating gravity wave is

$$w_1(x, z, t) = \tilde{w}e^{z/2H_s}e^{i(kx-mz-\Omega t)}.$$

The factor $e^{z/2H_s}$, which accounts for the exponential decrease of atmospheric density with height, can become large when $z > H_s$. This results in exponentially increasing wave amplitude w_1, and since $u_1 \propto \partial w_1/\partial z$, u_1 also increases exponentially. We also note that if either m or Ω is complex, wave amplitudes can grow as either $e^{m_I z}$ or $e^{\Omega_I t}$. It was shown in Section 5.4, that as a gravity wave approaches a critical level the amplitude of the horizontal wave perturbation velocity, u_1, grows as $|z - z_c|^{-1/2}$ where z_c is the height of the critical level. From Section 3.2, we saw that over a surface corrugation

$$u_1 = u_0 H m_s \sin(k_s x + m_s z),$$

where k_s is the wavenumber of the corrugation and $m_s = (N^2/u_0^2 - k_s^2)^{1/2}$. If we consider u_0 to be height varying,[1] then its amplitude is

[1] While this is inconsistent with our assumption of constant wind speed, these simple models are useful approximations to numerical calculations with height-varying winds.

FIGURE 8.1 Breaking waves at the shore. Only the top portion of the wave breaks, and continues to break as the wave continues to grow. Broken line in top photo depicts average wave height.

$$\frac{\partial |u_1|}{\partial z} = u_0 H m_s^2 \approx \frac{HN^2}{u_0}, \tag{8.1}$$

if the waves are hydrostatic. Thus $|u_1|$ increases with increasing height. This is especially the case when a mountain wave approaches a critical level, $u_0 \rightarrow 0$ where. Thus, we see that within the linear theory the height displacement of a wave streamline increases seemingly without bound. However, this is never seen in nature. Instead, as the wave height grows it reaches a point where it begins to overturn or breaks down into turbulence in much the same fashion as a surface wave at a beach as illustrated in Fig. 8.1. But the wave does not entirely collapse. Instead, only a portion of the wave breaks. *Wave saturation theory* (see, for example, Lindzen, 1981; Holton, 1982; Fritts, 1984; Baines, 1995) assumes that waves do not entirely collapse, but rather their amplitudes decrease just enough so that the wave field remains convectively stable. The term "saturation" originally described the damping of exponential growth of an instability as environmental conditions forced a return to finite amplitude. Wave saturation limits wave growth. This can be seen in Fig. 8.1a where the broken white line shows that the wave height is nearly constant even over the breaking portion. This height represents the convectively

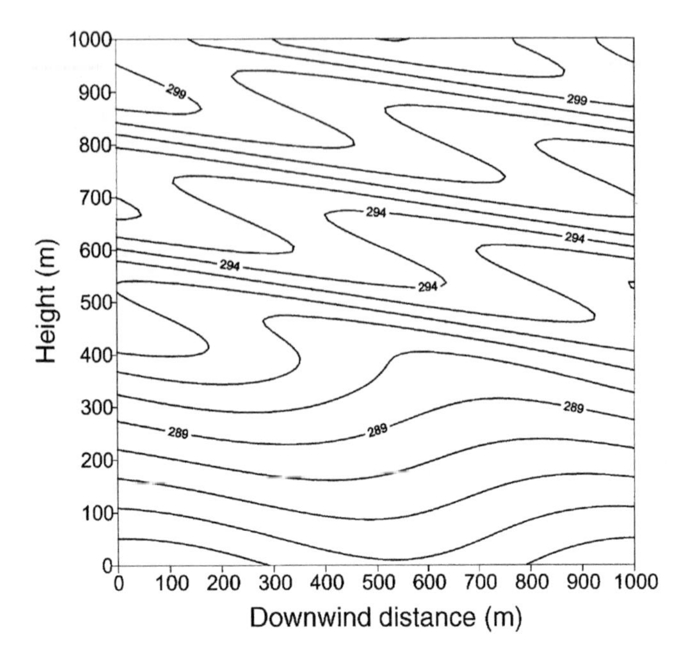

FIGURE 8.2 Isotherms over one cycle of a corrugated surface of wave length 1000 m and amplitude 60 m. $N = 0.022\ \mathrm{s}^{-1}$. Background wind speed profile is shown in Fig. 8.3. Regions of convective instability are seen above 400 m height. In reality, these regions would be unstable collapse.

stable portion of the wave. As the wave comes closer to the shoreline, it continues to grow, and the increasing wave growth continues to break down into turbulence. Videos of waves and wave saturation can be seen in the EURL\Ch7\WAVE-BREAKING. For breaking upward propagating gravity waves, the wave stress decreases with height because momentum is being lost to the environment.

In the linear theory, wave breaking cannot occur. Such a process is time dependent and therefore not possible in the linear theory. Figure 8.2 shows the streamlines (isotherms) over one cycle of a surface corrugation calculated with a numerical model. In these calculations, N was constant, but the wind profile is hyperbolic-tangent as shown in Fig. 8.3. In Fig. 8.2, we see that above about 350 m[2] the streamlines take on a nonphysical aspect. Specifically, there are regions where colder air lies above regions of warmer air. This represents a convectively unstable stratification, but because the linear theory is not a dynamic theory it cannot deal with instability. Such a condition represents a "break down" of the linear theory.

The Eliassen and Palm theory says that unless an upward propagating gravity wave is dissipated, the stress transported by the wave is constant. This is the case shown in Fig. 8.2. Even though there are regions of convective instability there

[2] This height corresponds to the inflection point of the velocity profile.

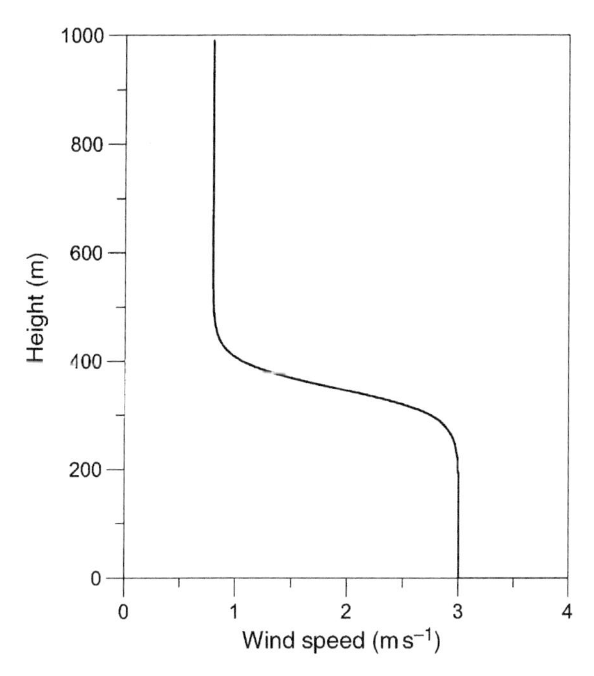

FIGURE 8.3 Hyperbolic-tangent wind profile used in the calculation of the wave field shown in Fig. 8.2.

is no wave dissipation and so the stress is constant. Following Nappo, Chun, and Lee (2004), the horizontal equation of motion for a one-dimensional non-rotating atmosphere including gravity wave stress is:

$$\frac{\partial}{\partial t}(\rho_0 u) = -\rho_0 w \frac{\partial u}{\partial z} + \frac{\partial}{\partial z}\left(D\frac{\partial u}{\partial z}\right) + \frac{\partial}{\partial z}\tau_x, \tag{8.2}$$

where D is the eddy diffusivity created by the breaking wave (see, for example, Holton, 1982) and $\tau_x = -\rho_0\overline{u_1 w_1}$ is the wave stress in the x-direction. If $\tau_x(z)$ is constant, then the wave stress does not affect the flow. Note also that if τ_x is constant, then the wave is not breaking and consequently the diffusivity D must be zero. However, if $\tau_x(z)$ decreases with height *i.e., convergence*, the wave stress will act to slow down $u_0(z)$. If gravity wave stress is to be meaningful in atmospheric dynamics, then we must find a way to "adjust" the linear theory so that wave stress convergence and diffusivity can be parameterized.

8.1.2 WAVE BREAKING HEIGHTS

In developing a wave saturation parameterization, the first thing we must do is determine the conditions for wave breaking. Consider a section of a streamline of constant potential temperature with value $\Theta(x, z_\Theta)$, where z_Θ is a function of x.

This stream line is illustrated in Fig. 8.4.

$$\theta_0(z_\Theta) + \theta_1(x, z_\Theta) = \Theta(x, z_\Theta) = \text{constant}, \tag{8.3}$$

where $\theta_0(z)$ is the background potential temperature; $\theta_1(x, z_\Theta)$ is a perturbation potential temperature, and z_Θ is the height of the streamline. The total differential of Θ is

$$\frac{\partial \Theta}{\partial x} + \frac{\partial \Theta}{\partial z_\Theta}\frac{\partial z_\Theta}{\partial x} = 0. \tag{8.4}$$

Using (8.3) in (8.4) gives

$$\frac{\partial \theta_1}{\partial x} + \frac{\partial z_\Theta}{\partial x}\frac{\partial}{\partial z_\Theta}(\theta_0 + \theta_1) = 0. \tag{8.5}$$

The condition for convective stability is

$$\frac{\partial}{\partial z_\Theta}(\theta_0 + \theta_1) \geqslant 0. \tag{8.6}$$

Now because the two terms in (8.5) are independent, and because $\partial\theta_1/\partial x$ in (8.5) need not be zero,

$$\lim_{\frac{\partial}{\partial z}(\theta_0+\theta_1)\to 0} \frac{\partial z_\Theta}{\partial x} \to \infty. \tag{8.7}$$

Thus, the onset of convective instability begins when the tangent to the isotherm becomes vertical as shown in Fig. 8.4. Using the continuity equation (2.8), we can define a stream function $\psi(x, z_\psi)$, where z_ψ is the height of the line along which ψ is constant. Then,

$$w_1 = \frac{\partial \psi}{\partial x}, \tag{8.8}$$

$$u_0 + u_1 = -\frac{\partial \psi}{\partial z_\psi}, \tag{8.9}$$

where we have taken u_0 to be horizontally uniform. Proceeding as above, if we take the total differential of ψ and use (8.8) and (8.9), we get

$$w_1 - \frac{\partial z_\psi}{\partial x}(u_0 + u_1) = 0. \tag{8.10}$$

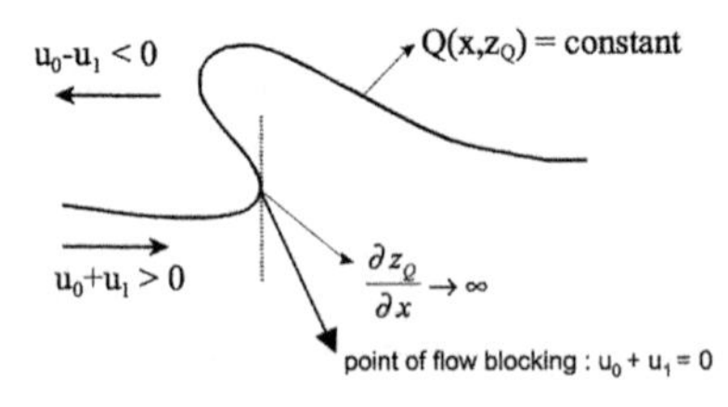

FIGURE 8.4 Section of an isotherm which is a streamline in the linear theory. When the slope of the streamline becomes vertical, the flow becomes convectively unstable. This is also the condition for flow blocking where $u_0 + u_1 = 0$.

Flow blocking occurs when $u_0 + u_1 = 0$, and since $w_1 \neq 0$ then

$$\lim_{(u_0+u_1)\to 0} \frac{\partial z_\psi}{\partial x} \to \infty, \tag{8.11}$$

and we see that convective instability and flow blocking occur simultaneously as illustrated in Fig. 8.4.

The convective stability limit (8.6) can be put into a more useful form by noting that the vertical displacement, ζ, of a flow streamline is related to the vertical velocity by

$$\frac{\mathrm{D}\zeta}{\mathrm{D}t} = w. \tag{8.12}$$

Linearizing (8.12) to first order gives

$$\frac{\partial \zeta_1}{\partial t} + u_0 \frac{\partial \zeta_1}{\partial x} = w_1. \tag{8.13}$$

If we now assume wave solutions of the form (2.13) for ζ_1 and w_1, (8.13) becomes

$$-i\Omega\tilde{\zeta_1} = \tilde{w_1}. \tag{8.14}$$

Using the relation $\rho_1/\rho_0 = -\theta_1/\theta_0$ in (2.25), gives

$$\tilde{\theta_1} = i\frac{\tilde{w_1}}{\Omega}\frac{\partial \theta_0}{\partial z}, \tag{8.15}$$

and combining (8.14) with (8.15) gives

$$\theta_1 = -\zeta_1 \frac{\partial \theta_0}{\partial z}, \tag{8.16}$$

where we have dropped the tildes. If we now use (8.16) in (8.7) and assume that the background potential temperature gradient is slowly varying, the condition for convective stability is

$$\frac{\partial \theta_0}{\partial z}\left(1 - \frac{\partial \zeta_1}{\partial z}\right) \geqslant 0, \tag{8.17}$$

which implies

$$\frac{\partial \zeta_1}{\partial z} \leqslant 1, \tag{8.18}$$

i.e., any change in the height of the streamline displacement must be less than the change in z. This is important because if one is using a vertically-girded wave model with resolution Δz the flow stream lines will intersect if $\zeta_1 > \Delta z$. By definition streamlines must not intersect. The height z_b marks the beginning of wave breaking, *i.e.*, where

$$\left.\frac{\partial \zeta_1}{\partial z}\right|_{z_b} = 1. \tag{8.19}$$

8.2 WAVE-SATURATION PARAMETERIZATION

Wave saturation is now recognized as an essential feature of the general circulation especially in the middle and upper atmospheres, *i.e.*, from the lower stratosphere to the thermosphere (see, for example, Lindzen, 1981; Holton, 1982; Fritts, 1989; Alexander and Dunkerton, 1999; Campbell and Shepherd, 2005). If we had ideal models with great spatial and temporal resolutions, then the hydrodynamic equations would resolve gravity waves and their effects on the atmospheric flow. But of course, such a model may be a long way off. Thus, just as with atmospheric turbulence, gravity wave processes must be parameterized. However, unlike turbulence a linear theory of gravity waves exists, which greatly simplifies the parameterization process.

The parameterization process is diagramed in Fig. 8.5. In the first half of a time step, $t + \frac{1}{2}\Delta t$, the most recent profiles of wind speed and temperature are passed to a linear wave model. The linear wave model calculates the wave stress τ and the eddy diffusivity, D as functions of height. These data are then used to update the wind and temperature profiles, *i.e.*,

$$u_0(z,t) = u_0(z, t+\Delta t/2) + \left[\frac{1}{\rho}\frac{\partial \tau}{\partial z} + \frac{\partial}{\partial z}\left(D\frac{\partial u_0}{\partial z}\right)\right]\Delta t, \quad (8.20)$$

$$\theta_0(z,t) = \theta(z, t+\Delta t/2) + \frac{\partial}{\partial z}\left(D\frac{\partial \theta_0}{\partial z}\right)\Delta t. \quad (8.21)$$

These methods are sometimes referred to as off line.

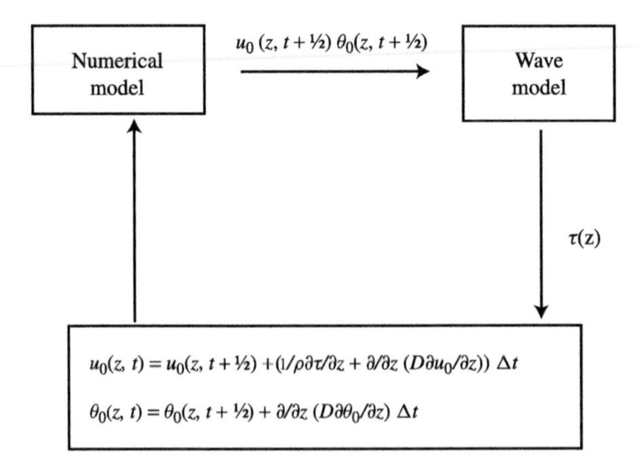

FIGURE 8.5 Schematic diagram of an off-line wave saturation parameterization.

8.3 PARAMETERIZATION METHODS

There exists today many methods for parameterizing wave dissipation. In this section, we will describe some of these methods as illustrations of the process.

8.3.1 SAWYER METHOD

One of the first parameterizations of wave stress was proposed by Sawyer (1959). Early numerical forecast models were based on the quasi-geostrophic approximation, and because of this gravity waves could not be simulated. Sawyer (1959) proposed the use of linear wave theory to evaluate the wave drag over a bell-shaped mountain. He assumed that this drag decreased with height in a manner similar to the decrease in the amplitudes of streamline displacements. Thus, he estimated that about half of the drag is exerted on the lowest 3 km of the atmosphere, and the remainder is distributed over the upper troposphere and stratosphere. He proposed the introduction of a force into the equation of motion directed against the low-level wind and decreasing linearly from a maximum near the ground surface to zero at the tropopause. We know now that such a scheme would fail because the wave drag is constant with height, unless wave breaking occurs.

8.3.2 LINDZEN–HOLTON METHOD

Wave saturation theory assumes that wave breaking results in turbulence. But turbulence is not passive; it is dispersive. The possibility of turbulence production by breaking gravity waves was posed by Hines (1963), Hodges (1967), and Lindzen (1967). Hodges (1969) suggested that this turbulence could be represented by an *eddy diffusivity*. Thus, vertical diffusion is associated with wave breaking, and we must add this effect to the dynamical equations. Following Holton (1982), we can write

$$\frac{Du_0}{Dt} = -\frac{1}{\rho_0}\frac{\partial}{\partial z}(\rho_0\overline{u_1 w_1}) + \frac{\partial}{\partial z}\left(D_e\frac{\partial u_0}{\partial z}\right), \tag{8.22}$$

where D_e is an eddy diffusivity produced by the wave breaking. The first term on the right-hand-side of (8.22) represents the drag produced by wave stress convergence, and the second term represents the effects of the Reynolds stress, *i.e.*, the turbulence stress. The eddy diffusion will also affect the background temperature, *i.e.*,

$$\frac{\partial \theta_0}{\partial t} = \frac{\partial}{\partial z}\left(D_e\frac{\partial \theta_0}{\partial z}\right). \tag{8.23}$$

Lindzen (1981) and Holton (1982) investigated this problem, and developed expressions for the wave stress convergence and eddy diffusivity, D_e, in wave breaking regions. In the discussion here, we rely much on the review paper by Fritts (1984) which presented Lindzen's model in greater detail than the original. Also, note that the notation in those papers has been changed to conform with that used in this book. The zone of application is the middle and upper atmosphere where the WKB approximation is most valid. We make the following assumptions: N^2 constant, u_0'' small, H_s large, and $\Omega = k(c - u_0) \ll kN$. Then, the Taylor–Goldstein equation reduces to the hydrostatic form,

$$\frac{\partial^2 \hat{w}_1}{\partial z^2} + \frac{N^2}{(c - u_0)^2} \hat{w}_1 = 0, \tag{8.24}$$

with WKB solution

$$\hat{w}_1 = A m^{-1/2} \mathrm{e}^{i \int_0^z m \mathrm{d}z}, \tag{8.25}$$

where A is an unknown amplitude, and

$$m = \frac{N}{c - u_0} \tag{8.26}$$

is slowly varying with height. Note that (8.24) is the equation for freely propagating waves ($c \neq 0$). In applications to the mountain wave problem, we must set $c = 0$. The complete solution is

$$w_1(x, z, t) = A m^{-1/2} \mathrm{e}^{z/2H_s} \mathrm{e}^{i(kx + mz - \omega t)}. \tag{8.27}$$

Using polarization (2.24) and (2.25) along with (1.66) we get

$$u_1(x, z, t) = -\frac{m}{k} w_1(x, z, t) \tag{8.28}$$

and

$$\theta_1(x, z, t) = -\frac{i}{k(c - u_0)} \frac{\partial \theta_0}{\partial z} w_1(x, z, t). \tag{8.29}$$

Lindzen (1981) assumed that the vertical variation of θ_1 is dominated by the perturbations e^{imz} in (8.27) so that

$$\frac{\partial \theta_1}{\partial z} = \frac{m}{k(c - u_0)} \frac{\partial \theta_0}{\partial z} w_1. \tag{8.30}$$

If we use (8.26) and (8.27) in (8.30), then the magnitude of $\partial \theta_1 / \partial z$ is primarily

$$\left| \frac{\partial \theta_1}{\partial z} \right| \propto m^{3/2} \mathrm{e}^{z/2H_s}. \tag{8.31}$$

The condition for convective stability is given by (8.7), and we see that this condition will be governed by the change of $\partial \theta_1 / \partial z$ with height since we have assumed that N is constant, and by implication $\partial \theta_0 / \partial z$ is either constant or changes very slowly with height. The factor $m^{3/2}$ in (8.31) causes growth if $|c - u_0| \to 0$, and causes decay if $|c - u_0|$ increases. If we express $m^{3/2}$ as an exponential, say e^{az}, then

$$\left| \frac{\partial \theta_1}{\partial z} \right| \propto \mathrm{e}^{(a + 1/2H_s)z}. \tag{8.32}$$

In the case $|c - u_0| \to 0$, which does not imply a critical level is present, $m^{3/2}$ adds to the $1/2H_s$ exponential growth rate an amount

$$a = \frac{1}{m^{3/2}} \frac{\partial m^{3/2}}{\partial z} = \frac{3/2}{(c - u_0)} \frac{\partial u_0}{\partial z}, \tag{8.33}$$

so that

$$\left| \frac{\partial \theta_1}{\partial z} \right| \propto e^{\left[\frac{3/2}{(c-u_0)} \frac{\partial u_0}{\partial z} + (1/2H_s)\right] z}. \tag{8.34}$$

Therefore, the eddy diffusion created by wave breaking must be such as to cancel the exponential growth with a rate equal to

$$\frac{3/2}{(c - u_0)} \frac{\partial u_0}{\partial z} + \frac{1}{2H_s}. \tag{8.35}$$

Using (8.28) in (8.30) gives

$$\frac{\partial \theta_1}{\partial z} = -\frac{u_1}{c - u_0} \frac{\partial \theta_0}{\partial z}, \tag{8.36}$$

and using this in (8.7) gives the condition for convective stability,

$$\frac{\partial}{\partial z}(\theta_0 + \theta_1) = \frac{\partial \theta_0}{\partial z} \left(1 - \frac{u_1}{(c - u_0)} \right) \geqslant 0. \tag{8.37}$$

Instability occurs when $u_1 > c - u_0$ or

$$u_0 + u_1 > c. \tag{8.38}$$

Thus, wave breaking occurs when the total velocity exceeds the wave phase velocity. For the case of terrain-generated waves, $c = 0$, and we see that wave breaking occurs with flow blocking as discussed in Section 8.1. Lindzen (1981) assumed that convectively unstable regions in the wave field result in the production of turbulence and just that level of eddy diffusion necessary to keep wave amplitudes to values near neutral stability. This is the saturation hypothesis. Wave amplitudes are constrained to values such that the slopes of the flow streamlines are not greater than $\pi/2$ or, equivalently, there are no flow reversals of the type illustrated in Fig. 8.4. It is also assumed that wave saturation does not otherwise affect wave propagation or wave characteristics.

Let wave saturation begin at height z_s. Then for breaking we require

$$\left| \frac{\partial \theta_1}{\partial z} \right| = \left| \frac{\partial \theta_0}{\partial z} \right|. \tag{8.39}$$

Using (8.39) in (8.30) and noting that $|w_1| = Am^{-1/2} e^{z_s/2H_s}$, we get

$$1 = \frac{m^{1/2}}{k(c - u_0)} A e^{z_s/2H_s}. \tag{8.40}$$

Using (8.26) in (8.40) gives

$$z_s = 2H_s \ln \left[\frac{k(c - u_0)^{3/2}}{AN^{1/2}} \right]. \tag{8.41}$$

The unknown wave amplitude A makes this equation problematical. However, we can apply (8.41) to a mountain wave where $c = 0$ and A can be evaluated. As an example, let's consider the case of constant flow over a surface corrugation of amplitude H. Now when considering the general case the proper form of (3.1) is

$$w_1(x, z) = -iu_0 H k_{\mathrm{s}} \mathrm{e}^{-i(k_{\mathrm{s}} x + m_{\mathrm{s}} z)} \mathrm{e}^{z/2H_{\mathrm{s}}}. \tag{8.42}$$

Then equating (8.42) with (8.27) gives

$$u_0 k H = A m^{1/2}. \tag{8.43}$$

Because we are using the WKB approximation, $m^{1/2} = N/u_0$ and solving (8.43) for A gives

$$A = (u_0 N)^{1/2} K H. \tag{8.44}$$

Then the breaking height for the surface corrugation waves is then

$$z_{\mathrm{s}} = 2H_{\mathrm{s}} \ln \left[\frac{u_0}{NH} \right]. \tag{8.45}$$

In Section 5.6 we identified u_0/NH as an internal Froude number. Then we can write

$$z_{\mathrm{s}} = 2H_{\mathrm{s}} \ln F. \tag{8.46}$$

For $z > z_{\mathrm{s}}$, the amplitude of the horizontal perturbation velocity is constrained by (8.38) to be

$$|u_{\mathrm{s}}| = |c - u_0|, \tag{8.47}$$

where u_{s} is the value of u_1 in the saturation zone, say between z_{s} and a critical level at height z_{c}. Figure 8.6, taken from Fritts (1984), illustrates the behavior of u_1 and the wave stress in the saturation zone. From (8.28), the amplitude of the vertical velocity is

$$w_{\mathrm{s}} = -\frac{k}{m} u_{\mathrm{s}} = -\frac{k}{N} (c - u_0)^2, \tag{8.48}$$

where we have used (8.26). In the absence of wave saturation, wave amplitudes grow in response to conservation of wave stress in regions of decreasing density, as discussed following (2.47). Following Fritts (1984) we represent the departure of the wave from conservative growth as

$$w_{\mathrm{s}} = w_1 \mathrm{e}^{-m_{\mathrm{I}}(z - z_{\mathrm{s}})}, \tag{8.49}$$

where m_{I} is the imaginary part of a complex vertical wavenumber, and w_1 is given by (8.16). Because the difference between w_{s} and w_1 is not large, we can write

$$\ln \left(\frac{w_s}{w_1} \right) \approx \frac{w_{\mathrm{s}}}{w_1} - 1, \tag{8.50}$$

and therefore

$$m_{\mathrm{I}} = \frac{1}{\delta z} \left[1 - \frac{w_{\mathrm{s}}(z_{\mathrm{s}} + \delta z)}{w_1(z_{\mathrm{s}} + \delta z)} \right], \tag{8.51}$$

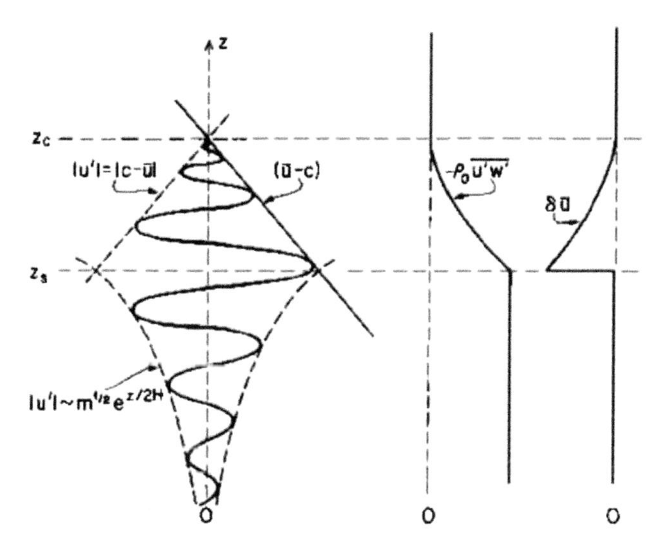

FIGURE 8.6 Illustration of wave growth below and dissipation below a critical level. Wave damping results in flux divergence. (Taken from Fritts (1984).)

where $\delta z = z - z_s$. Expanding $w_s(z + \delta z)$ and $w_1(z + \delta z)$ to first order in δz leads to

$$m_I = \frac{1}{w_1(z_s)} \left[\left. \frac{\partial w_1}{\partial z} \right|_{z_s} - \left. \frac{\partial w_s}{\partial z} \right|_{z_s} \right]. \tag{8.52}$$

Using (8.16) and (8.48) in (8.52) gives

$$m_I = \frac{1}{2H_s} + \frac{3}{2} \frac{1}{(c - u_0)} \frac{\partial u_0}{\partial z}. \tag{8.53}$$

Note that m_I is identical to (8.35).

Let us now consider only the diffusion part of (8.12), *i.e.*,

$$\frac{\mathrm{D}u_0}{\mathrm{D}t} = D_e \frac{\partial^2 u_0}{\partial z^2}, \tag{8.54}$$

where we have taken D_e to be slowly varying with height. Lindzen (1981) related eddy diffusion to a linear dampening. We can show this by first setting $u_0 = \bar{u} + u_1$, and then linearizing (8.54) to get:

$$\frac{\mathrm{D}u_1}{\mathrm{D}t} = D_e \frac{\partial^2 u_1}{\partial z^2}. \tag{8.55}$$

Setting $u_1 = A\mathrm{e}^{i(kx+mz-\omega t)}$ we get

$$D_e \frac{\partial^2 u_1}{\partial z^2} = -D_e m^2 u_1. \tag{8.56}$$

With a similar equation for θ_1. Using (8.56) in (8.55) gives

$$ik(c - u_0)u_1 + w_1 \frac{\partial u_0}{\partial z} = -m^2 D_e u_1, \tag{8.57}$$

Now assume a complex phase velocity, $c = c_R + ic_I$. Then the real part of (8.57) is

$$kc_I = -m^2 D_e. \tag{8.58}$$

From (8.26) we have

$$m_R + im_I = \frac{N}{c - u_0}, \tag{8.59}$$

and the imaginary part of (8.59) is,

$$m_I \approx -\frac{Nc_I}{(c - u_0)^2}. \tag{8.60}$$

Using (8.53), (8.58), and (8.60), the eddy diffusivity in the saturation zone is

$$D_e = \frac{k}{N^3}(c - u_0)^4 \left[\frac{1}{2H_s} + \frac{3}{2}\frac{1}{c - u_0}\frac{\partial u_0}{\partial z}\right]. \tag{8.61}$$

Note that D_e must be positive. It is easily shown that in the breaking zone $\frac{\partial u_0/\partial z}{(c-u_0)}$ is always positive, and $D_e \to 0$ as $z \to z_c$ as illustrated in Fig. 8.6.

We now turn our attention to the wave drag. Using (8.47) and (8.48), the wave stress in the saturation zone is

$$-\rho_0\overline{u_s w_s} = \frac{1}{2}\rho_0\Re(u_s w_s^*) = -\frac{1}{2}\rho_0\frac{k}{N}(c - u_0)^3. \tag{8.62}$$

Using (8.62) in (8.1) and assuming that N is constant or slowly varying gives

$$\frac{\partial u_0}{\partial t} = -\frac{k}{N}(c - u_0)^3 \left[\frac{1}{2H_s} + \frac{3}{2}\frac{1}{(c - u_0)}\frac{\partial u_0}{\partial z}\right]. \tag{8.63}$$

The first term in brackets comes about as a response to exponential growth, and the second term balances the $m^{3/2}$ growth term and is a function of whether the mean wind shear is positive or negative. In the breaking zone, the induced acceleration is governed by $(c - u_0)$. If a critical level exists, and if the wave is approaching it from below, then

$$\frac{\partial u_0}{\partial z} \approx \frac{c - u_0}{z_c - z}. \tag{8.64}$$

Using this in (8.63) we see that as the critical level is approached the induced acceleration is proportional to $(c - u_0)^3$, and is always negative. Comparing (8.61) with (8.63) we see that the induced flow acceleration due to wave drag is:

$$\frac{\partial u_0}{\partial t} = -\frac{N^2}{c - u_0}D_e. \tag{8.65}$$

As pointed out by Fritts (1984), (8.65) would suggest that the induced acceleration produced by saturation and the eddy diffusion responsible for wave dissipation are related; however, Lindzen (1981) points out that these two effects addressed by the simple linear model while related are separate manifestations of gravity wave saturation.

To scale the induced acceleration, Fritts (1984) gives the following example. We consider uniform flow at mesospheric heights with $H_s = 6$ km, $N = 0.02$ s^{-1}, $c - u_0 = 30$ ms^{-1}, and $\lambda_x = 200$ km. From (8.61) the eddy diffusivity $D_e = 265$ m^{2}s^{-1}. Using (8.65), the induced flow acceleration is

$$\frac{\partial u_0}{\partial t} = -305 \text{ ms}^{-1}/\text{day}. \tag{8.66}$$

This is a substantial reduction in the zonal wind, and such a reduction in velocity has not been observed. However, if the gravity waves were generated by terrain, then this deceleration would be confined to regions directly above the mountain. But even if limited to a small region, it is reasonable that such a deceleration would have some impact on the global scale. Therefore it is reasonable to think that this effect must be compensated by an acceleration of similar magnitude. A likely possibility proposed by Lindzen (1981) is the *Coriolis acceleration* produced by the meridional wind, v_0, *i.e.*,

$$\frac{\partial u_0}{\partial t} = 2\Omega_E \sin \Phi v_0 = -305 \text{ ms}^{-1}/\text{day}, \tag{8.67}$$

where Ω_E is the angular velocity of the Earth, Φ is latitude, and v_0 is the meridional wind. Solving for v_0 gives

$$v_0 \approx -\frac{24 \text{ ms}^{-1}}{\sin \Phi}. \tag{8.68}$$

Nastrom, Balsley, and Carter (1982) showed observations of mean meridional wind near the *mesopause* at Poker Flat, Alaska (65°N) for the Summer season. Values of v_0 ranged between approximately -10 to -30 ms^{-1}. From (8.68), v_0 at 65°N is about -26.5 ms^{-1}. This agreement between the theoretical estimates and observations seems to support the validity of the wave saturation mechanism.

8.3.3 THE PALMER METHOD

In order to improve the performance of the Meteorological Office 15-layer operational forecast model and 11-layer general circulation model, Palmer, Shutts, and Swinbank (1986) accounted for subgrid-scale terrain effects by implementing wave saturation theory. Using (3.33) and (8.36) the local Richardson number (5.117) is

$$R_i = \frac{g}{\theta_0} \frac{\frac{\partial \theta_0}{\partial z}\left(1 - \frac{\partial \zeta_1}{\partial z}\right)}{\left(\frac{\partial u_0}{\partial z}\right)^2 \left[1 + \frac{\partial u_1/\partial z}{\partial u_0/\partial z}\right]^2}, \tag{8.69}$$

where ζ_1 is the vertical displacement of a flow streamline defined by (8.24). Defining $R_{i,0}$ as the background Richardson number, we can write (8.69) as

$$R_i = R_{i,0} \frac{\left(1 - \frac{\partial \zeta_1}{\partial z}\right)}{\left(1 + \frac{R_{i,0}^{1/2}}{N} \frac{\partial u_1}{\partial z}\right)^2}. \tag{8.70}$$

Using (8.19) and (3.33) it is easily shown that

$$u_1 = -u_0 \frac{\partial \zeta_1}{\partial z} = -i u_0 m \zeta_1. \tag{8.71}$$

From (8.70) we see that the local or wave modulated R_i is a non-linear function of wave phase, and from Fig. 5.18 we see that instability occurs over only a limited range of the wave field. Palmer, Shutts, and Swinbank (1986) argued that ultimately the parameterization seeks to account for the ensemble subgrid-scale effects of terrain features on various space scales. The effects of these "phase-incoherent subgrid-scale gravity waves" are parameterized by ignoring the phase differences in (5.131). Then using (8.71) in (8.70) gives

$$R_{i,\min} = R_{i,0} \frac{1 - N|\zeta_1|/u_0}{\left[1 + R_0^{1/2}(N|\zeta_1|/u_0)\right]^2}, \tag{8.72}$$

where $R_{i,\min}$ represents the smallest value of the Richardson number that can be realized under the action of gravity waves. The free parameter in (8.72) is the wave displacement $|\zeta_1|$. Thus, (8.72) provides the mechanism for parameterizing wave saturation. This is done by controlling the value of $|\zeta_1|$ so the $R_{i,\min} \geqslant 1/4$. For example, if an upward moving gravity wave enters a region of decreasing wind speed and/or increasing stratification, then the quantity $N|\zeta_1|/u_0$ will increase and $R_{i,\min}$ will decrease. Decreasing $|\zeta_1|$ will balance the increase in N/u_0 so that $R_{i,\min} \geqslant 1/4$; but this adjustment also leads to a decrease in wave stress. Thus, the effects of wave saturation are parameterized. Note that Palmer, Shutts, and Swinbank (1986) used the dynamic stability condition, $R_{i,\min} \geqslant 1/4$ for the wave saturation criterion. This limit was chosen to account for large isotropic displacements which could induce Kelvin–Helmholtz instability.

Under the WKB approximation, the wave stress over sinusoidal topography is given by (3.83). Palmer, Shutts, and Swinbank (1986) assumed that the stress in the surface layer of the model, τ_s, is directed against the surface-layer wind, and is given by

$$\tau_s = \rho_0 \kappa N u_0 \sigma_H^2, \tag{8.73}$$

where σ_H is the rms value of the subgrid-scale topography, and κ is a "tunable" parameter which accounts for the fact that subgrid-scale topography is not represented by a single wavenumber. They use $\kappa = 2.5 \times 10^{-5}\ \mathrm{m}^{-1}$ which corresponds to a wave length of 250 km if the topography were sinusoidal. In the absence of wave dissipation, $\tau(z) = \tau_s$. In the case of wave dissipation, it is assumed that the direction of the wave stress will still be against the surface layer wind. The wave stress at any level above the surface layer is given by

$$\tau(z) = \rho_0 \kappa N U_0 |\zeta_1|^2, \tag{8.74}$$

where U_0 is the component of the background wind parallel to the surface-layer wind.

The parameterization begins with the calculations of surface layer stress (8.73) in each model grid cell. At the next layer, $R_{i,0}$ is calculated using the values of N, u_0, and v_0 at adjacent vertical levels. Then assuming that $\tau = \tau_s$, $|\zeta_1|$ is calculated using (8.74). This value is then used in (8.72) to evaluate $R_{i,\min}$. If $R_{i,\min} \gtrsim 1/4$, then τ is unchanged, and we proceed to the next upper layer. The preceding steps are repeated layer-by-layer, always initially setting the stress equal to the stress in the adjacent lower layer. If $R_{i,\min}$ remains about 1/4, then the wave stress remains constant and at its surface value. However, if this criterion is not met at a certain layer, then $R_{i,\min}$ is set equal to 1/4, and (8.72) is used to calculate a new displacement which we call $|\zeta_1|_{\text{sat}}$. The wave stress for the layer is then calculated using (8.74) but now with $|\zeta_1|_{\text{sat}}$ used in place of $|\zeta_1|$. This new value of stress is called τ_{sat}. We now go to the next upper layer, and proceed as before,with the initial estimate of displacement calculated using (8.74) but now using τ_{sat} calculated for the adjacent lower layer. When the top model layer has been reached, we have a vertical profile of wave stress which can then be used to update the velocity fields in each grid cell.

8.3.4 THE MCFARLANE METHOD

Mcfarlane (1987) developed a wave drag parameterization based on Lindzen's wave saturation theory (Lindzen, 1981). McFarlane's intent was to improve the climate simulations of the Northern Hemisphere wintertime circulations in the troposphere and lower stratosphere by accounting for the effects of orographically excited gravity waves. The wave drag parameterization was applied to the Canadian Climate Centre general circulation model. The linear wave equations used were,

$$\overline{U}\frac{\partial u}{\partial z} + w\frac{\partial \overline{U}}{\partial z} = -\frac{\partial \Pi}{\partial z}, \tag{8.75}$$

$$\frac{\partial \Pi}{\partial z} = \frac{g\theta}{\bar{\theta}}, \tag{8.76}$$

$$\frac{\partial u}{\partial x} + \frac{1}{\bar{\rho}}\frac{\partial}{\partial z}(\bar{\rho} w) = 0, \tag{8.77}$$

$$\overline{U}\frac{\partial \theta}{\partial x} + w\frac{\partial \bar{\theta}}{\partial z} = 0, \tag{8.78}$$

where $\Pi = (p/P)^{R/c_p}$ and P is a reference pressure. The analysis was done in terms of the first-order streamline displacement function $\psi(x, z)$, *i.e.*,

$$w_1 = \overline{U}\frac{\partial \psi}{\partial x}, \tag{8.79}$$

$$u_1 = -\frac{1}{\bar{\rho}}\frac{\partial}{\partial z}(\bar{\rho}\overline{U}\psi), \tag{8.80}$$

$$\theta_1 = -\psi\frac{\partial \bar{\theta}}{\partial z}. \tag{8.81}$$

The topography was represented by a corrugation with amplitude h, and horizontal wavenumber μ,

$$\psi(x, 0) = h\cos\mu x, \tag{8.82}$$

which is also the bottom boundary condition. The radiation condition was used at the top of the model. Using (8.79)–(8.81) the equation for ψ is

$$\frac{\partial}{\partial z}\left[\frac{\overline{U}^2}{\bar{\rho}}\frac{\partial}{\partial z}(\bar{\rho}\psi)\right] + N^2\psi = 0. \tag{8.83}$$

Away from critical levels the approximate WKB solution has the form

$$\psi(z, x) = A(z)\cos\left[\mu x + \int_0^z \phi(z')\mathrm{d}z'\right], \tag{8.84}$$

where amplitude $A(z)$ and phase function $\phi(z)$ change vertically on the scale of the mean flow. Substituting ψ from (8.84) into (8.83), equating the coefficients of the resulting trigonometric functions and neglecting second-order terms, one gets

$$\phi = \frac{N}{\overline{U}}, \tag{8.85}$$

$$A = h\left[\frac{\bar{\rho}(0)N(0)\overline{U}(0)}{\bar{\rho}N\overline{U}}\right]^{1/2}. \tag{8.86}$$

Wave breaking begins when

$$\frac{\partial \psi}{\partial z} \approx -F(z)\sin(\mu x + \phi) > 1, \tag{8.87}$$

where

$$F(z) = \frac{Nh}{\overline{U}}\left[\frac{\bar{\rho}(0)N(0)\overline{U}(0)}{\bar{\rho}N\overline{U}}\right]^{1/2} = A\phi. \tag{8.88}$$

Mcfarlane (1987) refers to F as the local Froude number.

McFarlane represents wave dissipation in the breaking zone by applying a dampening of the form

$$\psi = \psi_I \exp\left[-\int_0^z D(z')\mathrm{d}z'\right], \tag{8.89}$$

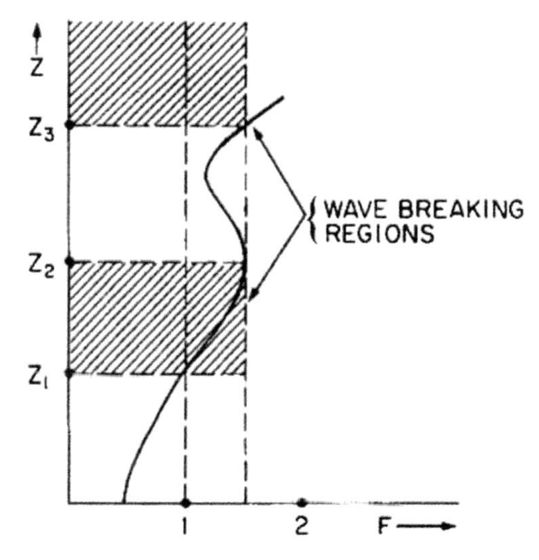

FIGURE 8.7 Schematic of vertical structure of the local Froude number showing regions of wave breaking and saturation. (Taken from Mcfarlane (1987).)

where ψ_1 is given by (8.84)–(8.86), and

$$D(z) = \frac{N^3 K}{\mu \overline{U}^4}. \tag{8.90}$$

It is assumed that wave dissipation by turbulence is represented by an eddy diffusivity, K, defined by

$$\left[\frac{N^2}{\mu \overline{U}^3}\right] K \leqslant \mathrm{O}\left(\frac{\overline{U}}{NH}\right). \tag{8.91}$$

The saturation hypothesis is satisfied if

$$\left|\frac{\partial \psi}{\partial z}\right| = F(z) \exp\left[-\int_0^z D(z')\mathrm{d}z'\right] \leqslant 1. \tag{8.92}$$

Figure 8.7, taken from Mcfarlane (1987), illustrates a possible vertical profile of F. Below height z_1, $F < 1$. Between z_1 and z_2, the local Froude number is greater than 1 and increases with height; z_1 marks the breaking height. Equation (8.92) can be satisfied using

$$D(z) = \frac{1}{F}\frac{\mathrm{d}F}{\mathrm{d}z}. \tag{8.93}$$

The wave stress is now given by

$$\tau = \frac{\tau_0}{F^2} = -\frac{1}{2}\frac{\bar{\rho}\mu \overline{U}^3}{N}, \tag{8.94}$$

where

$$\tau_0 \approx \frac{\mu h^2}{2} \bar{\rho}(0) N(0) \overline{U}(0). \tag{8.95}$$

At level z_2, the wave amplitude has been reduced so that wave breaking stops even though the Froude number is greater than unity. At z_2, $D(z)$ vanishes (because $\mathrm{d}F/\mathrm{d}z = 0$ in (8.93)) and remains zero between z_2 and z_3. Above z_3, convective instability begins again and $D(z)$ becomes nonzero to maintain a saturated, *i.e.*, neutral state. To represent the vertical variation of the wave stress the dampening factor is included in the definition of the wave amplitude, then

$$\tau = \tau(0) \frac{A^2 \bar{\rho} N \overline{U}}{A^2(0) \bar{\rho}(0) N(0) \overline{U}(0)}, \tag{8.96}$$

where $A(z)$ in independent of height except in saturation regions where

$$A(z) = \frac{\overline{U}}{N}. \tag{8.97}$$

The implied mean flow change is given by

$$\frac{\partial \overline{U}}{\partial t} = -\frac{1}{\bar{\rho}} \frac{\partial \tau}{\partial z} = -\frac{\mu}{2} \frac{\overline{U}^3}{N} \max \left[\frac{\mathrm{d}(\ln F^2)}{\mathrm{d}z}, 0 \right]. \tag{8.98}$$

In representing the wave drag effects in the general circulation model Mcfarlane (1987) denotes the large-scale flow conditions by a zero subscript. The wave momentum flux is then

$$\tau_0 = -\left(E \frac{\mu_\mathrm{e}}{2} h_\mathrm{e}^2 \right) \rho_0 N_0 \vec{V}_0, \tag{8.99}$$

where μ_e and h_e are the effective horizontal wavenumber and amplitude respectively, and E is an efficiency parameter which is taken to be less than unity. The modeled wave stress is constant with height except in saturation zones. Analogous to (8.96) the wave stress is given by

$$\tau(z) = \tau_0 \left[\frac{A(z)}{h_\mathrm{e}} \right]^2 \frac{\rho N U}{\rho_0 N_0 U_0}, \tag{8.100}$$

where $A(z)$ is the local amplitude. Above some reference height, which could be the first model level above the ground surface, the wind speed $U(z)$ corresponds to that component of the flow which is parallel to that at the reference level, *i.e.*,

$$U = \vec{V} \bullet \frac{\vec{V}_0}{\left|\vec{V}_0\right|}. \tag{8.101}$$

By implication, the reference level stress vector is directed opposite to the reference level velocity vector.

At any model level above the reference level, $A(z)$ is first estimated in terms of the amplitude at the next lower level, thus conserving wave stress. This new value is adjusted to prevent convective instability, *i.e.*,

$$A(z) \leqslant \frac{U}{N} F_c, \tag{8.102}$$

where F_c is a critical Froude number. For a linear monochromatic wave $F_c \approx 1$. The effective height in (8.100) is given by

$$h_e = \min\left(2S_d, \frac{F_c U_0}{N_0}\right), \tag{8.103}$$

where S_d is the standard deviation of that part of the orography most likely to generate waves. The factor $E\mu_e/2$ in (8.100) is a tunable parameter. In McFarlane's demonstration he used $F_c^2 = 0.5$ and $E\mu_e/2 = 8 \times 10^{-6}$ m^{-1}.

8.3.5 THE SCHOEBERL METHOD

The parameterization schemes discussed thus far assume that wave growth stops when the saturation condition is satisfied. However, Schoeberl (1988) citing the work of Fritts and Dunkerton (1984) argues that wave growth does not immediately stop with the onset of wave breaking. As we discussed with regards to Fig. 5.19, dynamic and convective instability first occur over a limited range of the wave field. Thus, the Reynolds stress produced by breaking waves may be different than that calculated with the assumption of constant wave amplitude. As an alternative procedure, Schoeberl (1988) presented a scheme where the flow streamlines are adjusted locally rather than globally, *i.e.*, over the whole wave. The WKB solution to the Taylor-Goldstein equation along with (3.33) is

$$\hat{\zeta}_1(z, k) = \hat{\zeta}_1(0)\left(\frac{m(0)}{m(z)}\right)^{1/2} e^{i\int m \mathrm{d}z}, \tag{8.104}$$

where $\hat{\zeta}_1(0)$ is the Fourier-transformed streamline displacement at the ground surface, and $m(0) = N(0)/u_0(0)$ is the vertical wavenumber at the ground surface. The saturation parameterization scheme consists of evaluating (8.104) upward layer-by-layer with the constraint that the flow be convectively stable. The WKB solution is constructed at each model level according to

$$\hat{\zeta}_1(z + \Delta z, k) = \hat{\zeta}_1(z, k)\left(\frac{m(z)}{m(z + \Delta z)}\right)^{1/2} e^{i\int_z^{z+\Delta z} m \mathrm{d}z}. \tag{8.105}$$

This so-called *streamline adjustment* algorithm begins at the ground surface or, equivalently, at the model surface layer where $\hat{\zeta}_1(0)$ is evaluated. Then (8.105) is used to carry the solution to the next upper level, z_1, where $\hat{\zeta}_1(z_1) = \hat{\zeta}_1(0)$. Next, the inverse Fourier transform of $\hat{\zeta}_1(k, z)$ gives the streamline displacement in physical space, $\zeta_1(x, z)$. This value is used in (8.29) to test the stability in the subgrid domain. If the layer is stable, then the solution is carried to the next layer.

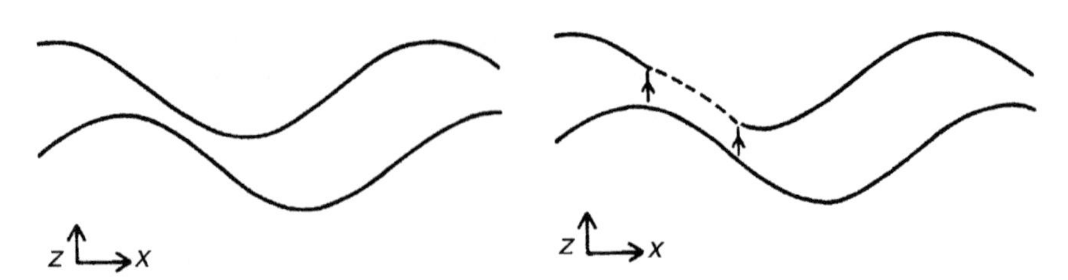

FIGURE 8.8 Schematic illustration of the convective adjustment process. Only a part of the wave is adjusted so that. (Taken from Schoeberl (1988).)

If the layer is unstable, then $\partial\zeta_1/\partial z$ is set to unity, and the new displacement is given by

$$\zeta_1(x, z + \Delta z) = \zeta_1(x, z) + \Delta z. \tag{8.106}$$

When all of the points in the wave field have been adjusted, the displacement is Fourier-transformed into wave space, and (8.105) is used to carry the solution to the next layer. Figure 8.8, taken from Schoeberl (1988), illustrates the adjustment process, and we see that convective adjustment is done only to part of the wave field, and does not initially limit wave growth as shown in Fig. 8.9 which displays potential temperature isotherms as function of altitude. In Fig. 8.9, we see that in regions of convective adjustment, the slopes of the isotherms become vertical as indicated by (8.8). We also see that wave amplitude continues to grow even when convective adjustment is occurring. If the wave encounters a region of the atmosphere where saturation does not occur, then wave amplitude growth ends; however, the vertically oriented slopes of the wave remain unchanged. Figure 8.10 plots the variation of displacement amplitude with height for the case shown in Fig. 8.9. For conventional saturation theory (Hodges, 1967) displacement amplitude is constant after the start of wave breaking; however, convective adjustment results in displacements which continue to grow after wave breaking, but at a decreasing rate. The dashed line in Fig. 8.10 shows the result without wave saturation. The continued growth of wave amplitudes after wave breaking is referred to as *supersaturation* (Lindzen, 1988).

8.3.6 THE TERRAIN-HEIGHT ADJUSTMENT SCHEME

The preceding wave saturation parameterization schemes made use of the WKB approximation which is applicable in the middle and upper atmosphere, *i.e.*, where the scale of vertical variation of the background variables is small compared with the vertical wave length of a gravity wave. However, in lower troposphere and especially in the stable PBL, vertical wave lengths are not necessarily small compared to background variations. For example, consider the vertical wavelength

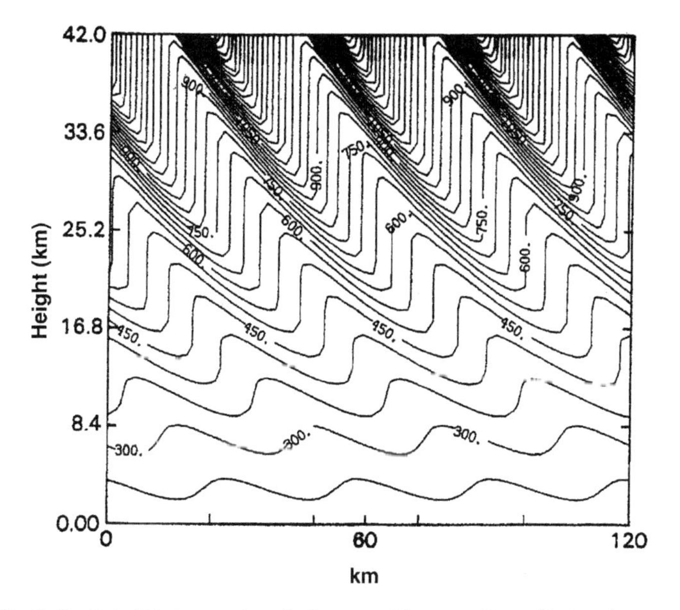

FIGURE 8.9 Potential temperature isotherms with convective adjustment over a surface corrugation. Wave breaking begins at about 14 km. (Taken from Schoeberl (1988).)

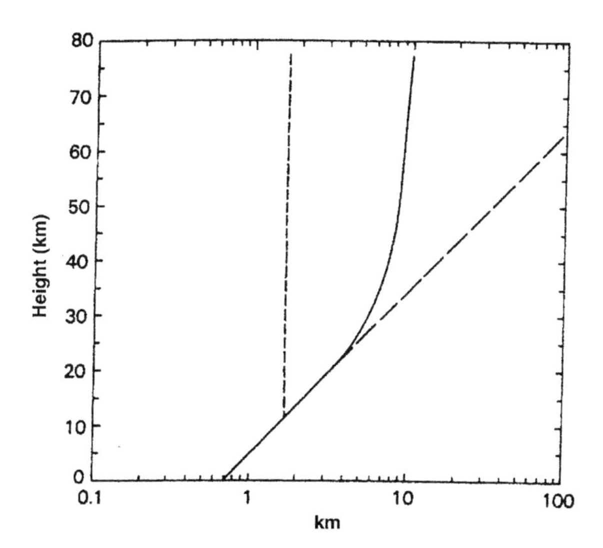

FIGURE 8.10 Displacement amplitude as a function of height for the experiment shown in Fig. 8.9. Dotted line shows results using saturation theory, solid line shows the results using streamline adjustment saturation parameterization, and dashed line shows result of no adjustments. (Taken from Schoeberl (1988).)

of a terrain generated wave when the background wind and stratification are constant

$$\lambda_z = \frac{2\pi}{m} = 2\pi\left(\frac{N^2}{u_0^2} - k^2\right)^{-1/2}. \tag{8.107}$$

While (8.107) applies only for the case where N and u_0 are constant, we can still use it for scaling. Considering now orographic features on the mesoscale, let a terrain feature have a horizontal width, b such that $k = 1/b$. Then, with $u_0 = 5\ \mathrm{ms}^{-1}$, $N = 0.03\ \mathrm{s}^{-1}$, and $b = 1000$ m, $\lambda_z \approx 1060$ m. An appropriate scale for the vertical variation of flow in the tropospheric surface layer is the Ekman-layer depth, z_E given by (4.75). Using the typical boundary-layer values of 5 ms^{-2} for the eddy coefficient of friction and $f = 10^{-4}\ \mathrm{s}^{-1}$, $z_\mathrm{E} \approx 1000$ m which is on the same order as the vertical wave length of the terrain-generated wave. We also note that within the stable boundary layer $u_0 \propto \ln(z/z_0)$, where z_0 is the aerodynamic surface roughness (see, for example Stull, 1988; Foken, 2008). Then as u_0 increases $m = (N_2/u_0^2 - k^2)^{1/2}$ decreases. Thus, gravity waves generated in the lower troposphere can be expected to have reflection levels, which cannot be considered under the WKB assumption. Because of the Ekman spiral, we can anticipate that there will be critical levels in the boundary layer (Nappo and Chimonas, 1992). Critical levels can also result from frontal systems, density currents, thunderstorm gust fronts, *etc*. For example, Fig. 8.11 taken from Mahrt (1985) shows the wind and temperature fields in the nighttime planetary boundary layer constructed from instrumented aircraft flights over moderately complex topography in central Oklahoma. During these flights, a northerly surface flow undercut a less stable southerly flow. In Fig. 8.11, we see regions of high turbulence (regions of vertical velocity variance greater than 0.01 $\mathrm{m}^2\ \mathrm{s}^{-2}$) centered in regions of wind reversals. For terrain-generated flows, critical levels occur at wind reversals, *i.e.*, where $u_0 = 0$. However, strictly speaking the WKB is not applicable near a critical level. Thus, streamline adjustment parameterization schemes are not expected to work well, if at all, in the lower troposphere.

In an effort to account for wave saturation effects in the planetary boundary layer over complex terrain, Nappo and Andren (1995) developed a method of saturation adjustment which could be used in a non-hydrostatic model to calculate wave drag. This parameterization was extended by Nappo and Physick (2000) to calculate wave stress and enhanced vertical mixing due to wave breaking. The method is quite simple. For all mountain waves, u_1 and w_1 are each proportional to the maximum terrain-height H, and hence $\tau \propto H^2$. In wave breaking regions, wave amplitudes decrease with height, and this decrease is simulated by *virtually* decreasing H. In application (Nappo, Chun, and Lee, 2004; Nappo and Svensson, 2008), the wave field in physical space, $u_1(x_i, z_j)$, $w_1(x_i, z_j)$, is calculated, where x_i is the i'th horizontal point of M subgrid points and z_j is the j'th vertical grid

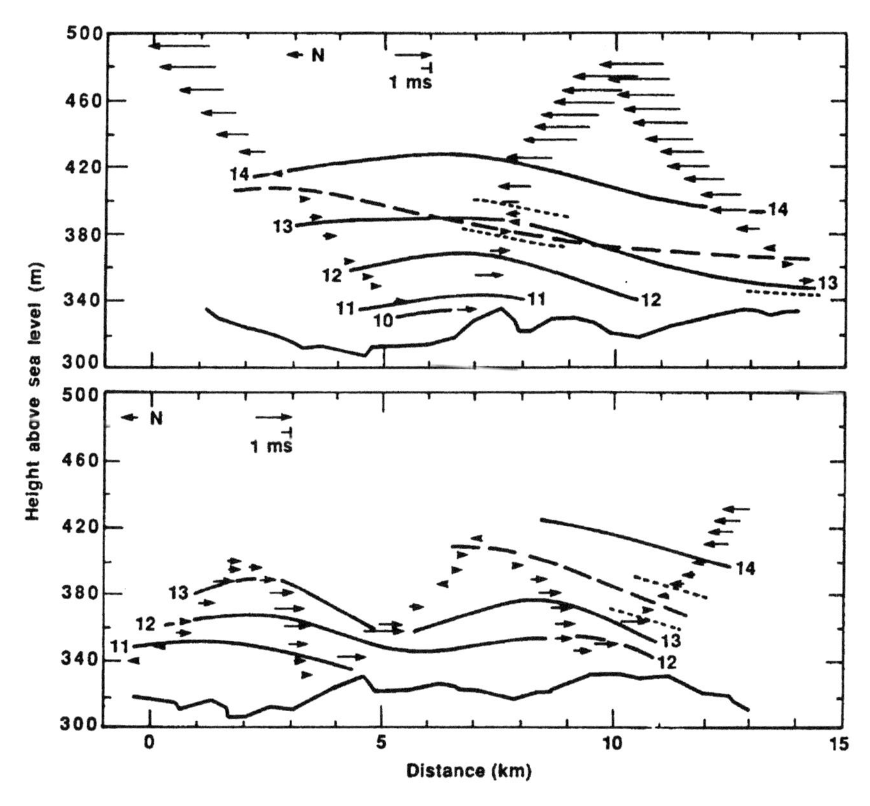

FIGURE 8.11 Two cross sections of low-level airflow constructed from aircraft data taken with two similar undulating flight tracks. The lower solid line in each figure marks the terrain surface. Solid lines are isotherms of potential temperature (°C); dotted lines enclose regions of vertical velocity variance greater than 0.01 $\mathrm{m}^2\ \mathrm{s}^{-2}$. (Taken from Mahrt (1985)).

level of the wave model. For each x_i, wave breaking is tested using the criterion,

$$\left|\frac{u_1(x_i, z_j)}{u_0(z_j)}\right| \leqslant 1. \tag{8.108}$$

Equation (8.108) is the criterion for no flow blocking. If (8.108) is satisfied for all x, then the calculations proceed to the next higher level, $j+1$. The above procedure is repeated, but if at some point (8.108) is not satisfied, H is decreased a certain percentage, usually 10% and $u_1(x_i, z_{j+1})$ and $w_1(x_i, z_{j+1})$ are recalculated. If (8.108) is still not satisfied, H is again reduced and u_1 and w_1 are again calculated. If now flow blocking does not occur, the calculations proceed to the next higher level, and the calculation-testing process is repeated. As each wave breaking point is adjusted to neutral stratification (no flow blocking) H becomes smaller and τ decreases. Figure 8.12 illustrates the changes in wave stress and disturbance amplitude for the case shown in Fig. 8.2 with wind profile shown in Fig. 8.3,

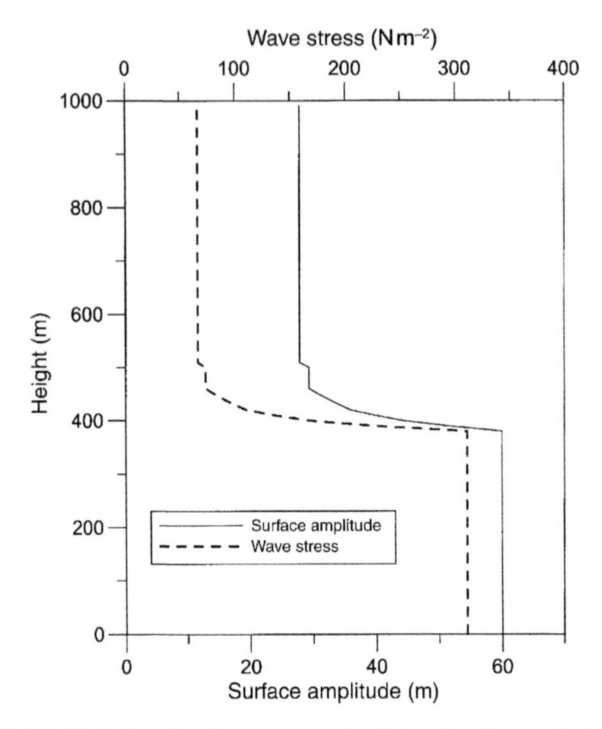

FIGURE 8.12 Vertical profile of wave stress and variation of surface amplitude resulting from the terrain-height adjustment wave saturation parameterization scheme used in the experiment shown in Fig. 7.12.

and the adjusted wave field is shown Fig. 8.13. We see that the wave field is now convectively stable at all heights. From Fig. 8.12, the wave stress decreases from about 310 N m^{-2} at about 380 m to about 70 N m^{-2} at about 510 m. This stress convergence results in a deceleration of about -0.5×10^{-3} ms^{-2} or a slowing down of the flow between 500 and 100 m of about 1.8 ms^{-1}.

The terrain-height adjustment scheme has been used by Tjernstrom *et al.* (2009) to explain observed turbulence in the residual layer over the CASES-99 field site. This turbulence is believed due to wave breaking below critical levels created by directional shear of the background wind.

8.4 SATURATION LIMITS AND OTHER PROBLEMS

Linear saturation theory and the parameterization schemes discussed in this chapter have limitations and omissions. Fritts (1984) listed several shortcomings, for example, neglect of wave superposition and interaction, wave transience and horizontal localization, quasi-linear mean flow accelerations, and the detailed

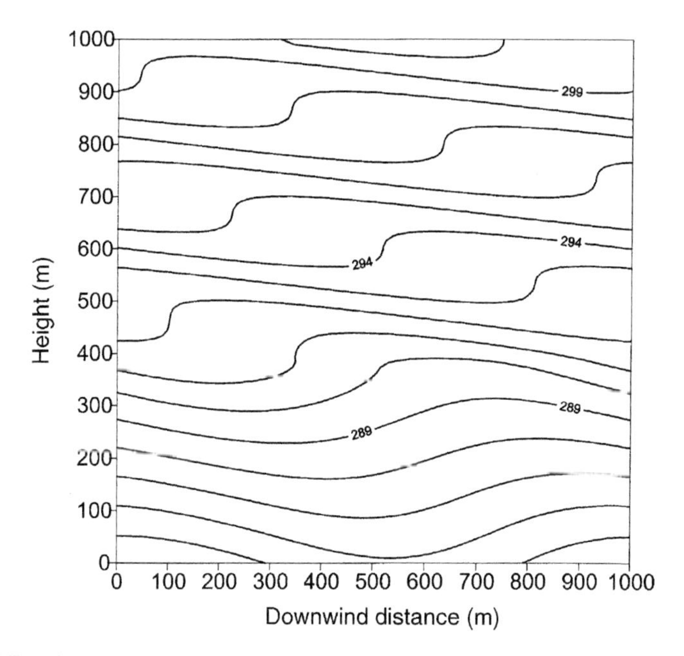

FIGURE 8.13 The same experiment as shown in Fig. 7.2, but now with terrain-height adjustment saturation parameterization.

nature of the saturation process may not be justified in general. Mcfarlane (1987) noted several deficiencies in parameterization schemes including the failure to account for the orientation of the surface-layer winds relative to the terrain, and the neglect of nonlinearity. Miller, Palmer, and Swinbank (1989) enhanced the Palmer, Shutts, and Swinbank (1986) scheme by introducing directionally-dependent subgrid-scale orographic variance, and this lead to significant improvements in forecast skill, and reduction in-model systematic errors. As shown by Smith (1977), nonlinearity due to advection terms and nonlinearity in the bottom boundary condition can lead to enhanced wave steeping in some vertical regions while suppressing steepening in others. This can have a pronounced effect on wave saturation which would be missed in a linear theory. Mcfarlane (1987) also pointed out that the failure to account for wave reflection because of the use of the WKB approximation negates the possibility of simulating wave-resonance and wave amplification effects observed by Lilly and Zipser (1972) and Lilly (1978) and numerically studied by Peltier and Clark (1979). Schoeberl (1988) noted that the schemes developed for general circulation models by Palmer, Shutts, and Swinbank (1986) and Mcfarlane (1987) should not be used to discriminate between various subgrid-scale processes because of the parameterization's crude estimates of breaking heights and induced acceleration. He pointed out that these schemes are not accurate enough representations for even idealized definitive conclusions about the influence of topography on the general circulation.

Saturation parameterization schemes developed for general circulation and forecast models usually represent subgrid orography in terms of a sinusoidal function. However, stratified flows over real mountains are much more complex (Baines, 1995). Kim and Mahrt (1992) compared calculated with observed wave stress over the coastal mountain range of in Croatia. They pointed out that stable flow over two-dimensional mountains are influenced by terrain- induced disturbances such as the blocking of low-level flow, the development of turbulent boundary layers, and the formation of stagnant cold-air pools in topographic depressions. These effects act to limit the vertical displacements of streamlines near the ground surface so that the *effective height* of the obstacle is less than its real height. Stern and Pierrehumbert (1988) proposed that the effective height, H_{e}, is

$$H_{\mathrm{e}} = \min\left[H, c\frac{u_0(0)}{N(0)}\right], \tag{8.109}$$

where H is the real terrain-height, $u_0(0)$ and $N(0)$ are surface layer values of wind speed and Brunt–Väisälä frequency, and c is a constant estimated to be between 0.4 and 0.8. Lott and Miller (1997) developed an orographic drag parameterization for subgrid-scale three-dimensional objects based on the ideas presented by Baines and Palmer (1990). Figure 8.14 taken from Lott and Miller (1997) illustrates the low-level flow behavior over an elliptically-shaped mountain. The non-dimensional height of the mountain, H_{n}, is taken to be

$$H_{\mathrm{n}} = \frac{HN}{|U|}, \tag{8.110}$$

where U is the scale speed of the incident flow. H_{n} corresponds to an inverse internal Froude number. For small H_{n}, all the flow goes over the mountain, and the effective height for gravity wave generation is H. However, for large H_{n} there is insufficient energy of the incident flow to overcome the buoyancy, and part of the low-level flow goes around the mountain. Then, as illustrated in Figure 8.14, the effective height of the mountain is $H - z_{\mathrm{b}}$ where z_{b} is the depth of the blocked

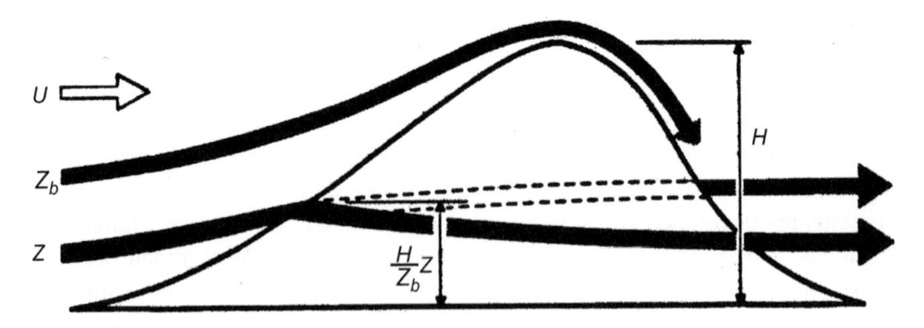

FIGURE 8.14 Schematic representation of the low-level flow behavior over a three-dimensional obstacle. (Taken from Lott and Miller (1997).)

layer. For constant U and N,

$$z_b = H\max\left(0, \frac{H_n - H_{nc}}{H_n}\right), \tag{8.111}$$

where H_{nc} is a critical non-dimensional height of order unity.

A currently important problem is the parameterization of stress carried by convectively-generated gravity waves in general circulation models (see, for example, Chun *et al.*, 2001 and references there in). The problem is complicated because it requires a cumulus cloud parameterization, and these schemes require the model atmosphere to be conditionally unstable. Thus, a difficulty arises in attempting to account for internal gravity waves in an unstable flow. Chun and Baik (1998) pointed out that above the cloud-top height internal gravity waves can propagate and affect the larger scale flow. They proposed that convection-induced gravity wave momentum flux be considered in the region above the cloud-top height, but not considered inside the cloud region.

8.5 PROBLEMS

1. What is the principle cause of wave breaking in the middle and upper atmosphere?
2. Explain the fact that wave breaking does not occur in a linear theory.
3. Is a critical level a necessary or a sufficient condition for wave breaking?
4. Can wave breaking lead to wave reflection? If so, then how?
5. What are the two effects of wave breaking in the saturation zone?
6. Explain the basic flaw in Sawyers parameterization of wave stress.
7. From (8.31) show that if $|c - u_0| \to 0$ then $\alpha > 0$, and if $|c - u_0| \to \infty$ then $\alpha < 0$.
8. Explain why eddy diffusion created by wave breaking must be such as to cancel exponential growth.
9. Using (8.59) to derive (8.60).
10. Show that. $F(z)\exp\left[-\int_0^z D(z')\mathrm{d}z'\right] \leqslant 1$ if $D = \frac{1}{F}\frac{\mathrm{d}F}{\mathrm{d}z}$

here [illegible] constant C_s and A_s [illegible]

$$[illegible] = H_{max}\left(0, \frac{H_{max} - H_{m}}{H_{m}}\right) \quad [illegible]$$

where H_{m} [illegible] some prescribed height of order unity.

[illegible] the parameterization [illegible] a more consistent [illegible] family of cumulus models [illegible] of convection model [illegible] example [illegible] are numerous. The [illegible] schemes [illegible] cumulus cloud parameterization, and these schemes require the model [illegible] numerically unstable [illegible] [illegible] in the cloud height and strong [illegible] [illegible] [illegible]

9

Observations and Measurements of Gravity Waves

9.1 INTRODUCTION

Thus far, this book has dealt mostly with theoretical issues. However, this information is meaningful only if it can be used in some real sense. The first step in this process is the ability to observe gravity waves and measure their characteristics, *i.e.*, linearity, amplitude, wavenumber, frequency, and dispersion. However, from the start we have a problem because we cannot see gravity waves any more than we can see the wind. We can only sense gravity waves by observing their effects on the

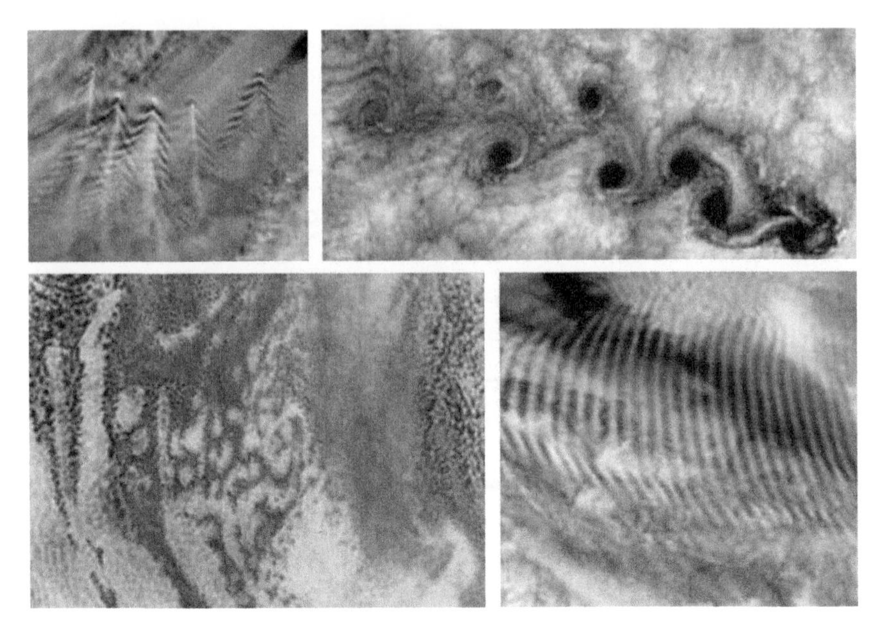

FIGURE 9.1 Cloud images of wave-like disturbances in the troposphere. Photographs from NASA.

atmosphere. It is only through analyses of the measurements of these effects can we estimate the characteristics of waves. These data analyses are based on what we know about wave behavior, and for the most part this is limited to the linear theory. It often happens that gravity waves, especially in the lower troposphere, do not appear linear; often they show neither constant frequencies nor amplitudes, or may not appear as waves (see, for example Figs. 1.3 and 9.1). In these cases, we often assume that the waves are nonlinear. However, it may be that the apparent nonlinearity is in fact due to errors in the observations. Thus, the study of real waves in the atmosphere is a daunting task.

In Chapter 2, we saw that a linear wave field is specified to within an undetermined amplitude when we know the dispersion relation, $\omega(\vec{\kappa})$. Thus, the objectives of all gravity wave observations are the estimations of wave frequency and wavenumber or, equivalently, phase speed and wavelength. While this specification seems at first glance straightforward, in practice it is not. The problem lies in the fact that we can observe only perturbations of wind speed and direction, temperature, density, pressure, trace gas and aerosol concentrations, and streamline displacements; however, since the advent of satellites and high-altitude aircraft images of complex wave cloud systems (Fig. 9.1) are common. The observations that can be made are determined by the characteristics of the instrument and its platform. The instrument may measure *in situ* or *remotely*, and the platform may be stationary or moving. *In situ* implies that measurements are made in the immediate surroundings of the measuring instrument. For example, a weather balloon measures the temperature, pressure, and humidity where the balloon is at any time. However, if wind speed and wind direction are calculated by the vector displacements of the ballon with time, then these are remotely made measurements. The analyses of meteorological data are limited by how the measurements are made. For example, a radiosonde or an upward-looking radar or sonar can be used to estimate the vertical structure of a wave field, but the horizontal characteristics of the wave, *i.e.*, wavelength and phase velocity can be estimated only by indirect means. A horizontal array of sensors can be used to estimate horizontal phase speed and wavenumber directly, but the vertical structure of the wave cannot be determined. Generally, obtaining the entire wave field from a single measuring technique is probably not possible; however, a gravity wave field (speed, direction, temperature, and pressure) can be sampled by instrumented aircraft (see, for example, Lilly, 1971; Mahrt, 1985; Dörnbrack *et al.*, 2001; Fritts *et al.*, 2003).

Gravity wave characteristics must be estimated by measuring perturbations to the mean atmospheric flow. Measurements include: (1) ground-based meteorological observations of wind speed, wind direction, temperature, and pressure; (2) vertical profiles of wind speed and temperature derived from tall towers, captured balloons, kites, sounding balloons, rocket sondes, radars, sonars, and satellites; (3) horizontal fields of vertical velocity and temperatures derived from instrumented aircraft, constant volume balloons, and remote sensors; (4) temperature soundings using GPS radio occultation techniques; (4) images of clouds captured by high-altitude aircraft flights and satellites, and (5) images of ionospheric wave perturbations revealed by nighttime airglow. All of these measuring techniques have been used in gravity waves observations.

9.2 GROUND-BASED MEASUREMENTS

Gravity waves with perturbations observable at the ground surface are confined mostly to the troposphere. Operational ground-based meteorological measurements are generally made in support of weather forecasting, airport operations, fire hazard, *etc.* These observations usually include horizontal wind speed and direction, temperature, relative humidity, surface pressure, precipitation, visibility, *etc.* Except for large-amplitude solitary-wave disturbances and low-frequency bores (see, for example, Cheung and Little, 1990; Rottman and Einaudi, 1993; Sun *et al.*, 2004; Coleman *et al.*, 2009) or mesoscale gravity waves (see, for example, Koch and Golus, 1988; Fiorino and Correia, 2002; Wang and Zhang, 2007), gravity wave perturbations are generally not detected by these types of measurements because the instrument sensitivities and reporting frequency are too low. Measurements of the more usual gravity waves require research-grade instrumentation, and the sampling rates should scale with the frequency of the waves under consideration. For high-frequency waves in the atmospheric boundary layer, sampling rates of at least 0.1 Hz should be used.

For lower-frequency mesoscale gravity waves, the sampling rates can be much lower. These data should include static pressure with a precision of at least 10 Pa, and wind speeds with a precision of 0.1 $\mathrm{m\,s^{-1}}$. Ground-based meteorological observations of gravity waves can be found, for example in Bosart and Sanders (1983), Koch and Golus (1988), Einaudi *et al.* (1989), Hauf *et al.* (1996), Rees, Staszewski, and Winkler (2001), Sun *et al.* (2004), and Nappo, Miller, and Hiscox (2008). Observations made on tall towers can be used to estimate the vertical structure of the wave field, see for example, Caughey and Readings (1975), Finnigan and Einaudi (1981), De Baas and Driedonks (1985), Blumen *et al.* (2001), Newsom and Banta (2003), Yague *et al.* (2007). In Chapter 10 we shall discuss how these observations can be used to estimate wave parameters.

9.2.1 PRESSURE

Although several techniques for detecting gravity waves near the ground surface have been tried, monitoring surface pressure is the one most commonly used. Reasons for this include, there are no moving parts; they are relatively cheap to make, and require little care once in operation. The pressure perturbations associated with most waves will be maximum at the ground surface, and because the background wind speed is generally small at the ground surface, there will be little Doppler shifting of the wave. Surface pressure perturbation is the most often used atmospheric variable for tracking gravity waves (see, for example, Herron and Tolstoy, 1969; Stobie, Einaudi, and Uccellini, 1983; Bedard, Canavero, and Einaudi, 1986; Einaudi *et al.*, 1989; Hauf *et al.*, 1996; Rees *et al.*, 2000; Poulos *et al.*, 2001; Yague *et al.*, 2007; Nappo, Miller, and Hiscox, 2008).

Gravity wave pressure perturbations observed near the ground surface range from $\sim$0.5 to $\sim$100 Pa (Cunning, 1974; Koch and Golus, 1988; Einaudi *et al.*, 1989; Hauf *et al.*, 1996; Grivet-Talocia *et al.*, 1999; Rees *et al.*, 2000). More typical values range from $\sim$1 to $\sim$10 Pa. Because absolute surface pressures $\sim$1000 hPa, a barometer would need a sensitivity of 1 in 25×10^4 to detect a 25 Pa signal. As an analogy, the average depth of the oceans is about 4.3 km and the corresponding surface wave amplitude would be about 0.5 m. Electronic barometers are commercially available, (for example, Furness Controls, Inc., Paroscientific, Setra, Validyne, and Väisälä), which can measure absolute atmospheric pressure with an accuracy of about $\pm$10 Pa. For high sensitivity, a differential pressure gauge in needed (see Wilczak and Bedard, 2004). Again using the analogy of surface waves on the sea, measuring absolute pressure is equivalent to measuring the total depth of water. But we are not interested in the depth of the sea, only the amplitude of the waves as illustrated in Fig. 9.2. In this application, accuracy is not as important as sensitivity. Thus, accuracy of absolute pressure measurement is not as important as the sensitivity of the measuring system. Because the output of the pressure sensor is usually a continuous voltage, sensitivity is a function of the precision of the conversion from analog to digital

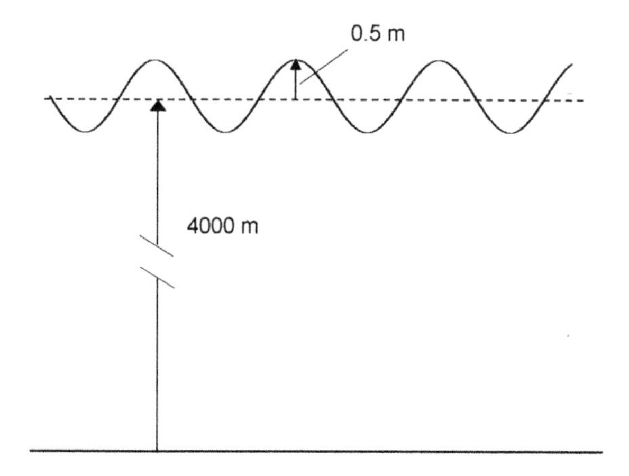

FIGURE 9.2 Analogy of microbarograph sensitivity, *i.e.*, measuring shallow waves on a deep sea.

format. Thus, the sensitivity is a function of the data logger as well as the barometer. We shall refer to any *electronic* instrument that can measure atmospheric pressures associated with gravity waves a *microbarometer*.[1]

9.2.1.1 Static Pressure Ports

The most challenging problem in measuring air pressure perturbations is the means of sampling the pressure, *i.e.*, getting a true pressure signal to the microbarometer. In meteorology, pressure is understood to be the *static* pressure, *i.e.*, the pressure of the atmosphere at rest. However, barometers cannot distinguish between static and *dynamic* pressure. Thus, if the input to the microbarometer is an open pipe, then because of the Bernoulli effect one will be measuring wind speed and turbulence as well as pressure. The *wind noise* must be filtered from the pressure signal. Many field studies, laboratory experiments, and theoretical analyses have addressed this sampling problem as early as the 1930s (see, for example, Prandtl and Teitjens, 1937; Sinclair, 1937; Wyngaard, Seigel, and Wilczak, 1994). Much of the interest in wind-noise filters comes from the study of infrasound (see for example, Georges and Greene, 1975; Bedard, 1978; Evers, 2005; Le Pichon, Blanc, and Hauchecorne, 2010; and references therein). Infrasound is a low-frequency (~0.002 to 20 Hz) longitudinal acoustic wave with pressure amplitudes ~0.01 to 100 Pa. Because infrasound amplitudes are generally less than gravity wave amplitudes, methods to measure the former are applicable to the latter. Walker and Hedlin (2010) give a comprehensive review of infrasound noise filters.

[1] Note that this is a distinction from the *microbarograph* which mechanically records pressure perturbations on a rotating chart (see, for example, Symons, 1890; Jacobs, 1937; Clark, 1950).

There are three basic types of noise filters for gravity waves; buried, mechanical, and aerodynamic.

Buried: If a microbarometer is buried beneath the ground surface, it is sheltered from turbulence and wind gusts and gives an accurate approximation of the static pressure. Gossard and Munk (1954) located a microbarograph in a flush pit at the shore end of a peer. Hauf *et al.* (1996) put their sensor in a plastic cylinder 1.5 m below the ground surface. The top of the container was covered with a solid piece of Styrofoam, and atop this is a metal plate with a small hole in its center. Rees *et al.* (2000) measured gravity waves over the Brunt Ice Shelf in Antarctica with pressure heads about 1 m under the snow.

Summater: A *summater* is an apparatus constructed to obtain a pseudo static pressure (see, for example, Burridge, 1971; Bedard, 1978). Basically, a solid pipe or a rubber hose has ports (holes) placed along its surface. Pressure perturbations with length scales less than that of the pipe or hose will destructively interfere resulting in a coherent perturbation along the length of the tube. In a sense, the input signal to the microbarometer has been spatially averaged. Sometimes three or four pipes are connected to a manifold. Daniels (1959) used metal pipes of width ranging from 1.27 cm at the front to 14.24 cm at the end with 100 sampling ports. Fig. 9.3 show an image of Daniel's summater; the total length of the pipe was about 604 m. Walker and Hedlin (2010) and Georges and Greene (1975) used a 61 m rubber hose with ports inserted about every 1.5 m; Bedard (1978) used a similar method. More recently, Walker and Hedlin (2010) discussed microporous or irrigation soaker hoses as shown in Fig. 9.4. Evers (2008) used a star-shaped array of six 5 m-long soaker hoses.

Aerodynamic: Aerodynamic filters make use of the fluid dynamics of the air, specifically the relation between wind speed and pressure, for example the Bernoulli equation. As suggested by Prandtl and Teitjens (1937), the most simple design for measuring static pressure is a flat circular disk with a small sampling hole in its middle as illustrated in Fig. 9.5. An internal boundary layer develops over the plate, and at the plate's surface, the wind speed is assumed zero and accordingly the air pressure must be the static pressure. Bedard and Ramzy (1983) used this design. The angle the wind makes with the plate is important. Sinclair (1937) used a similar but much smaller disk which consisted of a metal disk 3.1 mm thick, 10.1 cm in diameter, and supported by a 6.35 cm diameter metal tube. A modification of this design was made by Akyuz, Henry, and Horst (1991) and tested in a wind tunnel. Fig. 9.6, taken from Akyuz, Henry, and Horst (1991), is a schematic of the port. For wind direction angles less than 5° from the plate, they found that the error in pressure measurements is zero. They conclude that except for high winds the error resulting from the port is less than the rms error of the pressure transducer. Elliott (1972) using the experimental results of Willmarth and Wooldridge (1962) designed a disk-type pressure port with separate static holes on either side of the sampling port. The port, illustrated in Fig. 9.7 taken from Elliott (1972), was made of two circular disks each 0.127 cm thick and 4.5 cm in diameter. When joined together, the probe was about 0.228 cm thick with separate

FIGURE 9.3 Pipe pressure summater 604 m long with 100 evenly spaced holes. (Taken from Daniels (1959).)

sampling tubes for each of the static holes. These channels join at some distance from the probe and connect with the transducer. The probe is mounted horizontally. A vertically traveling turbulence eddy will result in a positive pressure perturbation on one face and negative pressure perturbation on the other, thus summing to zero dynamic pressure perturbation at the transducer input. A similar design was used by Donelan *et al.* (1999). Miksad (1976) designed a static pressure port consisting of two stacked disks with sampling holes on the inside of the stack as illustrated in Fig. 9.8, taken from Miksad (1976), This design was further modified and made commercially available from Väisälä. A typical field installation of this pressure port is shown in Fig. 9.9.

9.2.1.2 Noise Filtering

In applications where great sensitivity is required, a differential pressure sensor or *manometer* is used. These electronic manometers measure the difference in pressure between the atmospheric static pressure and a constant or slowly varying reference pressure. This allows the separation of the low-frequency large-

FIGURE 9.4 Commercial 5 m long pours (soaker) hose used as a pressure port. (Taken from Walker and Hedlin (2010).)

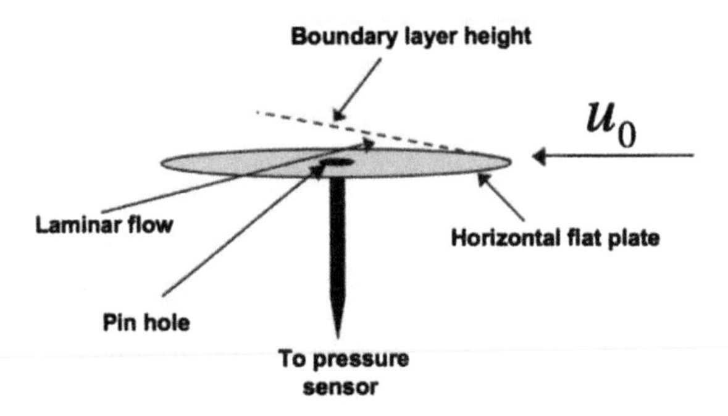

FIGURE 9.5 Aerodynamic pressure plate port. Friction at the plat surface results in zero wind speed so that only static pressure is sensed.

amplitude pressure changes from the high-frequency low-amplitude perturbations. Figure 9.10 illustrates the filtering setup. The manometer contains a piezoelectric diaphragm which responds electronically to the strains produced by pressure differences between the reference chamber and the atmosphere. If the reference chamber has a slow leak, then the chamber pressure will slowly adjust to low-frequency atmospheric pressure changes, and the signals of the high-frequency pressure changes will be enhanced. This is a high-pass filter, and the frequencies that are passed are controlled by the size of the leak. We refer to this type of filter

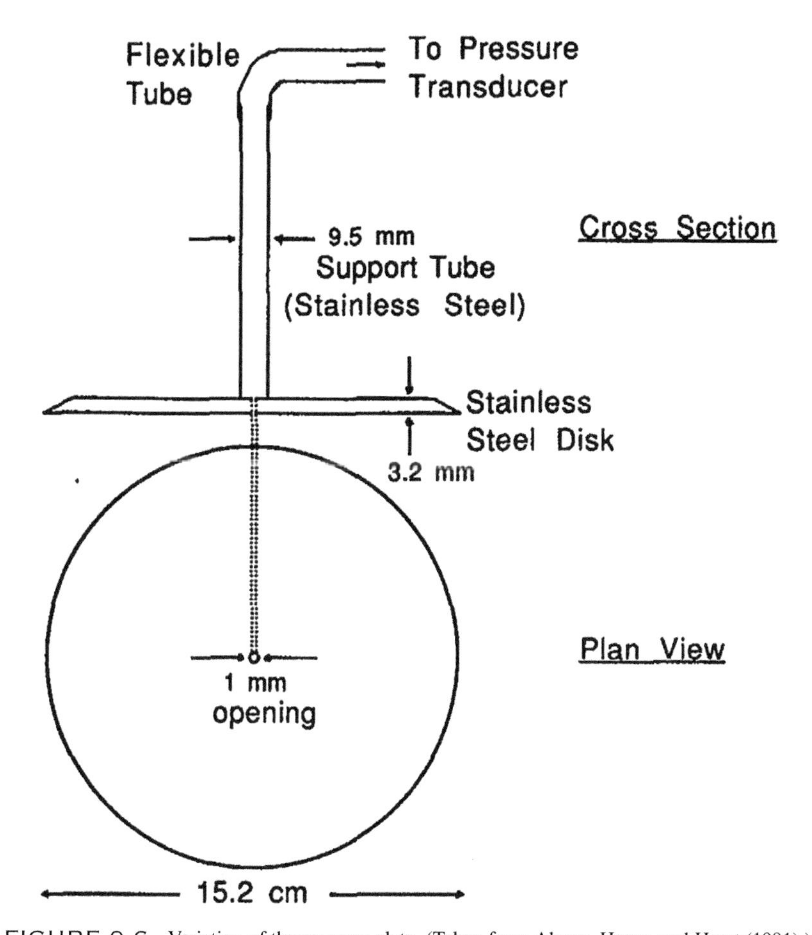

FIGURE 9.6 Variation of the pressure plate. (Taken from Akyuz, Henry, and Horst (1991).)

as *mechanical*. However, if the reference chamber is sealed, then all frequencies will be recorded. In this case, the low frequencies must be filtered numerically. We can refer to this type of filter as *digital*.

Mechanical noise filter: If there is a leak in the reference chamber, then the pressure, p_c, in the chamber changes according to

$$\frac{\mathrm{d}p_c}{\mathrm{d}t} = -\alpha\left[(p_c(t) - p_a(t))\right], \tag{9.1}$$

where p_a is the atmospheric pressure and α^{-1} is a time constant which is determined by the leak rate. The integration of (9.1) is straightforward if we multiply it by the integrating factor $e^{\alpha t}$ and note that $e^{\alpha t}\mathrm{d}p_c + \alpha p_c e^{\alpha t}\mathrm{d}t$ is an

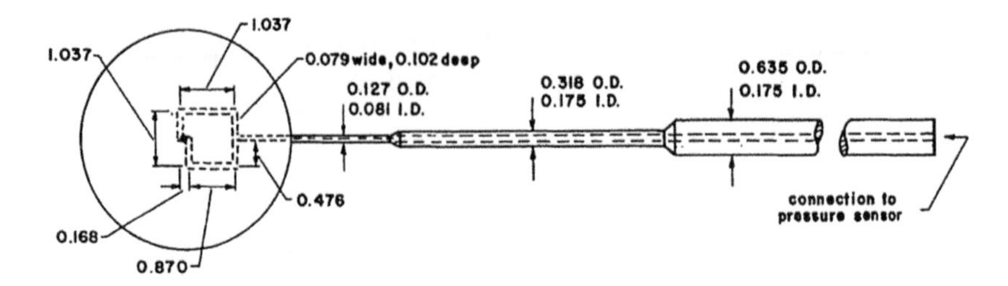

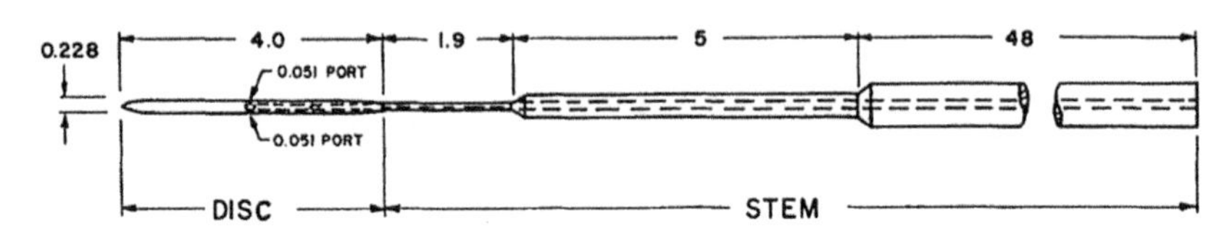

FIGURE 9.7 Details of the pressure probe designed by Elliott (1972).

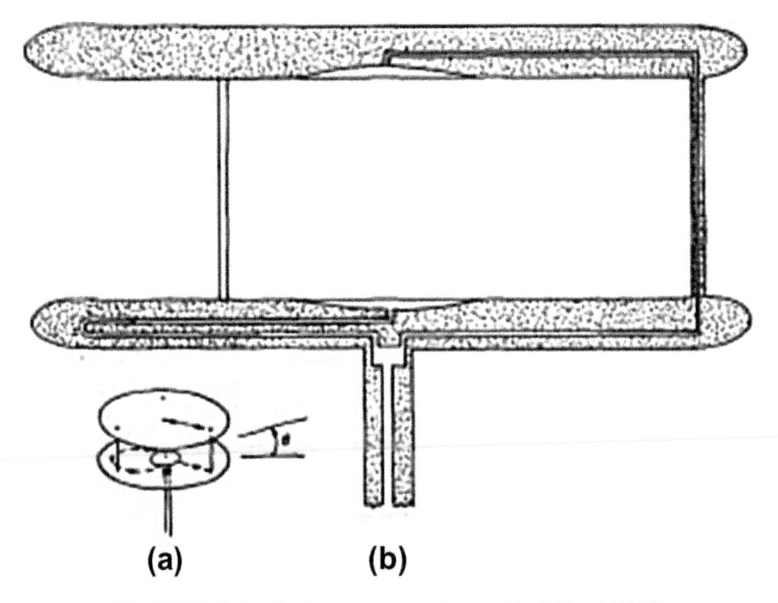

FIGURE 9.8 Static pressure port designed by Miksad (1976).

exact differential. We can then write (9.1) as

$$d\left[p_c e^{\alpha t}\right] = \alpha p_a e^{\alpha t} dt. \tag{9.2}$$

Integration of (9.2) gives the result,

$$p_c(t) = \alpha \int_0^t p_a(t') e^{-\alpha(t-t')} dt'. \tag{9.3}$$

FIGURE 9.9 Field installation using the Väisälä pressure port.

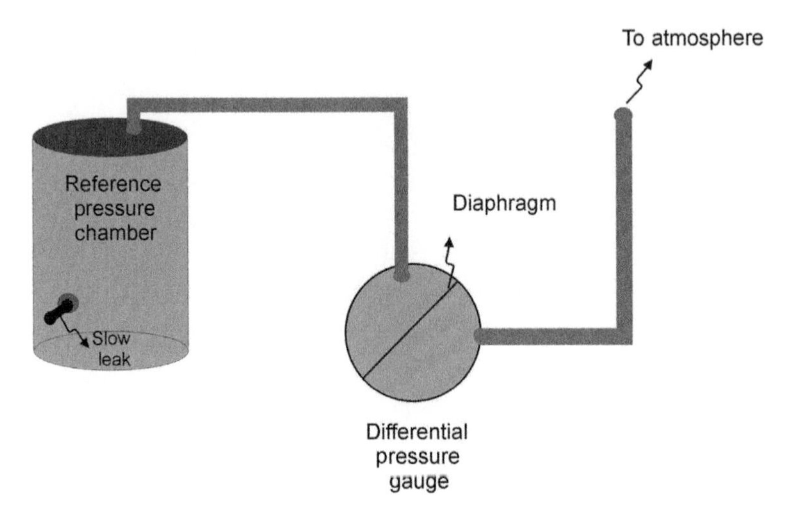

FIGURE 9.10 Schematic of an electronic-mechanical monometer. When the slow leak is sealed, the instrument becomes a digital monometer.

The differential pressure, Δp at time t is

$$\Delta p(t) = p_{\mathrm{a}}(t) - p_{\mathrm{c}}(t). \tag{9.4}$$

Using (9.3) in (9.4) gives

$$\Delta p(t) = p_{\mathrm{a}}(t) - \alpha \int_0^t p_{\mathrm{a}}(t')\mathrm{e}^{-\alpha(t-t')}\mathrm{d}t'. \tag{9.5}$$

Following, for example Bendat and Piersol (1971), a linear system which is stable and physically realizable, can be represented by

$$y(t) = \int_0^\infty h(\tau)x(t-\tau)\mathrm{d}\tau, \tag{9.6}$$

where $h(\tau)$ is a *weighting function* which relates the system output, $y(t)$, to the system input $x(t)$. Equating (9.6) with (9.5) we see that

$$h(\tau) = \delta(\tau) - \alpha H(\tau)\mathrm{e}^{-\alpha\tau}, \tag{9.7}$$

where $\delta(\tau)$ is the Dirac delta function, and $H(\tau)$ is the *Heaviside* function defined by

$$\begin{aligned} H(t-\tau) &= 1 \quad \mathrm{t} \geq \tau \\ &= 0 \quad \mathrm{t} < \tau. \end{aligned} \tag{9.8}$$

The dynamic characteristics of the system are given by the *frequency response function*, $\hat{h}(f)$, which is the Fourier transform of the weighting function. The Fourier transform of (9.7) is

$$\hat{h}(f) = \frac{f^2 + i\alpha f}{\alpha^2 + f^2}, \tag{9.9}$$

where f is the frequency. It is customary to express the frequency response in polar notation, *i.e.*,

$$\hat{h}(f) = G(f)\mathrm{e}^{-i\Phi(f)}, \tag{9.10}$$

where $G(f)$ is the *gain factor* of the system, and $\Phi(f)$ is the *phase factor* of the system. From (9.36) we see that

$$G(f) = \left(\frac{f^2}{\alpha^2 + f^2}\right)^{1/2} \tag{9.11}$$

and

$$\Phi(f) = \tan^{-1}\left(\frac{\alpha}{f}\right). \tag{9.12}$$

From (9.11), we see that frequencies less than α are attenuated while frequencies greater than α are less attenuated. Thus, the leak creates in effect a *high-pass filter*. The amplitude of a gravity wave with angular frequency a is reduced by a factor of 0.707, and the phase angle is shifted forward $\pi/4$ degrees. If the period of the wave is 15 min, then the wave phase is shifted forward about 112 s. These changes in wave amplitude and phase will affect statistics such as cross-correlations as well as the comparisons of pressure measurements with other measurements such as a velocity or temperature. Thus, it is necessary that the characteristics of the low pass filter be known so that the pressure data can be corrected before analysis.

As mentioned above, $1/\alpha$ is a time constant for the reference chamber pressure. This value depends on the frequencies of the wave of interest. For example, Jordan (1972) investigated atmospheric gravity waves generated by several mechanisms such as upper tropospheric winds, jet streams, weather fronts, thunderstorms, *etc.* He selected leak time constants which ranged from 50 to 100 s. Einaudi *et al.* (1989) examined gravity waves in the planetary boundary layer, and used a time constant of 7.5 s. Hauf *et al.* (1996) examined gravity waves on the mesoscale, and used a time constant of about 300 s.

Digital noise filter: If the pressure reference chamber is sealed, then the differential pressure given by (9.4) is

$$\Delta p(t) = p_{\mathrm{a}}(t) - p_{\mathrm{c}}, \tag{9.13}$$

where p_{c} is now a constant pressure. Such a system does not suffer degradation introduced by a high-pass filter. Indeed, pressure variations on all time scales can be detected if the instrument is operated long enough. Because all frequencies are present in the recorded data, digital filters must be used to isolate and separate desired frequency ranges. However, this enhanced frequency response comes at a cost. One of the most challenging problems is maintaining constant pressure in the reference chamber over the measuring time, which could range from minutes to days. Because the chamber is sealed and has constant volume (neglecting thermal expansion) the gas density is constant at least to first order. Then, from

Boyle's law

$$\frac{\delta p_c}{p_c} = \frac{\delta T_c}{T_c}, \tag{9.14}$$

where δp_c is the change in p_c due to a change δT_c in the reference chamber temperature, T_c. For typical values of surface pressure ($\sim$1000 hPa) and temperature ($\sim$300 K), a temperature change of 0.01 K results in $\delta p_c \approx 3.3$ Pa. This is a substantial pressure change, and is comparable to the amplitudes of high-frequency gravity waves in the planetary boundary layer. If T_c increase, then p_c increases, and from (9.13) it will appear as if Δp decreases. But this will be incorrectly interpreted as a decrease in atmospheric pressure. The reverse effect will occur if T_c decreases. If T_c is measured, then it would be possible to correct the data *post facto*. However, at this time it is not possible to measure temperature differences of 0.01 K or smaller. Thus, special care must be given to minimizing reference-chamber temperature variations. This can be done by insulating the reference chamber, increasing its thermal inertia, or keeping the reference chamber at near-constant temperature. Nappo *et al.* (1991) described a microbarograph system that insulated the reference chamber and used aluminum pellets in the chamber to increase thermal inertia in the chamber. Anderson *et al.* (1992) described a microbarograph system which used all three remedies to approach thermal stability. Note that these temperature effects will also effect the mechanical type of microbarograph, but not greatly if the temperature changes are slow. To alleviate these temperature problems, Hauf *et al.* (1996) installed their sensors in a plastic container placed 1.5 m below the ground surface. Another technique is to immerse the chamber in an ice bath. Wilczak and Bedard (2004) report on a newly-developed *state-of-the-art* high-precission microbarometer. This system can accurately measure Δp within $\pm$20 Pa. The system uses a unique combination of digital and mechanical filtering systems. The sealed thermally insulated reference chamber contains steel wool for thermal damping. Between the reference chamber and the output of the differential barometer there is a set of commercial ceramic flow resistors which act as a "slow leak." As in the case of mechanical filters, the output data must be corrected for phase shift and amplitude reduction.

Another difficulty with the digital instrument is saturation of the manometer. Commercial differential pressure sensors have limited operating ranges, say on the order of a few hundred hPa (Nappo *et al.*, 1991; Anderson *et al.*, 1992). The diurnal changes in atmospheric pressure, moving frontal systems, gust fronts, etc. can cause pressure changes of more than 100 hPa, and these changes can lead to instrument nonlinearity, saturation, and possible damage. To overcome these possibilities, the reference chamber pressure can be set to ambient pressure, *i.e.*, $\Delta p = 0$, by briefly opening the reference chamber using a solenoid-valve. For the Nappo *et al.* (1991) instrument, the "reset" valve is periodically opened; for the Anderson *et al.* (1992) instrument, the valve is opened whenever the differential pressure reaches a pre-determined value. Opening the valve sets $p_c = p_a(t_0)$ where

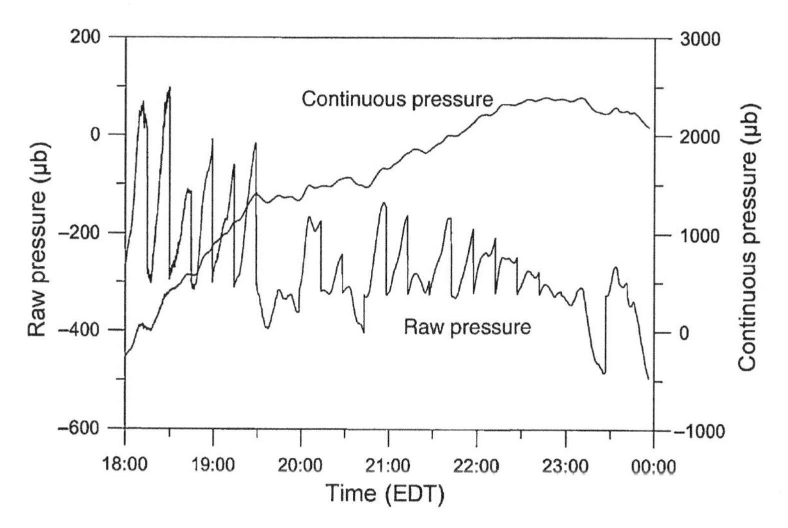

FIGURE 9.11 Time series of raw and post processed pressure data measured at Oak Ridge, TN on September, 2000.

t_0 is the time of pressure reset. Thus, (9.13) becomes

$$\Delta p(t) = p_a(t - t_0) - p_a(t_0), \tag{9.15}$$

and the graph of $\Delta p(t)$ takes the form of a series of ramp-like structures as shown in Fig. 9.11. The reference chamber was set to ambient pressure every 15 min. The differential pressure signal was sampled at 10 Hz from which 10 s averages were formed. Post-processing of the data requires the smooth joining of these ramps to form a continuous pressure time series. Fig. 9.11 shows the raw 15 min ramps and the reconstructed continuous pressure time series. Figure 9.12 shows the pressure perturbations observed between 18:00 and 21:00 EST for the case shown in Fig. 9.11. These data were obtained by band-pass filtering the post-processed data plotted in Fig. 9.11. The filter passed signals with periods between 1 and 30 min.

9.2.2 SAMPLING ARRAYS

Ideally, the horizontally moving component of a gravity wave is a *coherent* disturbance that moves with constant speed across an array of measuring instruments. In this case, coherence refers to a wave that does not change its shape or frequency as it moves. If the pressure perturbations associated with the wave are observed only at a single point, then all we can learn about the wave is its amplitude and frequency. If we have time series of pressures at two points, then we can estimate the component of the wave vector parallel to the line joining the two locations.[2] However, as illustrated in Fig. 9.13 there are

[2] How these estimates are made will be discussed in Section 10.2.4.

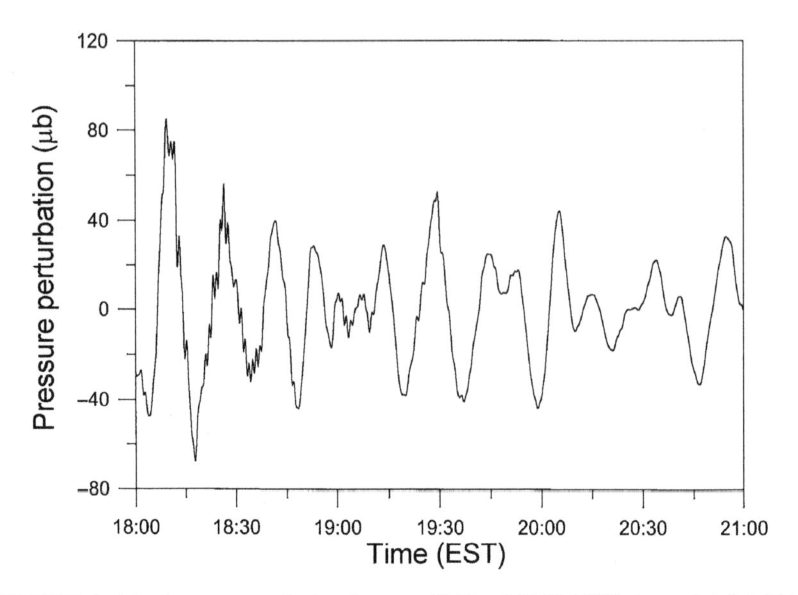

FIGURE 9.12 Pressure perturbations between 18:00 and 21:00 (EST) observed at Oak Ridge, Tennessee on September 1, 2000. Note the high-frequency activity between 18:00 and 18:45, and again around 19:00. These could be gravity waves or large-scale turbulence eddies from the previous convective period.

an infinite number of possible wave vectors with components parallel to a line joining two points. For example, let wave vector **AB** connect the points A and B in Fig. 9.13. However, wave vector **AC**, with wave fronts indicated by the dashed lines, and wave vector **AD** with wave fronts indicated by the solid lines also have components parallel to wave vector AB. Either of these wave vectors could be considered the true wave vector. Thus, the true wave vector cannot be determined with two measurements, and consequently we cannot estimate the true phase speed, and wavelength of the wave. As we shall see, at least three sensors are necessary to estimate wave characteristics; more sensors increases the accuracy of the estimates. The spatial distribution of the sensors describes an *array* that in effect acts as a radio-frequency antenna. The *response characteristics* of an antenna determines the range of frequencies and wavelengths of radio waves that can be received. Likewise, the response characteristics of the pressure sampling array determine the range of gravity waves that can be detected.

Fig. 9.14 is a schematic of an idealized 5-sensor array. Four sensors are placed on the corners of a square with sides 1 km, and one sensor is placed in the center. If the sampling period, *i.e.*, the time between observations, is Δt, and if the greatest distance between the stations is D, than the maximum speed of a disturbance crossing the array that can be detected is

$$c_{\max} = \frac{D}{2\Delta t}. \tag{9.16}$$

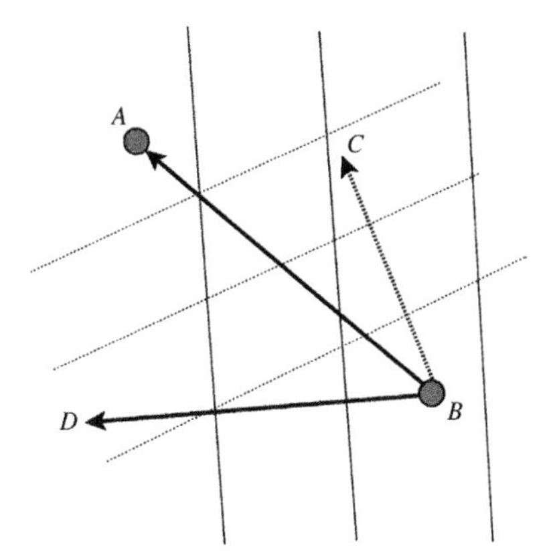

FIGURE 9.13 The apparent wave vector between points A and B is a component of an infinite number of possible wave vectors. Shown in the figure are two possible wave vectors that could have a component along the line joining stations A and B.

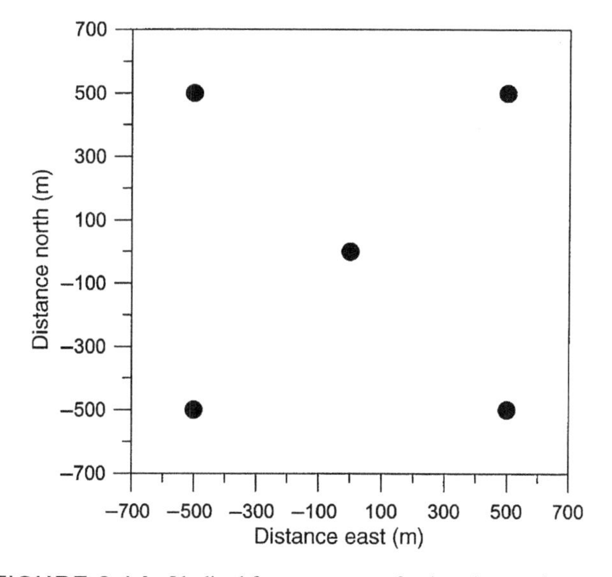

FIGURE 9.14 Idealized five-sensor array for detecting gravity waves.

The factor $2\Delta_t$, in (9.16) is the *Nyquist period*. It is the reciprocal of the *Nyquist frequency* which is the highest frequency content of a time series. Note that if the data are time averaged, for example 10 Hz data averaged to 0.1 Hz, then $\Delta t = 10$ s. Disturbances with periods less than $2\Delta t$ will not be detected by the array as a coherent signal. For the idealized array in Fig. 9.14, $D = 1000$ m. If $\Delta t = 10$ s, then $c_{\max} = 50$ ms^{-1}. A wave with phase speed greater than $c_{\max}$

will be incoherent in the sense that the wave shape observed at each sensor will be different. In this case, the wave energy will appear in waves of lower frequency, a processes known as *aliasing*. If the sampling period is 60 s, then $c_{max} = 8.3$ m s^{-1}. The minimum wavelength, λ_{min} that can be resolved by the array is comparable to the minimum spacing of the sensors. Waves with wavelengths less than the average spacing between station pairs will also suffer aliasing, and their energy will appear in waves with greater wavelengths. For the array illustrated in Fig. 9.14, $\lambda_{min} \approx 700$ m. The maximum resolvable wavenumber is $k_{max} = 2\pi/\lambda_{min}$, and for the array in Fig. 9.14 this is about 0.009 m^{-1}. The maximum detectable wavelength is given by

$$\lambda_{max} = c_{max} T_{max} = \frac{D}{2} \frac{T_{max}}{\Delta t}, \tag{9.17}$$

where we have used (9.16), and T_{max} is the maximum disturbance period that can be detected by the sensors. For the mechanical-filter microbarograph, T_{max} is determined by the filter time constant $1/\alpha$. From (9.11) we see that if the frequency of the wave is equal to α, then the amplitude of the wave is reduced by a factor of 0.707. We can take $T_{max} = 2\pi/\alpha$. If $1/\alpha = 300$ s and $c_{max} = 50$ ms^{-1}, then $\lambda_{max} \approx 90$ km. For the digital-filter microbarometer λ_{max} is determined by the lowest frequency retained after filtering. If that frequency has a period of 30 min, and if $c_{max} = 50$ m s^{-1}, then λ_{max} is also 90 km.

From the above, we see that the range of gravity waves that can be detected by a sampling array is a function of instrument spacing and sampling frequency. Studies of array design have been made, for example, by Barber (1959), Haubrich (1968), and Asten and Henstridge (1984). As Haubrich (1968) points out, the problem of array design is, (1) given N sensors, where should they be placed, and (2) what processing scheme should be applied to the array output. He also states that the solution for a design depends on the intended use of the array. Koch and Golus (1988) used a nested array of measurement systems to observe mesoscale gravity waves. They used an inner array of stations with an average spacing of 7 km, and a courser, regularly spaced array with an average spacing of 20 km. Grivet-Talocia *et al.* (1999) used eight digital barometers in an array with a diameter of about 50 km to compile a 4-year climatology of pressure disturbances over central Illinois, USA. They recovered wave disturbances with periods ranging from 30 min to 6 h. Fig. 9.15a shows the array used by Balachandran (1980) to observe thunderstorm-generated gravity waves. Fig. 9.15b shows the array used by Hauf *et al.* (1996) to observe gravity waves on the mesoscale. Fig. 9.15c shows the array used by Rees *et al.* (2000) to study gravity waves in the stable planetary boundary layer, and Fig. 9.15d shows the array used by Nappo *et al.* (2000) to study gravity waves in south-central Kansas during the CASES-99 field campaign. It is apparent from Fig. 9.15 that an overall "best" array design may not exist.

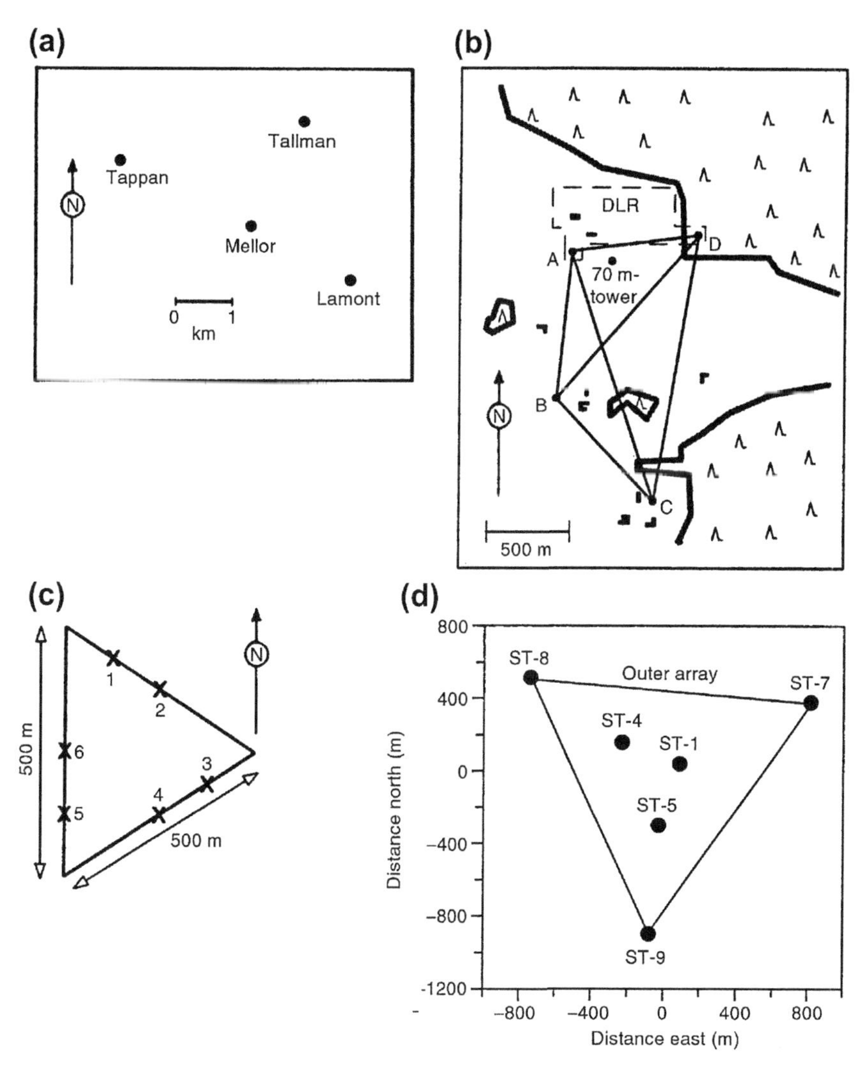

FIGURE 9.15 Microbarograph arrays. (a) used by Balachandran (1980); (b) used by Hauf *et al.* (1996); (c) used by Rees *et al.* (2000); and (d) used by Nappo *et al.* (2000).

9.3 FREE-BALLOON SOUNDINGS

Free-balloon soundings include *radiosondes* and *rawinsondes* and any other lifting measuring system device not attached to the ground surface. A radiosonde is a balloon-borne instrument for the simultaneous measurement and transmission of pressure, temperature, and humidity. A rawinsonde is a method of upper-air observation of wind speed, wind direction, temperature, pressure, relative humidity, and ozone by means of a balloon-borne radiosonde tracked by a radar or radio direction-finder. Today, many sondes use global positioning systems, (GPS),[3] (see, for example, Tsuda *et al.*, 2004; Tsuda and Hocke, 2004; Johny, Sarkar, and Punyasesdu, 2009). The GPS allows accurate tracking of the sonde balloon giving continuous wind speeds and wind directions. Corby (1957) suggested that observed periodic variations of a few meters per second of the vertical velocities of ascending radiosondes could be related to atmospheric gravity waves. Reid (1972) improved Corby's analysis method by relating departures of the vertical velocity of the airsonde from a mean value to changes in the gradient of the pressure trace which were automatically recorded on a paper chart. Typical changes in the vertical velocities were about 1 m s^{-1}. Lalas and Einaudi (1980) tested the ability of operational rawinsonde data to observe gravity waves in the troposphere. To couple the observed motion of the balloon with the gravity wave (observed by Doppler radar and a microbarograph array), the vertical position of the balloon, $z(t)$, was obtained by interpolation of its reported height which was once per minute, and the horizontal position was obtained by integrating the kinematic equation of motion,

$$\frac{\mathrm{d}x}{\mathrm{d}t} = u(z,t) = u_0\,[z(t)] + u_1\,[x(t), z(t), t{:}\phi_0]\,, \tag{9.18}$$

where ϕ_0 is the initial phase of the wave. The motivation for this study was in part to devise a method for determining the climatology of tropospheric gravity waves. Shutts, Kitchen, and Hoare (1988) analyzed a large amplitude gravity wave in the lower stratosphere detected by a radiosonde launched from Shanwell, UK. Fig. 9.16 shows the ascent profiles where we can clearly see pronounced wave activity between 15 and 22 km. They also used linear wave-saturation theory to argue that the disturbance was a quasi-stationary, terrain-generated gravity wave with a horizontal wavelength of about 16 km and a vertical wavelength of about 6 km. The wave temperature-perturbation amplitude was about 8 K, which they show is only slightly smaller than the critical amplitude for wave saturation. Shutts, Healey, and Mobbs (1994) described a pilot study to assess the potential of rapidly released radiosondes for studying terrain-generated gravity waves. The idea is that each sonde will be nearly simultaneously responding to a different part of the wave field. As they pointed out, sondes that are launched too close together in space and time provide no new information on the wave since then they are all experiencing the same part of the wave. Sondes which are separated over space scales greater than the wave scales are impossible to correlate. However, if the spacings of the sondes are between these limits, then this technique offers the possibility of determining not only the horizontal wavelengths but also the slopes of the constant phase lines. These values can then be used to infer wave drag. Gardner and Gardner (1993) and de la Torre and Alexander (1995) investigated the distortions to density and temperature spectra introduced by the characteristics of balloon motion and wave propagation. These studies sought methods to infer

[3] Examples of commercial radiosonde GPS systems are Meispi Electric Co. RS-06G GPS Radiosonde; GRAW Radiosonde DFM-09; Väisälä Radiosonde RS92-SGP.

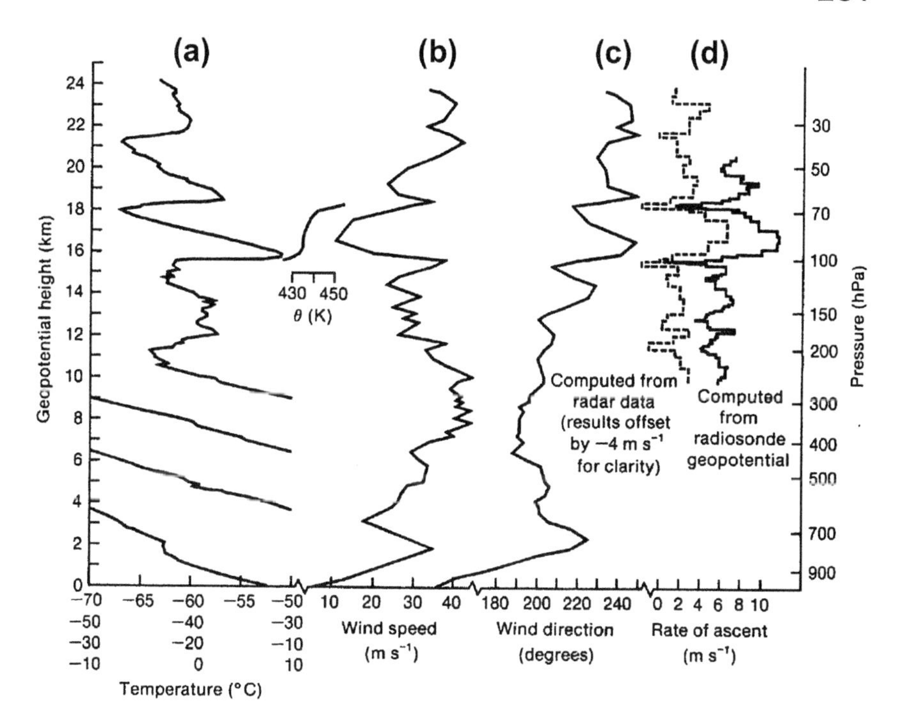

FIGURE 9.16 Radiosonde ascent profiles at Shanwell, 2318 GMT December 12 1986. (a) Temperature and potential temperature; (b) wind speed; (c) wind direction; (d) ascent rate, computed from the geopotential (solid line) and radar data (dashed line and ofset by 4 m s^1. From Shutts, Kitchen, and Hoare (1988).

the "real" wavelengths and frequencies from the "apparent" ones measured during soundings.

Tsuda *et al.* (2004) measured the wave characteristics of short vertical wavelength gravity waves (wavelengths < 3 km) in the troposphere and lower stratosphere using radiosondes launched every three hours for 120 h-long intensive observation periods (IOP). These launches were made during the Darwin Area Wave Experiment (DAWEX) (Hamilton, Vincent, and May, 2004). Of special interest was the wave kinetic energy, E_k, and wave potential energy, E_p, defined as

$$E_k = \frac{1}{2}\left[\overline{u'^2} + \overline{v'^2}\right], \tag{9.19}$$

$$E_p = \frac{1}{2}\left(\frac{g}{N}\right)^2 \overline{\left(\frac{T'}{\overline{T_0}}\right)^2}. \tag{9.20}$$

Three IOPs were made. Fig. 9.17, taken from Tsuda *et al.* (2004), shows the profiles of zonal and meridional wind speeds for the first IOP. A common wave characteristic was enhanced wave energy between 15 and 20 km and between

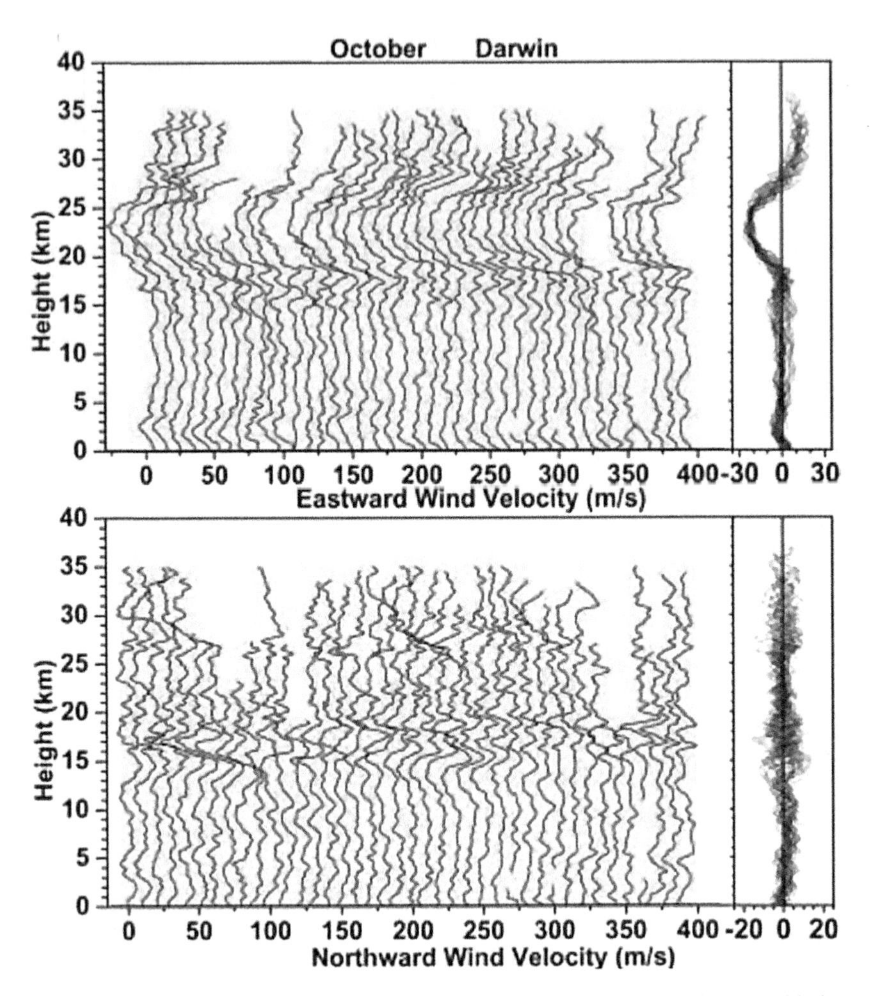

FIGURE 9.17 Profiles of (a) eastward and (b) northward wind velocity observed with three hourly radiosondes at Darwin during IOP1 (October 13–18, 2001). In the left panel, individual profiles are shifted every 10 m/s, according to the launch timings (every 3 h), while all results are superimposed in the right panel. (Taken from Tsuda *et al.* (2004).)

25 and 30 km and depressed wave energy between 20 and 25 km. These energy differences were attributed to different types of waves.

9.3.1 TETHERED LIFTING SYSTEMS

Tethered lifting systems (TLS) use an aerodynamic balloon attached to the ground surface with a line along which one or several instrument packages are raised and lowered or held at a fixed height. These systems can sample from the ground surface through 1 to 2 km. The system is limited to low wind conditions

FIGURE 9.18 TLS instrument package. Arrows on left picture show locations of instrument packages; instrument package shown in right picture. (Taken from Meillier, Jones, and Balsley (2008).)

for balloons and moderate to strong winds for kites. Fig. 9.18 is taken from Meillier, Jones, and Balsley (2008); the right panel shows the details of the instrument probe, and the left panel shows a package of five turbulence probes suspended at 6-m intervals along the tether. Details of the technology and a variety of results can be found in, for example, Balsley, Jensen, and Frehlich (1998), Balsley *et al.* (2003), Meillier, Jones, and Balsley (2008), Frehlich *et al.* (2003), Frehlich *et al.* (2004), and Muschinski *et al.* (2001). The TLS is able to provide relatively inexpensive, point-by-point, high-resolution measurements of winds, temperatures, and turbulence throughout the lower parts of the atmosphere. Fig. 9.19 taken from Meillier, Jones, and Balsley (2008) shows the horizontal wind speed and temperature wave perturbations (dashed line) observed with the TLS stationed at about 465 m AGL. These perturbations are associated with a ducted wave extending from about 200 to 900 m AGL. The solid lines in Fig. 9.19a–c are the results of a model calculation assuming a horizontal wavelength of about 17.1 km. Fig. 9.19d is the time series of turbulence kinetic energy eddy dissipation rate, ϵ, measured with a hot wire probe. The ability to simultaneously measure wave characteristics and in situ turbulence at arbitrary heights is an extremely useful resource.

9.3.2 SUPERPRESSURE BALLOONS

Balloons designed to drift along streamlines of constant density have been and continue to be used as tracers of atmospheric motions. Relatively inexpensive and easy to track with high accuracy using RADAR or GPS they provide ideal quasi-Lagrangian instrument platforms for tropospheric and stratospheric studies.

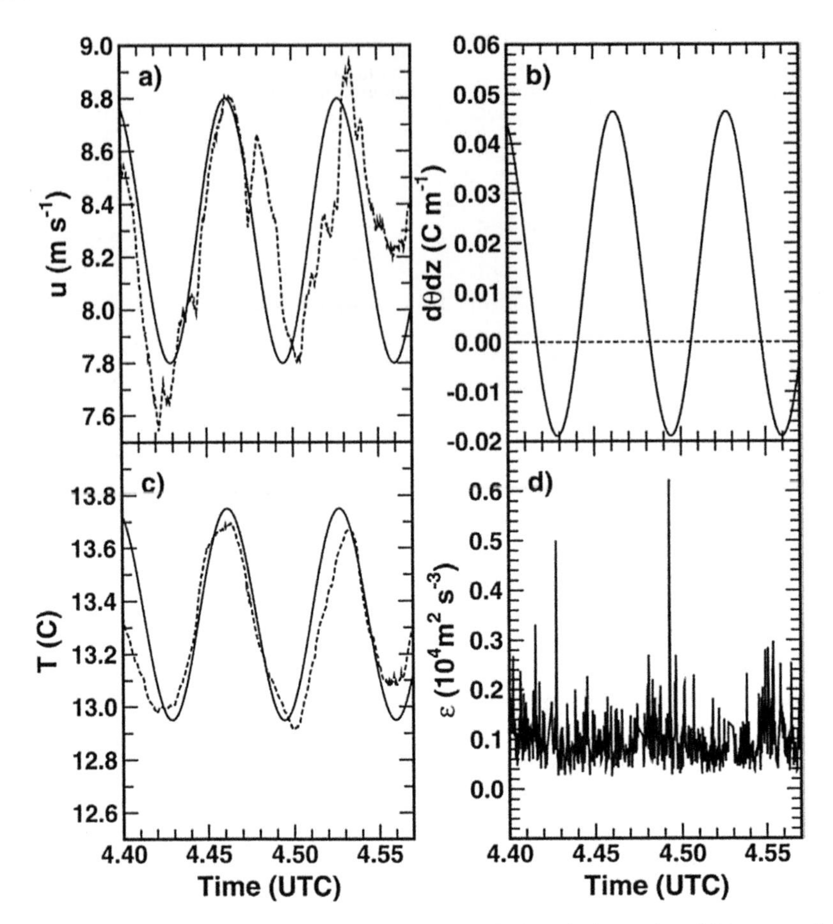

FIGURE 9.19 Computed (solid line) and measured (dashed) time series of (a) stream wise velocity and (c) temperature. (b) Computed time series of the wave-modulated, $\partial\theta/\partial/z$ and (d) time series of the kinetic energy dissipation rate measured by the TLS. (Taken from Meillier, Jones, and Balsley (2008).)

Isopynic surfaces are the most convenient streamlines for balloons to follow. This is because balloons are constructed to have constant volumes and can be weighted to obtain an arbitrary density. When released, the balloon seeks its equilibrium density surface (EDS). Buoyancy oscillations can occur about the EDS, and the EDS itself can be vertically displaced by large gravity waves and turbulence. It is commonly agreed that these special balloons quickly respond to the mean horizontal flow; however, vertical displacements of the balloon and their representations of wave-like structures is problematical. In this section, we look with some detail into how these vertical displacements have been treated.

A *superpressure balloon* (SPB) is made of a strong non-stretchable non-permeable material so that its volume is essentially constant. The balloon's density

FIGURE 9.20 Nearly constant level hydrogen balloon used by Japan during world war II (Taken from Mikesh (1990).)

is nearly constant, and when released it will seek its EDS. Buoyant forces will act to maintain the balloon's position on the isopycnic surface. The balloon can be displaced from this surface by vertical motions or changes in its density either by volume changes or gas leakage. Cadet (1978) gives an excellent history SPBs,

FIGURE 9.21 A tetroon with 0.32 m^3 volume. (Taken from Booker and Cooper (1965).)

fundamental technical issues, and reviews some applications. Perhaps the earliest use of unmanned balloons for horizontal transport came in 1944 when the Japanese used hydrogen balloons such as shown in Fig. 9.20, taken from Mikesh (1990), cleverly designed to drift over the north Pacific Ocean at an altitude between 9 and 12 km to deliver bombs over the northwest United States. The concept of using SPBs was introduced by Angell and Pack (1960). They described the construction of balloons using DuPont Mylar which could withstand superpressures as great as 150 mb. The construction of the balloons required a special heat sealing, and for purposes of economy and reliability it was desirable that these seals be straight lines. This requirement lead to a tetrahedron shape, and the balloons were called *tetroons*. Figure 9.21 shows a picture of a tetroon taken from Booker and Cooper (1965). Ideally, once released the tetroon (or other balloon shape) will drift with the wind along a surface of constant atmospheric density. By tracking the tetroon with radar or global positioning system, the wind field along the path of the tetroon can be resolved. However, this requires the balloon to behave as an air parcel. But generally this is not the case, especially for vertical motions, because the balloon's inertia is greater than that of a parcel of air.

Booker and Cooper (1965) pioneered the use of SPBs for estimating three-dimensional trajectories during mountain wave conditions. They point out that the only way an SPB can be displaced from its equilibrium height is by vertical air currents or a change of the mass of the balloon. The ideal SPB for air motion measurements would have zero restoring force, large drag, and small mass. In practice this is hard to achieve because, though small, Mylar stretches. Booker and Cooper (1965) experimentally determined that a 0.32 m^3 cubic meter SPB made of 2 mil ($\approx$0.05 mm) clear Mylar showed less than a 2.5% change in volume up to a superpressure of about 1.2 hPa. Because the buoyancy of the balloon is a function of its density, the balloon's change in volume would affect the restoring force, but this effect might be small. Ideally, a balloon's motion would mimic that of an air parcel. However, forces such as buoyancy, gravitation, and drag makes the inertia of the balloon greater than that of an air parcel. Thus, balloon motions will be retarded relative to true air motion. This effect will be most noticeable in the vertical velocity variations of the SPB. It was subsequently reasoned that if the balloon's vertical motion could be calculated theoretically, then these results could be used to estimate the true motions of an air parcel by "correcting" the observed motion of the SPB.

The first quantitative theory of the vertical motion of an SPB was developed by Booker and Cooper (1965) and reported on by Hirsh and Booker (1966). Booker and Cooper (1965) considered the response of an SPB to a steady sinusoidal variation of vertical velocity as a function of time. They considered a balance between the drag force D that would displace a balloon from its equilibrium height H_0 and a restoring force R that would return the balloon to it equilibrium position. Their equation for the vertical velocity $W_\mathrm{b}(t)$ of the SPB is

$$W_\mathrm{b}(t) = H_x \cos \frac{2\pi vt}{\lambda} \mathrm{d}t - \left(\frac{2R}{\rho_\mathrm{a} C_\mathrm{d} A} \right)^{1/2}, \tag{9.21}$$

where H_x is the vertical amplitude of the sinusoidal disturbance, v is the constant horizontal velocity, λ is the disturbance wavelength, R is the restoring force, ρ_a is the density of the free air, C_d is the drag coefficient, and A is the cross sectional area of the SPB exposed to vertical motions. The restoring force is

$$R = V(\rho_\mathrm{a} - \rho_\mathrm{g}) - W_\mathrm{b}, \tag{9.22}$$

where V is the volume of the balloon and ρ_g is the density of the inflation gas. Also considered in their computer program were the effects of adiabatic changes in ρ_a and the changes in V due to superpressure. A sample result is shown in Fig. 9.22 taken from Hirsh and Booker (1966). Typically the balloon trajectories under estimate the amplitude of the disturbance and lead the phase of the disturbance. They concluded that SPB trajectory error was greater for high wind speeds than for low wind speeds. The error is also greater for longer wavelengths and smaller amplitudes.

Vergeiner and Lilly (1970) used SPBs and instrumented aircraft in their study of lee waves in the Front Range of the Colorado Rocky Mountains. They used an equation similar to that of Hirsh and Booker (1966), *i.e.*,

$$-MW_B - Mg + V\rho g + C_D A\rho(W_A - W_B)|W_A - W_B| = 0 \tag{9.23}$$

to correct SPB trajectory errors; however, they did not assume an ideal steady sinusoidal disturbance. In (9.23), M is the total mass of the balloon + transponder, and the other symbols have the same meaning as in 9.22. The derived equation is

$$(W_A - W_B)|W_A - W_B| = C_1 \frac{\mathrm{d}^2 z_B}{\mathrm{d}z^2} + C_2(z_B - z_e), \tag{9.24}$$

where $C_1 \approx 2.67$ m and $C_2 = 0.0033$ ms^{-1}, and ρ_e is the equilibrium height of the SPB. Equation (9.24) was used in the form

$$\frac{\mathrm{d}}{\mathrm{d}z}(z_A - z_B) = \begin{cases} +\sqrt{+\text{right-hand side}} & \text{if RHS} > 0, \\ -\sqrt{-\text{right-hand side}} & \text{if RHS} < 0. \end{cases}$$

A typical result is shown in Fig. 3.4.

Hanna and Hoecker (1971) expanded the analyses of Booker and Cooper (1965) and Vergeiner and Lilly (1970) by including the effects of dynamic buoyancy and acceleration drag. Their vertical equation for the SPB has the form

$$M_{\mathrm{b}}\frac{\partial w_{\mathrm{b}}}{\partial t} = M_{\mathrm{a}}\frac{\partial w_a}{\partial t} + \frac{1}{2}M_a\left(\frac{\partial w_a}{\partial t} - \frac{\partial w_{\mathrm{b}}}{\partial t}\right) - M_{\mathrm{b}}g\left(\frac{\rho_{\mathrm{b}} - \rho_{\mathrm{a}}}{\rho_{\mathrm{b}}}\right) - \rho_{\mathrm{a}}A\frac{C_D}{2}(w_{\mathrm{b}} - w_a)|w_{\mathrm{b}} - w_a|, \quad (9.25)$$

where M_{b} and M_a are the masses of the balloon and displaced air respectively; w_{b} and w_a are the vertical speeds of the balloon and air respectively; the other symbols have the meanings as above, and the SPB is assumed to have a spherical shape. Defining the hydrostatic stability, s as

$$s = -g\frac{\partial}{\partial z}\left(\frac{\rho_{\mathrm{a}}}{\rho_{\mathrm{b}}}\right), \quad (9.26)$$

(9.25) takes the form

$$\frac{3}{2}\frac{\partial}{\partial t}(w_a - w_{\mathrm{b}}) = -g(1 - \mathrm{e}^{-sz/g}) - \frac{3C_D}{8R}(w_{\mathrm{b}} - w_a)|w_{\mathrm{b}} - w_a|, \quad (9.27)$$

where R is the radius of the balloon. Hanna and Hoecker (1971) used several combinations of values for s, C_{D}, and R and assumed a vertical wind speed of

$$w_a = W_a \sin\left(\frac{2\pi t}{T}\right), \quad (9.28)$$

where W_a is the magnitude of the wind speed, and T is the wave period. For $T \gtrsim 500$ s, the numerical results showed that the phase lead, ϕ, was a monotonically increasing function of the dimensionless parameter $\epsilon_H = (RTs)/(C_D W_a)$. (Note that ϵ_H does not appear in Hanna and Hoecker (1971).) We can interpret ϵ_H as the ratio of the buoyant force to the drag force. Fig. 9.23, taken from Hanna and Hoecker (1971) shows a plot of ϕ as a function of ϵ_H. A similar result (not shown) is obtained for the dimensionless magnitude $(1 - W_{\mathrm{b}}/W_a)$ when plotted against $sR/C)DW_a(T - 10s^{0.5})$. As an example for values of $R = 0.65$ m, $C_D = 0.8$, and $s = 0.57 \times 10^{-3}$ s^{-2}, they calculate a phase lead of the SPB $\phi \backsim 30°$ and the magnitude response $W_{\mathrm{b}}/W_a \backsim 0.95$ for $W_a = 0.5$ m s^{-1} and $T = 1000$ s. For W_a and T equal to 0.25 m s^{-1} and 2500 s respectively, $\phi \approx 65°$ and $W_{\mathrm{b}}/W_a \approx 0.50$. An over all conclusion is that better SPB response to sinusoidal variations of the vertical wind speed is obtained with small balloon and large drag coefficients.

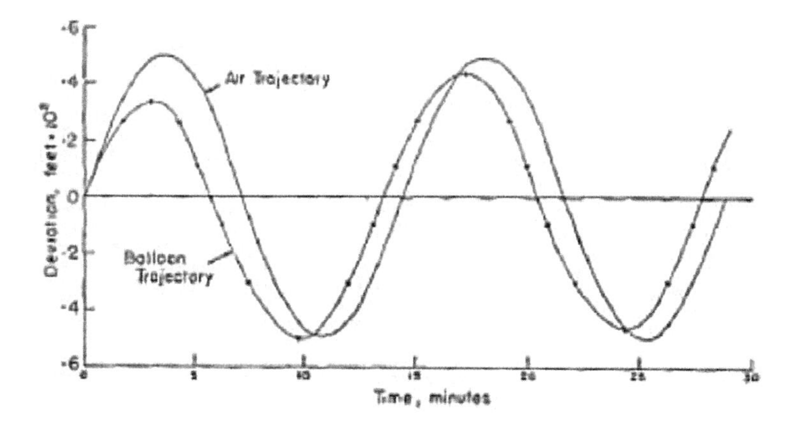

FIGURE 9.22 Theoretical comparison of an SPB's response to a wave perturbation of vertical velocity. (Taken from Hirsh and Booker (1966).)

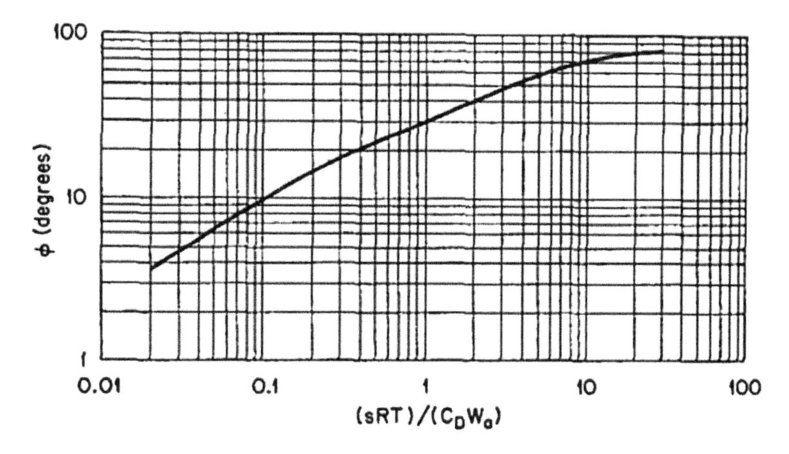

FIGURE 9.23 Theoretical calculation of SPB phase lag from the real vertical wind, θ as a function of the ratio of buoyant force to drag force. (Taken from Hanna and Hoecker (1971).)

Massman (1978) examined the oscillations of SPBs in the upper troposphere and lower stratosphere. He identified two possible modes of motion, neutral buoyant oscillations (NBO) and vertical motions of the balloon's EDS due to gravity waves and turbulence. Beginning with a three-dimensional version of the Hanna and Hoecker (1971) equation, Massman neglected shear and velocity terms, and after collecting terms his resulting equations were

$$\mathbf{v}_H \approx \mathbf{u}_H \tag{9.29}$$

and

$$\frac{d^2 z_b}{dt^2} = -\frac{2}{3} g \left(\frac{M_b - \rho_a V_b}{M_b} \right) - \frac{1}{3} \rho_a C_d A_b \left(\frac{dz_b}{dt} - w \right) \left| \frac{z_b}{dt} - w \right| + \frac{dw}{dt}, \quad (9.30)$$

where $\mathbf{u}_H$ is the horizontal velocity of the atmosphere, $\mathbf{v}_H$ is the horizontal velocity of the balloon, w is the vertical velocity of the atmosphere, M_b and V_b are the mass and volume of the balloon respectively and z_b is the altitude of the balloon; the other symbols have their usual meanings. The first term on the right-hand side of (9.30) is the buoyancy term. Levanon *et al.* (1974) show that this term can be written as $\omega_N^2(z_b - z_0)$, where z_0 is the height of the EDS and ω_N^2 is the balloon's NBO frequency. Ignoring slight variations in V_b

$$\omega_N = \left[\frac{2}{3} g \left(\frac{\Gamma_a - \gamma}{T_0} \right) \right]^{1/2}, \quad (9.31)$$

where Γ_a is the autoconvective lapse rate (g/R) and equals 34.2 K km^{-1}, γ is the environmental lapse rate and T_0 is the ambient temperature. During the Tropical Wind Energy Conservation and Reference Level Experiment (TWERLE) (Julian, Kellogg, and Suomi, 1977), 35 radar-tracked SPB flights and 67 instrumented SPB flights were launched at three tropical sites (American Samoa; Ascension Island and Accra, Ghana), and 23 mid-latitude SPB flights were launched from Christchurch, New Zealand. Flight levels were roughly 14.5 km in the upper tropical troposphere and 13.2 km in the mid-latitude lower stratosphere. For the tropical flights, the mode of the distribution of NBO periods was in the 3–4 min range, and for the mid-latitude flights the mode was in the 4–5 min range. Massman (1978) concluded that the NBO period at tested altitudes is most likely to be between 3 and 6 min. This oscillation is innate to the system and periodic motion much greater than about 6 min cannot be considered NBO. The SPB will not only respond to oscillation of the atmospheric vertical wind speeds and turbulence, but also the vertical motions of the EDS due to waves. Thus, the time dependence of z_0 must be explicitly included in (9.30). When the EDS is disturbed by a wave of the form $\Delta z_a \cos \omega t$ where Δz_a is the amplitude of the wave and ω is its frequency, and if $\omega \ll \omega_N$, *i.e.*, when the wave period is much greater than the NBO period, the departure of the balloon from its equilibrium surface is given by

$$[z_b(t) - \bar{z}_0] = \kappa \Delta z \cos \omega t \pm (1 - \kappa)^2 \epsilon \Delta z_a (\cos 2\omega t - 1), \quad (9.32)$$

where $\kappa = (\Gamma_d - \gamma)/(\Gamma_a - \gamma)$ and $\epsilon \propto \omega^2/\omega_N^2$. The parameter ϵ is roughly the inverse of ϵ_H. In (9.32), the $\pm$ is determined by the vertical velocity of the balloon relative to that of the EDS. When $\kappa \gg \epsilon$, *i.e.*, where the environmental lapse rate γ approaches the autoconvective lapse rate Γ_a and the period of the wave disturbance is much greater than the Brunt–Väisälä period, the SPB will tend to remain very nearly on its EDs and the departures from it will be very small. In regions where $\kappa \ll \epsilon$ the SPB oscillations are not due to the vertical wind but rather due to the motions of the balloons EDS. In these cases, the SPB is a true isopynic tracer.

The analyses of Hirsh and Booker (1966), Hanna and Hoecker (1971), and Massman (1978) considered only wave-like vertical variations of the EDS. In reality, periodic variations of streamline heights do not occur on their own. Instead, periodic height variations of streamlines are due to gravity waves. Thus, changes in wave-perturbation vertical velocity are correlated with changes in density, temperature, and pressure. Nastrom (1980) examined the response of an SPB to a single gravity wave including variations in both density and vertical wind speed, which from the polarization equations are in quadrature (2.31). Nastrom develops an equation of motion for a spherical balloon with constant density and assumes the mass of the air displaced by the balloon, M_{a}, equals the mass of the balloon, M_{B}, *i.e.*,

$$\frac{\partial^2 z}{\partial t^2} = -\omega^2 z + 2/3gR(t) - A\left(\frac{\partial z}{\partial t}\right)\left|\frac{\partial z}{\partial t} - w_a\right| + \frac{\partial w_a}{\partial t}, \tag{9.33}$$

where $\omega^2 = -2/3g(\partial M_a/\partial z)/M_B$, $R(t)$ is the percentage change of atmospheric density with time, $A = C_D/4r$, r is the radius of the balloon, and w_{a} is the vertical velocity of the air. Assuming a realistic density profile, it is found that ω^2 changes less than 1% per 100 m in the lower troposphere. Thus, ω^2 is considered constant. Defining $w_a = w \sin \nu$, $R = -\zeta \cos \nu t$ where ν is the wave frequency, an analytical solutions of (9.33) is obtained as

$$\frac{\partial^2 z}{\partial t^2} = -\omega^2 z - 2/3g\zeta \cos \nu t - A\left(\frac{\partial z}{\partial t} - w \sin \nu t\right) \times \left|\frac{\partial z}{\partial t} - w \sin \nu t\right| + w\nu \cos \nu t. \tag{9.34}$$

Nastrom points out that while (9.34) is simpler than (9.33) it is still nonlinear due to the drag term. He proposes to approximate this term by a sinusoidal function, *i.e.*,

$$A\left(\frac{\partial z}{\partial t} - w \sin \nu t\right)\left|\frac{\partial z}{\partial t} - w \sin \nu t\right| \approx A \sin \nu t \, |\sin \nu t| . \tag{9.35}$$

This is based on the numerical solutions of (9.33) and (9.34) which both show that only the odd harmonics have noticeable amplitudes, and the first harmonic is considerably larger than other harmonics. This is further justified by the Fourier expansion $F(x) = \sin x |\sin x|$, *i.e.*,

$$F(x) = \frac{8}{3\pi}\sin x - \frac{8}{15\pi}\sin 3x + \cdots + \frac{8}{n\pi(4-n^2)}\sin nx. \tag{9.36}$$

A solution of (9.34) of the form

$$z = -\frac{C}{\nu}\cos(\nu t - \theta) - \frac{D}{2\nu}\cos(2\nu t - \phi) - \cdots \tag{9.37}$$

is used which is approximated by its first harmonic and the nonlinear term is expanded in the Fourier series of the form (9.36). The resulting solutions for phase lead, θ, are complicated and the reader is referred to Nastrom (1980) for details. The prime conclusion of this analysis was that linearized equations can be solved analytically for steady wave conditions to render estimates of the amplitude and phase of an SPB relative to a sinusoidal vertical wind and atmospheric density variations provided they are in quadrature. In the presence of a horizontally propagating linear gravity wave, the amplitude response (C/w) increases with the amplitude of w_a and approaches linearity for large amplitude waves. For long period waves and high static stability, the SPB asymptotically approaches a true isopynic tracer with some phase shift, but this is not true for low static stability. Nastrom points out that in a vertically propagating gravity wave when phase is changing with height, a real SPB in a real gravity wave will encounter constantly changing phase as it oscillates vertically. For gravity wavelengths considered by Nastrom ($\backsim$ 1 km), this changing phase is unimportant.

A key component of the global circulation is the poleward transport of heat in the extratropical middle atmosphere. This transport is the result of gravity wave breaking in the stratosphere and mesosphere (Holton *et al.*, 1995), which in turn results in temperatures that greatly exceed equilibrium values (Andrews, Holton, and Leovy, 1987). The horizontal and vertical scales of these waves are respectively $\backsim$ 100–1000 km and $\backsim$ 100 m–10 km. These waves are generally too small to be realized in global-scale models, and while space borne techniques have provided considerable information on the global scale gravity wave potential energy, the global-scale momentum fluxes by vertically propagating gravity waves are poorly measured. Boccara *et al.* (2008) give a brief review of momentum flux measurements, and describe a methodology for estimating gravity wave characteristics from quasi-Lagrangian observations provided by long duration flights ($\backsim$ 2 to 4 months) of SPBs (see Hertzog *et al.*, 2008 for a description of these flights and results.) Boccara *et al.* (2008) take into account that SPBs do not perfectly follow isopynic surfaces, and assess this effect in the retrieval of gravity wave momentum fluxes. The Lagrangian pressure perturbations, p'_1 measured by the balloon are the sum of two components: (1) the Eulerian perturbation caused by the gravity wave, p'_{w}, and (2) a contribution associated with the vertical gradient of the background pressure, p'_{g}. Using the polarization equations developed by Fritts and Alexander (2003), the wave momentum flux is:

$$\Re(w'u'^{*}_{\parallel}) = -\frac{\Omega}{\overline{\rho} H N^2}\Im(p'_l u'^{*}_{\parallel}), \tag{9.38}$$

where w' is the wave perturbation vertical velocity, $u'_{\parallel}$ is the wave perturbation horizontal velocity in the direction of wave propagation, Ω is the intrinsic frequency, H is the atmospheric density scale height ($\backsim$6 km), and p'_l is the Lagrangian pressure perturbation. The phase shift, ϕ, of the balloon is given by

$$\phi = gH\frac{\Re(\rho'_l u'^{*}_{\parallel})}{\Im(p'_l u'^{*}_{\parallel})}, \tag{9.39}$$

where ρ'_l is the Lagrangian density perturbation. Once ϕ is known, the intrinsic phase speed is given by

$$c_I = \frac{2}{\overline{p}\delta\_u'^{2}_{\parallel}}\Re[p'_l(u'_{\parallel}\mathrm{e}^{-i\phi})^{*}], \tag{9.40}$$

where $\delta_{-} = 1 - f^2/\omega^2$ where f is the inertial frequency. Hertzog *et al.* (2008) applied the above analyses to the 27 long duration SPB flights launched from McMurdo, Antarctica (77.8S, 166.7E), between September 5 and October 28, 2005. The mean flight duration during the campaign was 59 days, and the longest flight lasted almost 4 months (109 days). The last flight terminated on 1 February 2006 (*i.e.*, during the SH summer).

9.3.3 AIRCRAFT

Aircraft have been used as moving platforms for remote sensors (see, for example, Blumen and Hart, 1988; Kwon and Gardner, 1990); however, aircraft have also been used to make *in situ* measurements from which wave parameters can be calculated. The majority of observations have been made over mountains. There are several reasons for this: mountain waves have large amplitudes and can be easily detected; because the waves are stationary, repeated observations of the same wave field are possible; and because the wave fronts tend to be located downwind of the obstacles that generate them, the lee waves are easy to locate. Perhaps the first observations of mountain waves using an aircraft were made by Radok (1954). He proposed repeated flights up- and down-wind along a number of tracks. During each flight, the airspeed and throttle settings are held constant corresponding to level flight. Changes in the recorded elevation of the aircraft as well as changes in temperature and pressure are related to vertical velocities associated with the standing waves. Vergeiner and Lilly (1970) used a Beech Queen Air 80 in their study of lee waves off the Colorado Rocky Mountains. The aircraft was equipped with sensors and digital conversion equipment to measure and record temperature, static pressure, air speed, heading, and ground speed using a Doppler navigation system. Most flights were made at constant power settings and pitch angle, and they assumed that the aircraft vertical velocity was equal to the atmospheric vertical velocity. They estimate the accuracy of this assumption to be about 1 m s^{-1}. Another source of error was uncertainty in the x,y-position of the aircraft which they estimate to be about ± 2 km. The Colorado Lee Wave Program (Lilly *et al.*, 1971; Lilly and Kennedy, 1973) was an ambitious flight program involving 4 aircraft, the NCAR DeHavilland Buffalo (turboprop) and three jets, the NCAR Sabreliner, the NOAA B-57B, and an Air Force B-57F stratospheric jet. A primary goal of that program was the evaluation of terrain-induced momentum flux and turbulence generated by breaking gravity waves as described by Lilly (1972). An interesting side result reported by Lilly and Kennedy (1973) was that gustprobe equipment is apparently not necessary for the direct aircraft measurement of wave momentum flux, but an inertial platform or similar stable attitude reference is essential. Smith (1976) used a much less ambitious program, *i.e.*, a lightweight aircraft (Bonanza F33A) to study lee waves generated by the Blue Ridge Mountain in the central Appalachians. The aircraft had no special meteorological instrumentation other than a laboratory thermometer located outside the cockpit window. Simple as it was, the aircraft had no difficulty in responding to the wave-induced vertical motions of the flow. Comparison of the observations with linear theory and laboratory experiments showed that the linear theory correctly predicted the wavelength, but seriously under predicted the wave amplitude. Turbulence kinetic energy and waves were measured by Karacostas and Marwitz (1980) using the NCAR Queen Air 304D over Elk Mountain, in Wyoming. Brown (1983) used the Hercules and Canberra aircraft of the UK Meteorological Research Flight to investigate mountain waves over the British Isles. As with the previous studies mentioned, Brown (1983) used linear wave theory to calculate wave parameters and momentum fluxes. Gravity waves in the upper atmosphere, between 60 and 140 km, have been analyzed by Fritts, Blanchard, and Coy (1989) and Fritts, Wang, and Blanchard (1993) using density fluctuations measured by high-resolution accelerometers on board the space shuttle during re-entries. Aircraft continue to be a primary tool for gravity wave research; recent examples include Moustaoui *et al.* (1999), Leutbecher and Volkert (2000), Lane *et al.* (2000), Poulos *et al.* (2001), and Dörnbrack *et al.* (2001).

9.4 REMOTE MEASUREMENTS

Remote sensing can be defined as measuring the characteristics and properties of the atmosphere in a region far removed from the sensing instrument. In this section we shall give only qualitative details of these instruments. The discussion of several technical issues are beyond the scope of this book. Examples of remote sensors include radar, lidar, and sodar. For example, Kjelaas *et al.* (1974) used a triangular array of sodars to estimate gravity waves characteristics. Observations taken with the sodar array were compared with those obtained with an array of microbarographs. The results from both measurement techniques were comparable when waves were observed by both methods. However, the sodar array was able to detect waves propagating in an elevated inversion above the convective boundary layer, but these waves were not detected by the microbarographs. A similar instance was reported by Nappo, Eckman, and Coulter (1992) who observed a Kelvin–Helmholtz wave at about 300 m AGL in the nighttime boundary layer, but the wave was not seen by a surface array of microbarometers Eymard and Weill (1979) used a triangular array of Doppler sodars to study gravity waves at two locations in France. Xing-sheng *et al.* (1983) used an optical triangle to calculate wave speeds and directions. The optical anemometers sense path-averaged instantaneous wind speeds which is perhaps a more acurate measure of wave perturbation velocity than observations made at three points. Carter *et al.* (1989) used three vertically-directed 50 MHz radar wind profilers spaced between about 5 and 6 km to detect vertical velocities associated with gravity waves during the ALPEX experiment in southern France. They concluded that "...monochromatic wave activity is a relatively rare occurrence and that most of the time a wide spectrum of waves influences the vertical velocities in an incoherent fashion." However, the somewhat

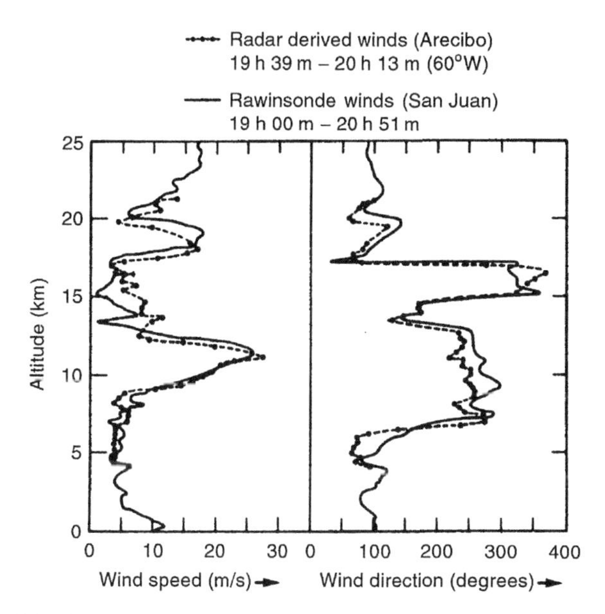

FIGURE 9.24 Comparison of vertical profiles of wind speed and wind direction observed by the Arecibo radar with nearly simultaneous rawinsonde observations from San Juan. (Taken from Farley *et al.* (1979).)

large horizontal scale of the array may be partly responsible for these incoherent disturbances. As we have seen in Section 9.2.2, if the average separation between stations in the array is large relative to a horizontal wavelength, then aliasing of the wave signal can occur. This can result in an incoherent wave signal.

9.4.1 RADAR

Radar has been extensively used to study gravity waves throughout the atmosphere. For example, using radar, Ottersten, Hardy, and Little (1973) described gravity wave observations in the planetary boundary layer; Gauge and Balsley (1978) described probings of the troposphere and stratosphere; Manson (1990) presented a climatology of wave structures in the mesosphere; and Fritts and Isler (1994) described wave motions in the mesosphere and lower thermosphere. Perhaps one of the first reported radar observations of a gravity wave was made by Hicks and Angell (1968) who described "...a hitherto undetected phenomenon observed on several occasions..." that "...is an apparent horizontally twisted, braided, or helical-appearing atmospheric structure ...". They conclude that the object in question is a breaking Kelvin–Helmholtz gravity wave. Since then, numerous studies of gravity waves have been performed using radar.

Perhaps the most widely used radar is the MST (mesosphere–stratosphere–troposphere) system. The MST technique (see, for example Balsley and Gage, 1980) uses ultrasensitive VHF (30–300 MHz) and UHF (300–3000 MHz) radars to study the weak backscattering arising from refractive index fluctuations in the neutral atmosphere and lower ionosphere. Figure 9.24, taken from Farley *et al.* (1979) shows plots of radar-derived wind profiles observed with the Arecibo radar in Puerto Rico compared with radiosonde wind profiles. The agreement between radar winds and airsonde winds is very good. The radiosonde was launched from San Juan, about 70 km east of Arecibo, and Farley *et al.* (1979) attributed the differences between the profiles to the physical separation between the observing points rather than to experimental uncertainties. Vincent and Reid (1983) used these techniques to measure gravity wave momentum fluxes in the mesosphere. At a height range of from 80 to 90 km, they calculate a westerly acceleration due to momentum flux convergence of about 20 m s^{-1}/day due to waves with a dominant wavelength of about 50 km and a phase speed of about 50 m s^{-1}. While this induced acceleration is almost an order of magnitude smaller than that predicted by Fritts, (1984) using saturation theory (see Section 8.3.2), it may well be correct since Vincent and Reid (1983) estimated the average wave stress over three days of observations. Tsuda *et al.* (1989) used MST radar techniques to compare wave spectra observed in the troposphere, stratosphere and mesosphere with the model predictions of Dewan and Good (1986) and Smith, Fritts, and VanZandt (1989). A further discussion of the basic concepts of radar and its applications to wave studies can be found in Chadwick and Gossard (1986).

Two types of radar are commonly used in atmospheric research, Doppler radar and frequency-modulated continuous-wave (FM-CW) radar.

9.4.2 DOPPLER RADAR

The principles of *Doppler radar* as used in atmospheric studies can be found, for example, in Gauge and Balsley (1978) and Kropfli (1986). Basically, pulses of radio frequency energy are transmitted via a directional antenna. These pulses are scattered by refractive inhomogeneities caused by temperature and humidity fluctuations on scales of half a wavelength of the pulse. These fluctuations are generally caused by turbulence or large gradients of temperature or humidity along the radar beam. Ultra-high-frequency (UHF) radars operate over frequency ranges of from 300 to 3000 MHz with wavelengths ranging from 0.1 to 1 m. Sensitive Doppler radars operate in the very-high-frequency (VHF) range from 30 to 300 MHz with wavelengths ranging from 1 to 10 m. The component of the scattered beam parallel to the incident beam is received by the antenna, and the time delay between transmitted and received pulses is a function of the distance to the scattering region. If the scattering region is moving, then the frequency of the returned pulse is Doppler shifted by an amount that is proportional to the velocity component in the direction of the radar beam. If the radar is directed vertically, then only the vertical velocity of the scattering region is observed. If, however, the beam is inclined to the vertical by angle θ, then the instantaneous Doppler velocity, v, at range R is

$$v(\theta, R, t) = w(\theta, R, t)\cos\theta + u_{\mathrm{h}}(\theta, R, t)\sin\theta, \tag{9.41}$$

where w and u_h are the instantaneous vertical and horizontal velocities respectively. If the tilted beam is aimed along orthogonal azimuth angles, and if it is assumed that the horizontal wind speed is much greater than the vertical wind speed, than as demonstrated by Farley *et al.* (1979) the instantaneous horizontal winds speeds u and v can be resolved. By making similar measurements at various range distances, the wind profile can be resolved.

9.4.3 FM-CW RADAR

Generally, Doppler radar returns are limited to temperature and humidity inhomogeneities on the scale of meters, and to ranges greater than several hundreds of meters above the ground surface. The first restriction is imposed by the radar frequency, and the latter by the fact that during the transmission of the radar pulse no signal can be received. However, to better understand the dynamics of the planetary boundary layer it is necessary to know the true thickness of thin radar backscatter layers and the processes generating these fine-scale refractive index inhomogeneities. The frequency-modulated continuous-wave (*FM-CW radar*) developed by Richter (1969) makes these observations possible. To eliminate the need for radar pulses which limits the range, a continuous microwave frequency is transmitted and simultaneously received by an identical antenna close bye. In itself, this mode of operation would provide high spatial resolution, but no range information. To get range information, the transmitted frequency is linearly modulated between two frequencies over a time T_M. The received signal will be Doppler shifted by moving scatters, and when the transmitted and received signals are combined in real time, a sinusoidal beat frequency, f_b is generated. The time delay, Δt, of the appearance of reflected signal is related to the distance of the scatter by

$$f_b = \frac{F}{T_M}\Delta t = \frac{2F}{cT_M}H, \tag{9.42}$$

where F is the frequency excursion, c is the velocity of propagation, and H is the height of the scatter (assuming the instrument is pointed upward). In the case of multiple returns, a spectrum analysis of the beat frequency allows the different targets to be resolved according to their range, and the amplitudes of the beat frequencies are measures of the reflection coefficients of the targets. Fig. 9.25 taken during the CASES-99 field program illustrates the fine wave-like structures that can be revealed by the FM-CW radar. The capability of the FM-CW radar was enhanced in 1976 when Doppler capability was added (Chadwick *et al.*, 1976; Strauch *et al.*, 1976). This was accomplished by using a digital Fourier transform that preserved the phase and amplitude of spectral density of the radar signal obtained during each sweep. Monitoring the change in phase from sweep to sweep provides the Doppler information needed to estimate radial velocities. Fig. 9.26 taken from Chadwick *et al.* (1976) compares FM-CW Doppler winds with winds observed using a tethered balloon and a rawinsonde. The FM-CW radar

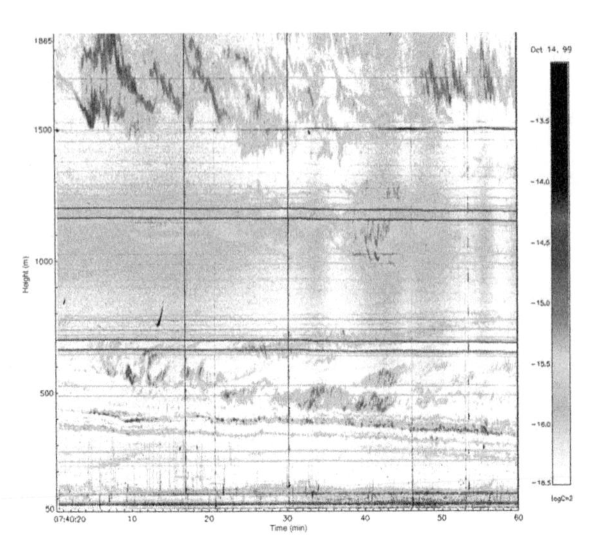

FIGURE 9.25 FM-CW radar images recorded on October 14, 1999. Record begins at 07:40:20 GMT. Kelvin–Helmholtz waves are between 1500 and 1800 m and between 500 and 600 m. Courtesy of Stephen Frasier, Univ. of Mass.

is becoming a standard instrument for boundary layer studies, and has proved especially useful in the studies of wave and turbulence in the stable boundary layer (see, for example, Eaton, McLaughlin, and Hines, 1995; De Silva *et al.*, 1996).

9.4.4 SODAR

Sodar, which stands for sound detection and ranging, was first described as a research tool for probing the lower atmosphere by Little (1969) and McAllister *et al.* (1969). The historical development of sodar and its principles of operation can be found in Beran, Hooke, and Clifford (1973) and the reviews by Brown and Hall (1978) and Neff and Coulter (1986). Basically, sodars are like radars but with sound energy replacing radio frequency (RF) energy. Little (1969) pointed out that the interactions of sound waves with the lower atmosphere is much stronger than interactions of electromagnetic waves. For example, the change in sonic refractive index for a 1 K fluctuation in temperature is about 1700 N-units, where one N-unit equals 1 part in 10^6. For RF wavelengths, the corresponding change in refractive index is about 1 N-unit. For fluctuations in wind speed, the difference is greater; for a 1 $\mathrm{m\,s^{-1}}$ variation in wind speed the change in the sonic refractive index is about 3000 N-units, whereas RF waves are essentially unaffected. Thus, sodar offers a much more sensitive means than radar for observing fine structure and turbulence in the PBL. An example of this ability is shown in Fig. 1.4, taken from Zamora (1983). The dark bands represent sound returns from regions of

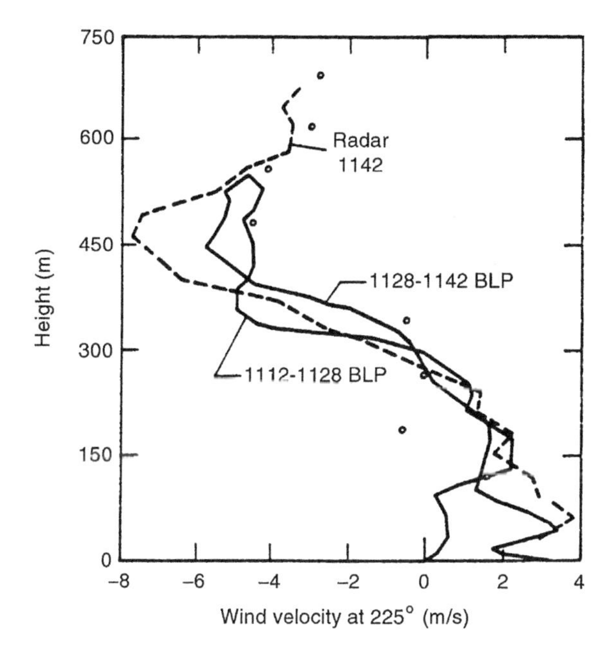

FIGURE 9.26 FM-CW Doppler radar wind speed profile (dashed line) compared with tethered balloon wind profiles (solid lines) and rawinsonde profile (circles). (Taken from Chadwick *et al.* (1976).)

turbulence and stable stratification. The light regions between the bands are either quiescent or neutrally stratified, *i.e.*, $\partial\theta_0/\partial z = 0$. The small-scale structures of the type shown in Fig. 1.4 are discussed by Chimonas (1999); they represent complex interactions between waves and turbulence in the stable PBL. Emmanuel *et al.* (1972), Hooke, Hall, and Gossard (1973), and Emmanuel (1973) were among the first to use sodar as a means of identifying regions of shear instability in the nocturnal inversion. Merrill (1977) made use of sodar to identify wave instabilities in one of the first comprehensive field studies of waves and turbulence. Since then, the sodar has been an integral part of most studies of the boundary layer. Neff and Coulter (1986) present several examples of these types of studies. The first reported application of Doppler technology to sodar was made by Beran, Little, and Willmarth (1971) who used it to measure vertical velocities in the convective boundary layer. Descriptions of Doppler sodar systems are given by Beran, Hooke, and Clifford (1973) and Brown and Hall (1978). A discussion of early studies of gravity waves and sodar is given by Beran, Hooke, and Clifford (1973). Since then, the Doppler capabilities have been greatly extended so that routine profiles of the three-dimensional wind field and its accompanying turbulence structure are possible.

9.4.5 LIDAR

Light detecting and ranging or *lidar* is the optical counterpart of radar. Lidar is the most recent of remote sensing techniques, and some aspects of the technology are still under development. Instead of RF pulses, lidars transmit pulses of light, and measure its backscatter as a function of time. A discussion of the basics of lidar can be found in Schwiesow (1986). The transmitted light is backscattered by aerosols and molecules depending on the frequency of the light. A lidar system consists basically of a laser, a telescope for transmission and reception, a detector, and of course a computer for signal control, signal processing, and data analysis. Because the absorption and scattering of the laser beam is proportional to the density of the absorbers and scatters, lidars have been often used to remotely measure concentrations of aerosols, trace gases, and atmospheric pollutants. The first reported use of a lidar for observing gravity waves made by Collis, Fernald, and Alder (1968). They used two pulsed ruby lidars to observe the structure of wave clouds in the lee of the Sierra Nevada mountains in California. Gardner and Shelton (1985) and Gardner and Voelz (1987) developed equations which relate variations of mesospheric sodium (Na) concentrations to gravity waves, and have used lidar measurements of Na to estimate wave characteristics in the mesosphere. Kwon and Gardner (1990) used airborne lidar measurements of mesospheric Na to estimate not only vertical and horizontal wavelengths, but also intrinsic frequency, phase speed, and propagation direction. Gardner and Taylor (1998) discuss the limits of remotely measuring middle atmosphere gravity wave parameters.

Because the light scatterers move with the mean flow, Doppler techniques can be used to remotely measure wind speeds. However, Eloranta, King, and Weinman (1975) and Sroga, Eloranta, and Barber (1980) demonstrated that three-dimensional boundary-layer flow fields could be measured without the Doppler technique by following the motions of naturally occurring aerosol density inhomogeneities. Examples of the early uses of Doppler lidar for studying winds in the boundary layer can be found in the reports by DiMarzio *et al.* (1979), Bilbro *et al.* (1984), and Köpp, Schwiesow, and Werner (1984). One of the first observations of gravity waves with a Doppler lidar was reported by Blumen and Hart (1988). They used an airborne Doppler lidar to study the flow fields in the wake of Mount Shasta in California. Ralph *et al.* (1997) reported on the use of Doppler lidar and other instruments in two studies of trapped gravity waves in the lee of the Rocky Mountains near Boulder, Colorado. The NOAA/ETL High Resolution Doppler Lidar (HRDL) which is described in Grund *et al.* (2001) has been used in many field experiments for studying gravity waves. The lidar can achieve 30 m range resolution, a working range of between 2 and 3 km, and a velocity precision of about 0.1 $\mathrm{m\,s^{-1}}$. Figure 9.27, taken from Newsom *et al.* (2000), shows contours of stream-wise velocity as a function or height and horizontal range. (A video of these return signals can be seen at EURL\Ch9\video-1.) A well-defined wave is seen between 40 and 70 m, and a low level jet with a maximum velocity of 10 $\mathrm{m\,s^{-1}}$ at a height of about 120 m. Eichinger *et al.* (1999) reported on the

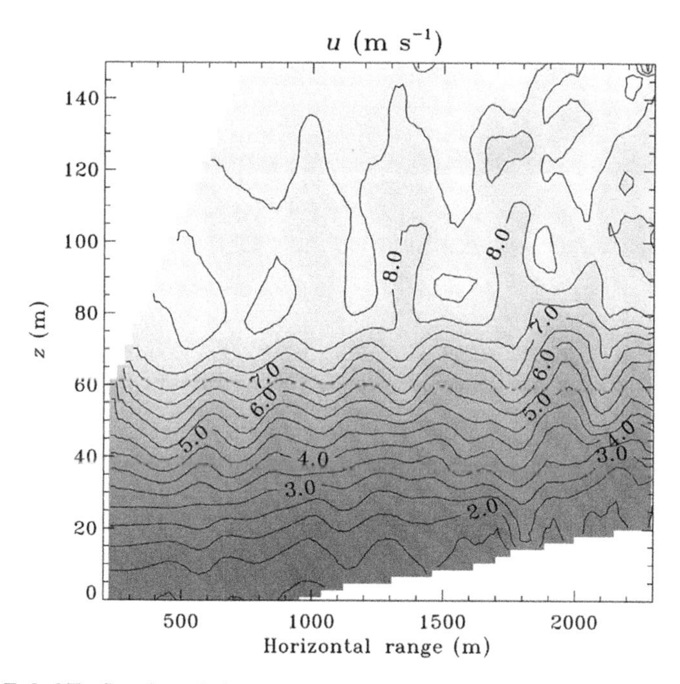

FIGURE 9.27 Sample vertical cross-section wind speed scan taken by the HRDL radar during the CASES-99 field program. Positive wind speeds are away from the radar. (Taken from Newsom *et al.* (2000).)

development of a scanning Raman water vapor lidar designed for boundary layer and tropospheric observations. Horizontal range while scanning is about 700 m, and the vertical range can be up to 12 km. Spatial resolutions range from 1.5 m in the near field to 75 m in the far field. Fig. 9.28 taken from Eichinger *et al.* (2001) shows wave activity during the night of October 13–14, 1999 during the CASES-99 field program. While these types of lidars do not give wind speed data directly, their fine spatial resolution is especially useful in estimating wavelengths and regions of wave activity. Also, by observing a wave over successive scans it is possible to estimate wave speed.

Temperature soundings from 1 to 105 km altitude with 0.2 to 1 km vertical resolution were made by Gerding, Rauthe, and Höffner (2004) at the Leibniz-Institute of Atmospheric Physics (IAP) in Kyhlungsborn, Germany. They used a combination of two different lidar-systems and three different methods of temperature measurements applied in four altitude regions from the lower troposphere up to the lower thermosphere. Nighttime temperature soundings of up to 16 h in duration were used to examine the temperature variations on scales greater than 10 min. Atmospheric waves have been identified with vertical wavelengths typically between 10 and 17 km and periods larger than 4.5 h. The waves amplitude often exceeds 20 K.

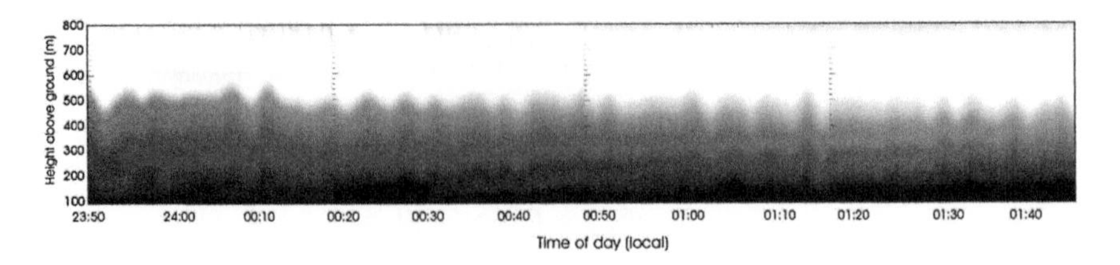

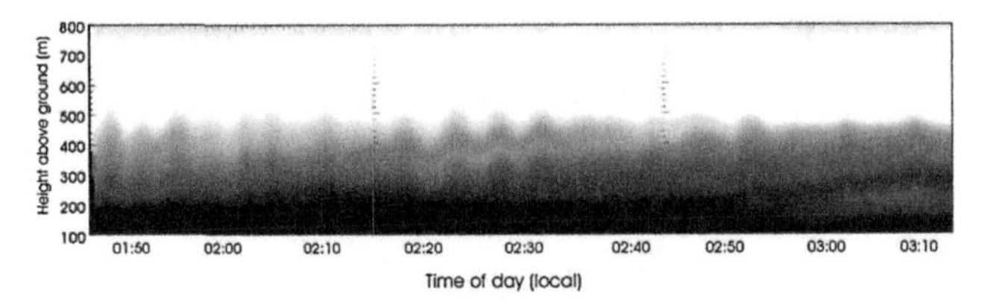

FIGURE 9.28 Boundary layer waves observed by vertically pointed Ramon lidar during the night of October 13–14, 1999 during the CASES-99 field program. Time is CDT. (Taken from Eichinger *et al.* (2001).)

9.4.6 AIRGLOW

Airglow is a quasi-steady faint photochemical luminescence occurring in the upper atmosphere between about 80 and 120 km over middle and low latitudes. (Links to videos of air glow are found at EURL\Ch9\AirGlow.) The near steadiness of airglow distinguishes it from the sporadic aurorae which occur over higher latitudes and at higher elevations. Although the emitted electromagnetic radiation results from photochemical reactions of many atmospheric trace gases, several sources of airglow radiation occur at specific wavelengths, *i.e.*, atomic oxygen at 558 nm (10^{-9} m), sodium at 589 nm, and the hydroxyl radical OH at 600–2000 nm. As examples, the excited OH arises from the reaction (Gardner and Taylor, 1998)

$$H + O_3 \rightarrow OH^* + O_2, \tag{9.43}$$

and the sodium emission is based on the Chapman mechanism (Molina, 1983), *i.e.*,

$$Na + O_3 \rightarrow NaO + O_2, \tag{9.44}$$

$$NaO + O \rightarrow Na^* + O_2, \tag{9.45}$$

$$Na^* \rightarrow Na + h\nu \ (589.3 \text{ nm}). \tag{9.46}$$

Each of these emissions occur at a particular height over a distance of from 10 to 20 km (see, for example, Swenson and Gardner, 1998). Although not visible to the naked eye, the airglow can be photographed under dark skies using highspeed films or charged-coupled-device (CCD) imagers. Peterson and Keiffaber (1973) used infrared film and a fast 35-mm camera to make a series of 15-min time exposures of a moonless night sky in New Mexico. The photographs all showed bright cloud-like structures which moved across the sky and varied in

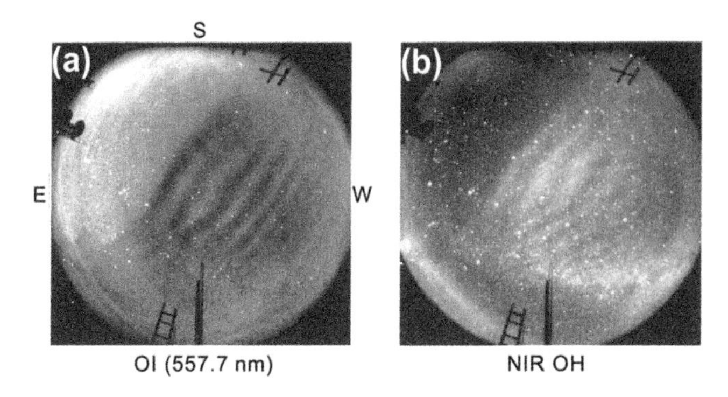

FIGURE 9.29 OH and OI airglow all-sky images from Haleakala, Hawaii on October 10, 1993, 1057 UTC. (Taken from Taylor, Turnbull, and Lowe (1995).)

brightness. They concluded that the bands were due to varying airglow emission intensity and not intervening atmospheric clouds moving against an otherwise uniform emission background. Assuming a height of 100 km, they calculated speeds of the structures at 20 $m\,s^{-1}$ and 43 $m\,s^{-1}$. Because the intensity of the photochemical emission is proportional to the local density and the temperature, variations in intensity can be related to variations in density and temperature. It is now widely accepted that these variations are due to gravity waves (see, for example, Molina, 1983; Hecht *et al.*, 1997; Fritts *et al.*, 1997; Nakamura *et al.*, 2003 and references contained therein). Fig. 9.29, taken from Taylor, Turnbull, and Lowe (1995) shows a striking example of these traveling waves. Indeed, a whole literature exists devoted to gravity waves in the upper atmosphere as revealed by airglow signatures and lidar sensing (see, for example, Swenson and Gardner, 1998). Of particular interest is the number of waves observed to be nearly monochromatic. Gravity waves launched in the troposphere propagate upward in the form of wave packets. Critical levels and wave reflections encountered as the packet moves upward erode the packet until only a small portion of the original wave spectrum exists. The result is a nearly monochromatic wave. Current thinking suggests that some of these waves are ducted modes in upper atmospheric wave guides (see, for example, Munasinghe *et al.*, 1998). Vargas *et al.* (2009) measured gravity waves and momentum fluxes in the mesosphere from OH airglow emissions. Horizontal wavelengths ranged from 10 to 150 km, vertical wavelengths ranged from $\sim$6 to 40 km, phase speeds ranged from 50 to 150 $m\,s^{-1}$, and horizontal momentum fluxes were as great as 5 $m^2\,s^{-2}$. Fig. 9.30, taken from Vargas *et al.* (2009), shows three OH air glow images taken at intervals of a few minutes. Cloud heights were between 80 and 100 km. Their results give horizontal wavelengths between 10 and 150 km, vertical wavelength between $\sim$6 and 40 km, periods between $\sim$6 and 60 min, and horizontal phase speeds between 50 and 150 $m\,s^{-1}$.

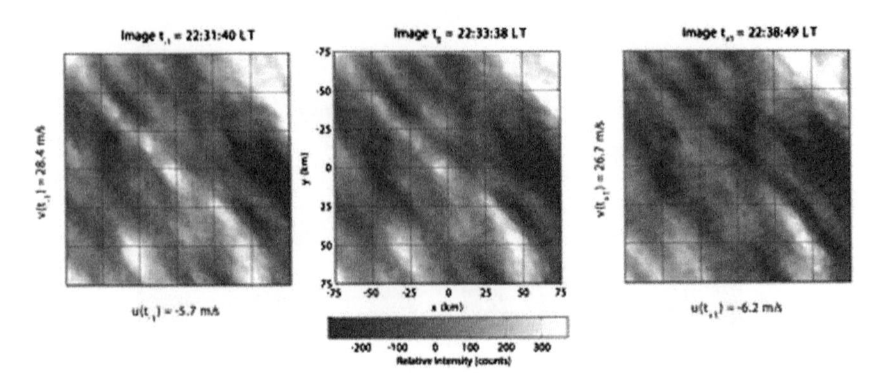

FIGURE 9.30 Three OH air glow images between 80 and 100 km taken at intervals of a few minutes. (Taken from Vargas *et al.* (2009).)

9.4.7 SATELLITES AND GLOBAL POSITIONING SYSTEMS

9.4.7.1 Photographic Analyses

Almost all analyses of gravity waves using satellite images are limited to wave clouds in the lee of mountains. Some of the first analyses were made by, for example, Döös (1962), Conover (1964), Fritz (1965), and Cohen and Doron (1967). Ernst (1976) analyzed infrared images of low level mountain waves taken by a geostationary satellite. He suggested that these data could be used for the early detection and identification of mountain waves and possible episodes of clear-air-turbulence. Satellite imagery continues to be used in mountain wave research, see for example, Ralph *et al.* (1997), Lane *et al.* (2000), Leutbecher and Volkert (2000), Smith *et al.* (2002), Jiang, Doyle, and Smith (2005), and Smith *et al.* (2007).

9.4.7.2 Sounding Rocket

Rockets have been used since the middle of the 20th century for measuring winds in the middle and upper atmospheres. Development of the National Aeronautics and Space Administration's (NASA) Aerobee liquid-propellant sounding rocket was begun in 1946 by the Aerojet Engineering Corporation (later Aerojet-General Corporation) under contract to the US Navy. The Applied Physics Laboratory (APL) of Johns Hopkins University was assigned technical direction of the project. James A. Van Allen, then Director of the project at APL, proposed the name "Aerobee." He took the "Aero" from Aerojet Engineering and the "bee" from Bumblebee, the name of the overall project to develop naval rockets that APL was monitoring for the Navy. The 18-k–thrust, two-stage Aerobee was designed to carry a 68-k payload to a 130-km altitude. The suite of Aerobee soundings rockets can reach altitudes between 150 and 2200 km and carry payloads ranging from 45 to 136 km. Ference *et al.* (1959) used sounding rockets to launch grenades for measuring sound transmission to determine air temperatures. Grenades were ejected from the rocket at predetermined altitudes during its ascent. Bartman *et al.* (1956) used the falling sphere technique to measure pressure, density, and temperature up to 100 km. The method consists of ejecting near the peak of the rocket's trajectory, an inflatable 1.2 m diameter nylon sphere contain pressure and temperature sensors, the data being returned by radio. Tracking the trajectory of the balloon allowed calculations of wind speed and direction.

Currently, the most often used rocket sounding system makes use of radar tracked "clouds" of metal coated foil chaff. Wu and Widdle (1992) obtained high-resolution wind profiles (25 m) between 85 and 103 km using a precision radar and a foil chaff that is about 1 μm thick. The use of very light chaff is necessary at very low atmospheric densities. About 6000 aluminized plastic foils were used for each sounding. Foil chaffs have also been used by, for example, Devarajan, Reddy, and Reddi (1985), Murayama *et al.* (1999), and Wu *et al.* (2001).

9.4.7.3 GPS Radio-Occultation Sounding

In the early days of the interplanetary space program, *i.e.*, early 1960s, scientists from the Stanford University and the Jet Propulsion Laboratory used radio signals from Mariner 3 and 4 spacecraft to study the atmosphere of Mars. As the planet occulted the spacecraft, radio signals passed through increasing depths of the Martian atmosphere resulting in refraction or bending of the radio signals. The bending causes an increase in the transmission path between Earth and Mars resulting in increasing phase shifts of the signals. Because the refraction is a function of temperature, pressure, and water vapor, the vertical distributions of these variables could be related to the increasing refraction as the spacecraft moved behind the planet. During the next 3 decades, subsequent developments and refinements of the technique have been used to study the celestial bodies containing an atmosphere. The historical development of these techniques is outlined in, for example, Melbourne *et al.* (1994) and Businger *et al.* (1996), and Yunck, Liu, and Ware (2000).

With the advent of the Global Positioning System (GPS), it became possible to make near-continuous soundings of the middle and upper atmospheres. This is made possible by the low Earth orbit (LEO) GPS receiver satellites. The LEOs orbit the Earth at a height of about 750 km while the GPS satellites orbit at a height of about 20,200 km. Fig. 9.31 is a schematic of the GPS radio occultation (RO) system, not drawn to scale. It is clear that the refracted path is longer than the direct unrefracted path, and this difference allows the calculation of the degree of refraction and the corresponding atmospheric temperature. As of 2009, there are 29 GPS satellites and 6 LEO satellites (Cucurull, 2009). Because refraction is strongly influenced by water vapor as well as temperature, temperature sounding is limited to heights from about 40 km to about 5–7 km a region where the atmosphere is considered dry (Ware *et al.*, 1996). The RO temperature profiles have vertical resolutions ranging from 200 to 1000 m (Tsuda and Hocke, 2004). Fig. 9.32 taken from Tsuda and Hocke (2004), shows a comparison of tropical temperature profiles, above Indonesia, measured with a radiosonde and RO system.

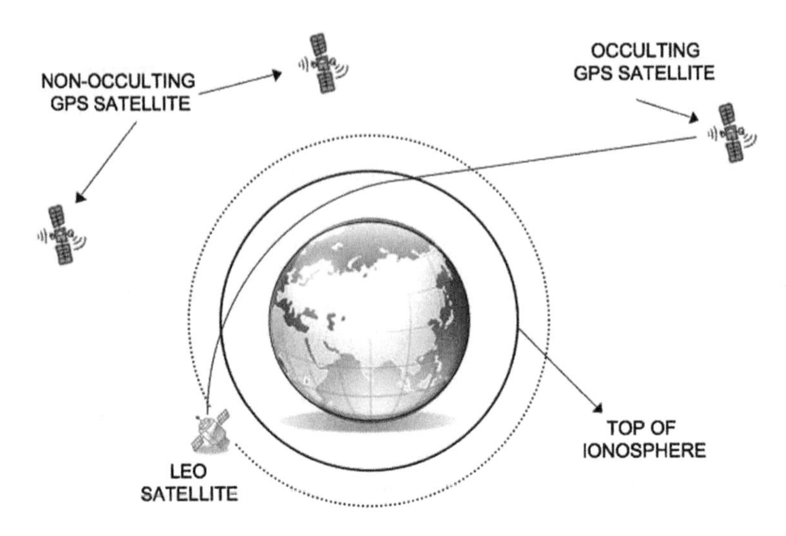

FIGURE 9.31 A schematic of the GPS Radio-Occultation system.

Below about 10 km, the RO profile deviates greatly from the radiosonde profile because the RO temperatures were calculated assuming a dry atmosphere. The expanded portion of Fig. 9.32 shows the RO sounding achieving a high-resolution measurement of the detailed temperature structure around the tropopause. The RO profile also shows that the wave activity above the tropopause continues far past the height where the radiosonde profile ends. GPS-RO sounding is now widely used for studying gravity waves in the middle and upper atmospheres.

9.4.7.4 Microwave Limb Sounding

Microwave limb sounding (MLS) is another satellite-observing technique for measuring the vertical variations of temperature and atmospheric trace gasses (Waters *et al.*, 1975). As the instrument-baring satellite passes over the Earth, millimeter- and submilleter-wavelength thermal emissions are scanned through the atmospheric limb. Development of the MLS experiments began at the Jet Propulsion Laboratory in the mid-1970s and included instruments deployed on aircraft and balloons prior to application of the technique from space. The measurement technique is described by, for example, Waters *et al.* (1975), Waters (1989), and Waters *et al.* (1999). MLS features include (1) the ability to measure many atmospheric gases, with emission from molecular oxygen providing temperature and pressure; (2) measurements that can be made reliably, even in the presence of heavy aerosol, cirrus, or polar stratospheric clouds that can degrade ultraviolet, visible, and infrared techniques; (3) the ability to make measurements at all times of day and night and provide near-global coverage on a daily basis; (4) the ability to spectrally resolve emission lines at all altitudes, which allows measurements of very weak lines in the presence of nearby strong

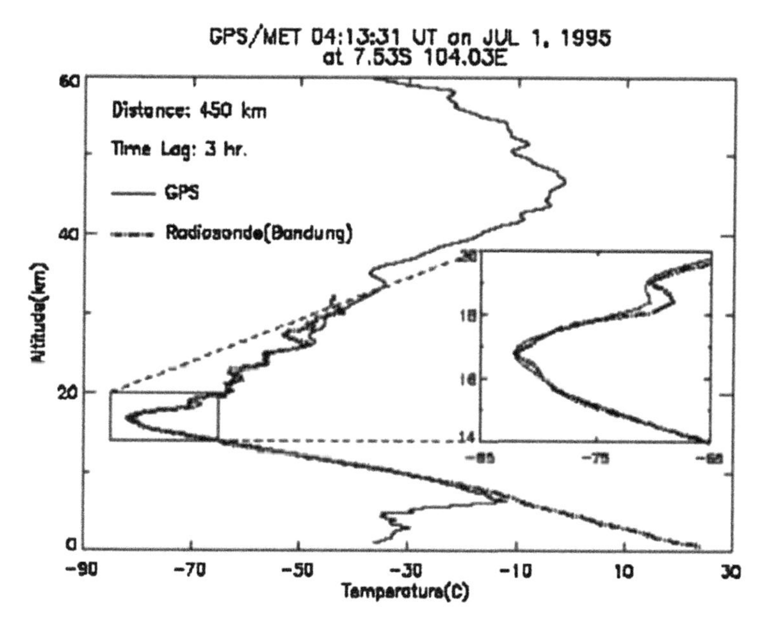

FIGURE 9.32 Comparison of temperature profiles made with radiosonde and GPS-RO system. (Taken from Tsuda and Hocke (2004).)

ones; (5) composition measurements that are relatively insensitive to uncertainties in atmospheric temperature; (6) a very accurate spectroscopic database; and (7) instrumentation that has very accurate and stable calibration, adequate sensitivity without necessarily requiring cooling, and provides good vertical resolution set by size of the antenna. In the MLS system, gravity waves are identified by the vertical and horizontal fluctuations of temperature. Because MLS is continuous in time and space, the horizontal-along-track wavelengths as well as the vertical wave-lengths can be estimated. Wu and Eckermann (2008) evaluated temperature variances obtained from the MLS on the Aura satellite at 12 pressure altitudes between 21 and 51 km. Modeling simulations show that these variances are controlled mostly by gravity waves with vertical wavelengths greater than 5 km and horizontal along-track wavelengths of 100–200 km.

PROBLEMS

1. Plot the path of a light aircraft flying over a surface corrugation of amplitude 200 m and wavelength 2000 m. The mean wind speed is constant at 5 m s^{-1} perpendicular to the corrugations, the Brunt–Väisälä frequency is 0.05 s^{-1}, and the average potential temperature is 280 K. The air craft flies with constant power at 40 m s^{-1} at a mean altitude of 2500 m.

2. A pilot balloon ascends at 3 $\mathrm{m\,s^{-1}}$. Plot the trajectory of a pilot balloon released at the crest of the corrugation as in Problem 1 with the same atmospheric conditions as in Problem 1.
3. A spherical superpressure balloon 2 m in diameter is inflated with helium so that its equilibrium surface density (ESD) if 500 hPa, and the drag coefficient of the SPB is 0.8. The atmosphere is isothermal with a temperature of 270 K and a surface pressure of 1000 hPa. (1) Calculate the lift of the balloon. (2) Calculate the vertical speed of the balloon at its EDS. (3) What is the maximum altitude reached by the SPB? (4) How long will it take for the SPB to reach an approximately constant height?
4. Calculate the power spectral peaks for Station 1 in the CASES-99 file P-192.DAT EURL\Ch9\DATA. Look for disturbances between 0030 to 0230 and 0530 to 0730 UTC. Use any method you wish.
5. A gravity wave with wavenumber $4.2 \times 10^{-4}\ \mathrm{m^{-1}}$ has a phase speed of 15 $\mathrm{m\,s^{-1}}$. Microbarometers are located at the vertexes of an equilateral triangle with side length L. What are the minimum values of sampling rate and side length for resolving this wave?
6. A mechanical-filter microbarometer has a response time constant of 120 s. Calculate the gain factor and the true frequency of a wave when the phase factor is 45°.
7. If the phase speed of a gravity wave is 5 $\mathrm{m\,s^{-1}}$ at what background wind speed do the wind and pressure wave perturbations become 180° out of phase?
8. The pressure, P_c, in a reference chamber is $1.001 P_a$, where P_a is the atmospheric pressure. At initial time $t = 0$, a valve in the chamber is opened resulting in a pressure leak. What is the mass flux of air from the chamber if pressure difference between the chamber and the atmosphere decreases by a factor of 1/e in 100 s?
9. Describe the trajectory of a rising balloon if the horizontal wind perturbations u_1 and v_1 are out of phase by $\pi/2$.
10. Ionospheric OH airglow images often show horizontally propagating gravity waves between the upper mesosphere and lower thermosphere. It is believed these are ducted waves. Explain what would cause ducting at these altitudes.

10

GRAVITY WAVE ANALYSES

10.1 INTRODUCTION

In this chapter, we examine basic techniques for identifying gravity waves and estimating their characteristics. Although the waves may be hydrostatic or non-hydrostatic, linear or nonlinear, in the lower or upper atmosphere, we use analysis techniques based on the linear theory. In the lower atmosphere, gravity waves can be found over a vast spectrum of horizontal wavelengths and amplitudes. This region is especially amenable to observation and analysis; however, the middle and upper atmospheres are less accessible to measurement. Wave analysis requires data over at least several cycles; this is especially true in the lower troposphere where many waves and turbulence, which can be considered "noise," are almost always present. However, in the higher atmosphere gravity waves can be more easily identified because there is less high frequency noise and wave amplitudes are greater then in the troposphere. However, these waves are generally of low frequency and so continuous measurements are problematical. The range of frequencies in the middle atmosphere for hydrostatic waves is $f < \Omega < \Omega_c << N$ where Ω_c is a cutoff frequency (7.81) which is a function of horizontal wavenumber. For wavelengths greater than about 100 km, $\Omega_c \sim N$; however for wavelengths greater than about 1000 km, $\Omega_c \sim N/2$. For $N = 0.02\ \mathrm{s}^{-1}$, 1000 km long waves will have intrinsic periods of about 1 h. Measuring these waves over several cycles is not possible at this time. We see that analyses of waves in the lower and upper atmospheres require different techniques.

10.2 ANALYSES OF TROPOSPHERIC GRAVITY WAVES

Because of nonlinear effects, turbulence, and modulation by mesoscale processes, gravity waves in the troposphere are more *wave-like* than wave. This is especially true of waves in the lower troposphere and planetary boundary layer. Figure 10.1 shows plots of vertical velocity observed with the University of Wyoming King-Air B200 research aircraft during the CASES-99 field campaign. The wave-like structures and patches of what appears to be turbulence seen in the plots often are observed in the stable troposphere. The spatial series shown in Fig. 10.1 extend vertically from about 160 m to about 1000 m above the ground surface, and cover about a 90 min period. Between about 600 m and 800 m we see waves and turbulence, but below 600 m we see mostly waves. The wave field is highly irregular suggesting the existence of many waves. As pointed out by Finnigan (1988), gravity waves in the boundary layer are mostly nonlinear with time-changing amplitudes and frequencies. It is only when we go far from the disturbing effects of the ground surface that gravity waves take on a more ideal, *i.e.*, linear structure.

The wave dynamicist faces a daunting task when analyzing gravity waves and their characteristics. Much time could be spent in describing these difficulties, but these will become evident when one embarks on this process. Our time is better spent in learning the steps in wave analysis rather than discussing frustrations. We shall see that while there are several methods for estimating wave characteristics such as frequency, wavelength, phase speed, direction, and amplitude these various methods might not give identical results. Indeed, some might not apply, and we must remember that in almost all cases our estimations are approximations. Although we shall study waves in the stable boundary layer, many of these methods are applicable throughout the atmosphere. In the following, we shall start with a 6-h sample of data and proceed steep-by-steep in the analysis, developing our tools along the way. The data are taken from the morning hours (UTC) of October 14, 1999 during the CASES-99 field campaign. The data consist of surface pressure measured on the six-sensor array diagramed in Fig. 9.15d. The data extend from

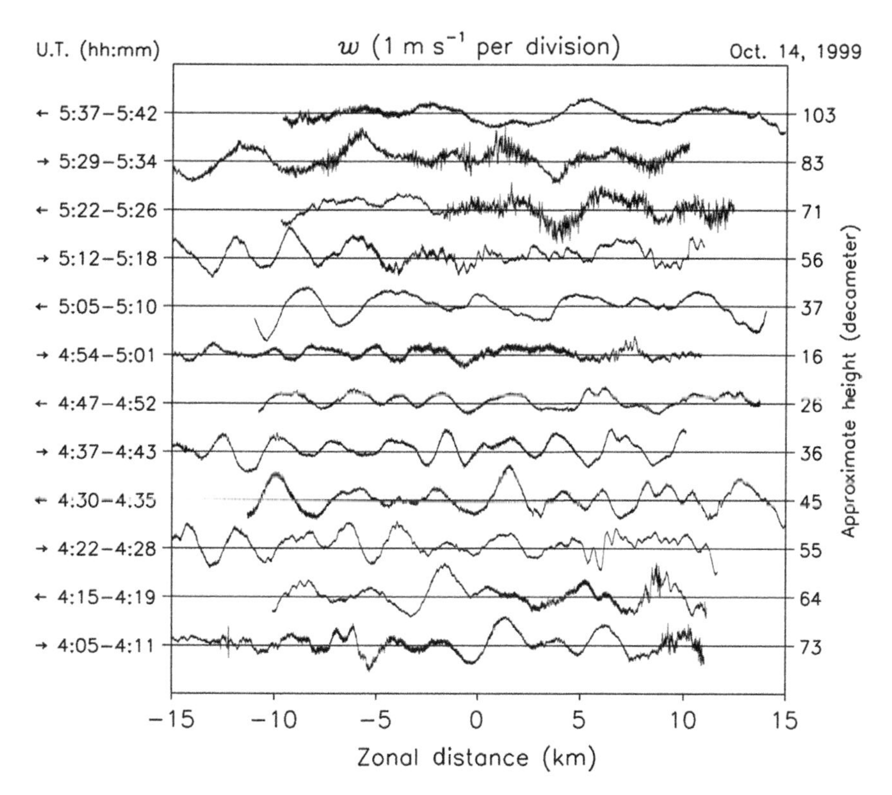

FIGURE 10.1 Vertical velocity recorded by the University of Wyoming King-Air B200 research aircraft on October 14, 1999 during the CASES-99 field campaign. The flights were in the east–west direction at various heights. The arrows on the left indicate the flight direction; the recording times are in universal time.

0300 to about 0900 UTC,[1] and are 10-s block averages of 10 Hz observations. These data are available at the website.

10.2.1 INITIAL STEEPS

The first step is always to plot the data and visually look for wave-like fluctuations; however, the raw data is on the order of about 1000 hPa while the wave-like perturbations are on the order of 10 Pa; thus we first remove the mean or background pressure p_0 leaving the pressure perturbation $p_1(t)$. The next step is to separate the perturbations from the long-term time trends of P_0. This is done by removing the *linear trend* from the pressure time series. The formula for detrending is

[1] Unless otherwise indicated, all times are UTC.

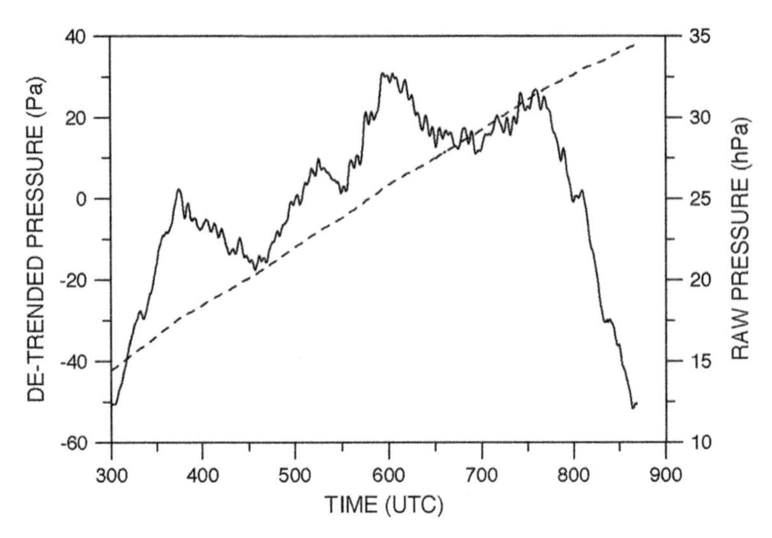

FIGURE 10.2 De-trended pressure perturbation (solid line) and raw pressure for October 14, 1999 at CASES-99.

$$p_\ell(t) = p(t) - \frac{p(t_f) - p(t_i)}{t_f - t_i}(t - t_i), \tag{10.1}$$

where $p_\ell(t)$ is the de-trended pressure, t_i is the beginning time of the data record and t_f is its ending time. The solid line in Fig. 10.2 is the de-trended pressure $p_\ell(t)$, and the dashed line is $P_0(t)$. For p_ℓ we can see at least two wave-like signals a small-amplitude wave with a period of several minutes superimposed and a large-amplitude wave with a period of about 2 h. A time series of several hours would be required to analyze this low frequency wave-like signal. There might be other wave-like signals in p_ℓ, but they are not immediately recognizable. We wish to isolate or separate the high frequency signals. A common way to do this is through spectral analysis.

10.2.2 SPECTRAL ANALYSIS

A time series of perturbations consists of wave-like and turbulence-like disturbances. The waves might differ in frequency and wave length, and might be moving in different directions. Although we may speak of a particular wave with a particular frequency, we must remember that what appears as a wave is in fact a packet of waves. The energy content of this disturbance is spread over a range of frequencies which can be either narrow or wide; however, for ease of discussion we shall use the term "wave." Spectral techniques have often been used to detect gravity waves in data records (see, for example, Caughey and Readings, 1975; De Baas and Driedonks, 1985; Hunt, Kaimal, and Gaynor, 1985; Koch and

Golus, 1988; Guest *et al.*, 2000; Newsom and Banta, 2003; Meillier *et al.*, 2008). Spectral methods generally work well when wave amplitudes are large and persist for many cycles. However, small-amplitude waves and waves of short duration can easily be over looked by a power spectrum. For example, if a wave persists for 15 min, and if the spectrum is calculated over an hour, the wave might not be detected. Sometimes two or three waves can exist each with different but not greatly different frequency. A spectrum calculation may merge these frequencies into a single broad frequency band, which might have relatively little scientific meaning. A short-period wave could sporadically appear for short durations over a few hour's time; however, in a spectrum analysis the wave might appear to have a much longer period. Thus, spectral methods may not always be useful in analyzing gravity waves. The Fourier decomposition of several large disturbances each several minutes long can result in spectral peaks which can be mistaken as due to a single wave. Also, we know nothing about the time evolution of the disturbances. Are they continuous or sporadic? How can we analyze these disturbances if we do not know when they occur? The Fourier transform can also be affected by large changes in disturbance amplitude. Thus, while spectral analysis is a good first view of the data, its usefulness in gravity wave analysis is limited.[2]

Since the advent of the Fast Fourier Transform (FFT), spectral analysis has become a standard and many software packages come with spectral algorithms. Figure 10.3 shows the normalized power spectrum ($F \bullet S_F$) for the de-trended signal in Fig. 10.2. We see two distinct peaks, one with a period of about 4.6 min and a lesser peak at about 12 min. The 4.6 min period wave is easily seen in the de-trended data in Fig. 10.2 however, a 12 min wave is less obvious. Is this real or an artifact of the Fast Fourier Transform? This spectrum extends over a 3 h period; however we have no idea as to whether or not the frequencies of these peaks correspond to continuous wave activity or several large disturbances.

10.2.3 WAVELET ANALYSIS

The *wavelet analysis* can be used to resolve disturbance energy simultaneously in frequency and time, and thus remove many of the ambiguities in spectral analysis. Wavelet transforms, which are the bases of the wavelet analysis, were formally introduced in the early 1980s. Historical backgrounds and introductions to the formalism can be found in, for example, Combes, Grossman, and Tchamitchian (1989), Chui (1992), Foufoula-Georgiou and Kumar (1994), Treviño and Andreas (1996), Torrence and Compo (1998), and Addison (2005). Examples of wavelet analysis used in gravity wave studies are found in Hauf *et al.* (1996), Rees, Staszewski, and Winkler (2001), Zink and Vincent (2001), Fritts *et al.* (2003), and Nappo, Miller, and Hiscox (2008). In the MUA wavelet analysis was used by Sato (1994), Sato and Yamada (1994), Zhang *et al.* (2001), Zink and Vincent

[2] We must note that in the study of turbulence and geophysical problems spectral analysis is extensively used.

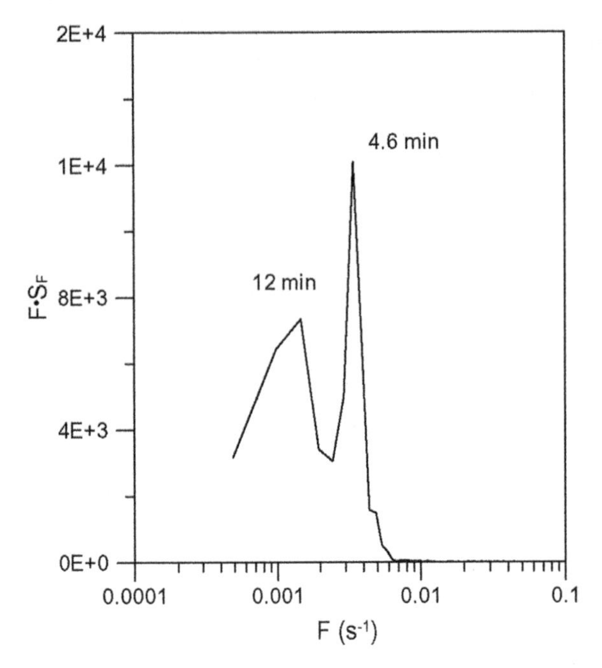

FIGURE 10.3 Normalized power spectrum ($F \bullet S_F$) for the time series shown in Fig. 10.2.

(2001), Boccara *et al.* (2008), and Simkhada *et al.* (2009). It is beyond the scope of this book to give a formal discussion of wavelet analysis. Instead we present a brief description of the technique, and demonstrate its application in our wave analysis. We also note that wavelet analysis capabilities are available in, for example, IDL, FORTRAN, and MATLAB. Thus, except for special circumstances it is unnecessary to write wavelet analysis codes.

The *wavelet coefficients*, $y(t, a)$, are defined by the continuous wavelet transform of a time series $x(t)$ as (see, for example, Meneveau, 1991; Farge, 1992)

$$y(t, a) = \int_{-\infty}^{\infty} g_a(t - t')x(t')\mathrm{d}t', \tag{10.2}$$

where

$$g_a(t) = \frac{1}{\sqrt{a}} g\left(\frac{t}{a}\right). \tag{10.3}$$

Equation (10.2) has the form of a *convolution integral* (see Appendix), *i.e.*,

$$\eta(t) = \int_{-\infty}^{\infty} h(\tau)\zeta(t - \tau)\mathrm{d}\tau, \tag{10.4}$$

where $h(\tau)$ is a weighting function. The dynamic characteristics of a linear system, $\zeta(t)$, can be described by a weighting function which is defined as the output of the system at any time to a unit impulse input applied a time τ before. That is, the value of the output $\eta(t)$ at any particular time is given as a weighted linear sum over the entire history of the input $\zeta(t)$. Thus, (10.4) is in essence a *filtering function* as, for example, a running mean or a band-pass filter. The characteristics of the filter are contained in the weighting function. In (10.4), $h(t)$ is a single parameter linear function of time. However, the weighting function, $g_a(t)$, is a two parameter function, *i.e.*, t and a. It is seen that the wavelet coefficients are functions that are filtered with a variable weighting function. This is the bases of wavelet analysis.

The wavelet weighting function is the *mother wavelet* $g(t/a)$, and a is a *dilation* time scale. The function g_a in (10.3) represents a time window of varying width, a, and the coefficients $y(t, a)$ represent the amplitude of $x(t)$ at dilation a and time t. The Fourier transform of the mother wavelet replaces the time window by a frequency window, *i.e.*,

$$\hat{g}(\omega) = \int_{-\infty}^{\infty} g(t)\mathrm{e}^{-i\omega t}\,\mathrm{d}t. \tag{10.5}$$

The windowing properties are determined by the mother wavelet. We want the window to decay quickly for large time and frequency which is necessary for Fourier transformations. The Fourier-transformed window function will have a central frequency ω_c and an effective width σ. Thus, $\hat{g}_a(\omega) = \hat{g}(a\omega)$ with central frequency ω_c/a and effective width σ/a.

We want a window that will give the best resolution of wave-like signals. The window most frequently used is the complex-valued Morlet wavelet

$$g(t) = \mathrm{e}^{i2\pi t}\,\mathrm{e}^{-t^2/2}, \tag{10.6}$$

which represents a sine wave with Gaussian-weighted amplitude. The family of dilated wavelets (10.3) is then

$$g_a(t) = \frac{1}{\sqrt{a}}\exp\left(i2\pi\frac{t}{a} - \frac{t^2}{2a^2}\right). \tag{10.7}$$

Figure 10.4 plots the real, *i.e.*, symmetric, part of the Morlet wavelet (10.7) in physical space for dilation values of 1, 2, and 3 s. We see that the amplitude of the window, g_a decreases with increasing dilation, a, but the window width increases with increasing a. The Fourier transform of (10.7) is the window function in frequency space, and is given by

$$\hat{g}_a(\omega) = \frac{1}{(2\pi)^2}\exp\left[-\frac{\omega - 2\pi/a}{2/a^2}\right]. \tag{10.8}$$

The center frequency of the window is $\omega_\mathrm{c} = 2\pi/a$ and the window width is $\sigma = 1/a$. The oscillation period of the center frequency is $T_a = 2\pi/\omega_\mathrm{c} = a$. Note that because the real part of g_a is symmetric, its Fourier transform is real. Figure 10.5 shows plots of $\hat{g}_a(\omega)$ for $a = 1$, 2, and 3 s. The functions g_a behave much like a filtering function in a band-pass filter, *i.e.*, admitting only frequencies bounded by the curves.

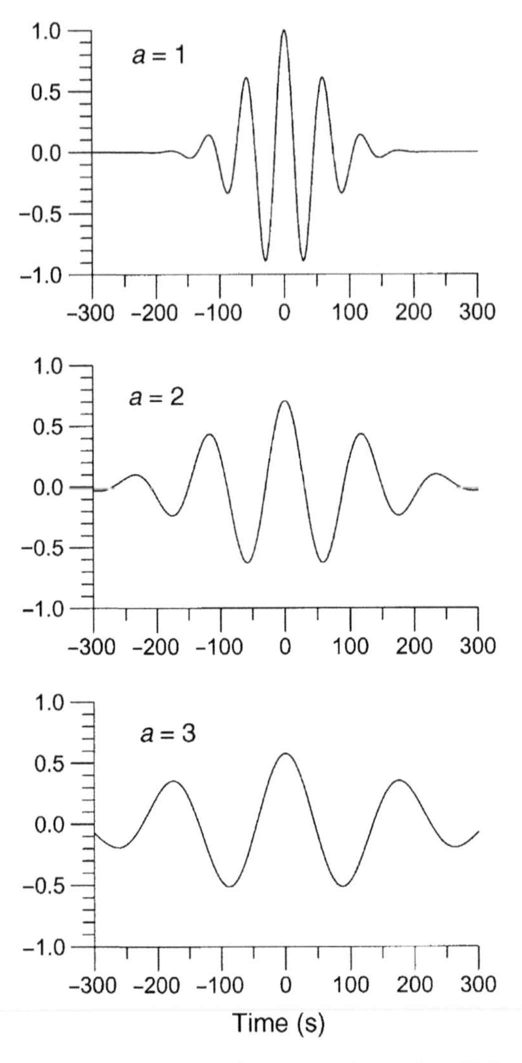

FIGURE 10.4 The Morlet wavelet for various values of time-dilation parameter, a.

The *wavelet energy density* is defined as

$$W(t, a) = \frac{1}{a^2} y(t, a) y^*(t, a). \tag{10.9}$$

A *wavelet analysis diagram* (Hauf *et al.*, 1996) or a *scalogram* (Addison, 2005) is created by contouring W in either time–frequency space $(t, 1/a)$ or time-period space (t, a). Figure 10.6 shows contours of wavelet energy density calculated in p_ℓ. Throughout the period 0430–0730, we see an intermittent disturbance with period ranging from 2 to 6 min. This disturbance is most likely a ducted gravity

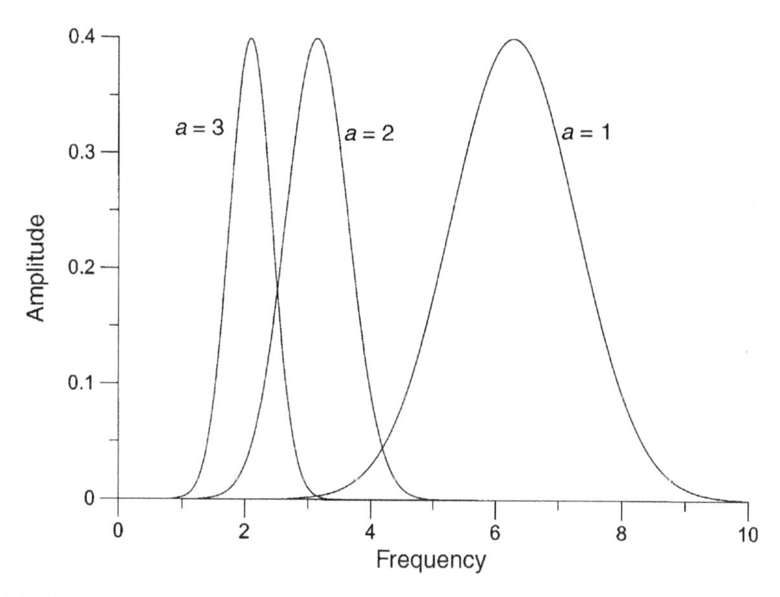

FIGURE 10.5 The Fourier-transformed Morlet wavelet for various values of time-dilation parameter, a.

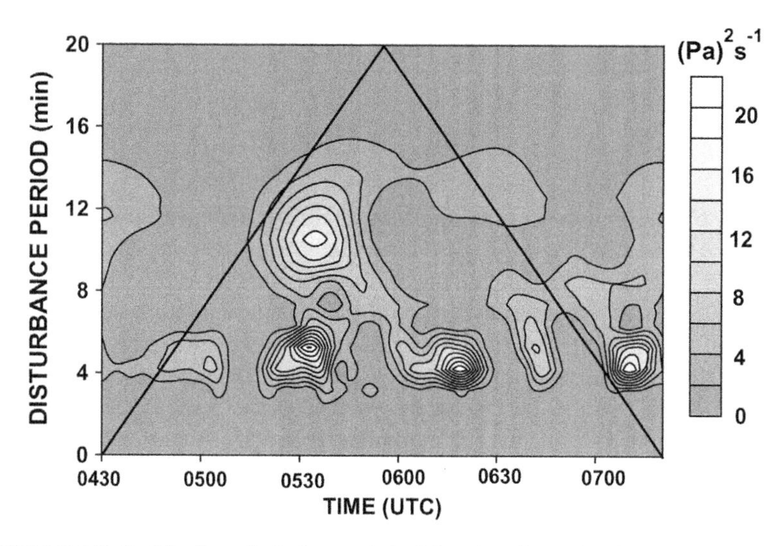

FIGURE 10.6 Wavelet analysis diagram derived from data observed on October 14, 1999 during the CASES-99 field campaign. The dark lines mark the cone of influence.

wave. A strong disturbance with periods centered at about 10 min is also seen. This disturbance is due to the second peak in Fig. 10.3. The solid lines define the *cone of influence* (Torrence and Compo, 1998). Values outside the cone of influence are not reliable because of edge effects at the ends of the time series. Note the evaluation of (10.2) is made using the Fourier Parseval formula,

$$\hat{y} = \hat{g}_a \hat{x}, \tag{10.10}$$

where the $\hat{y}$ indicates the Fourier transformation, *etc.* In (10.9) the inverse Fourier transform of (10.10) is used. The derivation of (10.10) is given in the Appendix.

10.2.4 BAND-PASS FILTERING

Wavelet analysis has identified in our data a fairly persistent disturbance with period ranging from 2 to 6 min. We wish to isolate or filter this signal for further analysis. This is most easily done using a band-pass filter as defined in (10.4). The weighting or more properly the *windowing function* $h(t)$ selects the range of frequencies to admit or pass through the filter. The simplest function is based on the unit top-hat defined by

$$h(t) = \begin{cases} 0 & \text{if } t < T_{\text{low}}, \\ 1 & \text{if } T_{\text{low}} \geqslant t \leqslant T_{\text{high}}, \\ 0 & \text{if } t \geqslant T_{\text{high}}, \end{cases} \tag{10.11}$$

where T is disturbance period, which in our case $T_{\text{low}} = 2$ min and $T_{\text{high}} = 6$ min. The implementation of the filtering is almost trivial. First the time series p_ℓ is Fourier transformed to $\hat{p}_\ell(f)$ using an FFT, where f is frequency. Next, a function $H(f)$ is constructed, *i.e.*,

$$H(f) = \begin{cases} 0 & \text{if } t < f_{\text{low}} = 1/T_{\text{high}}, \\ 1 & \text{if } f \geqslant f_{\text{low}} = 1/T_{\text{high}}, \\ 0 & \text{if } f \leqslant f_{\text{high}} = 1/T_{\text{low}}. \end{cases} \tag{10.12}$$

Note that $H(f)$ is not the Fourier transform $h(t)$. The filtered Fourier series, $\hat{p}_{1,\ell}$ is obtained by multiplying H by $\hat{p}_\ell$. The filtered perturbation $p_{1,(2\text{–}6)}$ is given by the inverse FFT of $\hat{p}_\ell$. The filtered pressures at CASES-99 Station 1 is plotted in Fig. 10.7. In Fig. 10.7a wave activity occurs with varying amplitude throughout the test period. The ringing of the amplitude indicates either wave packets as shown in Fig. 1.18 or intermittent wave activity. Figure 10.7b shows the wave signals at Stations 7 (filled triangle), 8 (block), and 9 (filled circle) respectively from about 0525 to about 0555. Note that wave amplitudes are nearly uniform during this period. Thus, by band-pass filtering the wave signal we can focus more clearly on the wave's motion.

We must point out a potential problem with band-pass filtering. Because we are dealing with a finite number of data points with a desecrate time resolution, there will be a finite number of frequencies in the desecrate Fourier transformation. If we select a frequency range between f_1 and f_2, we must be sure that a representative number of Fourier components will pass through the filter. Let the

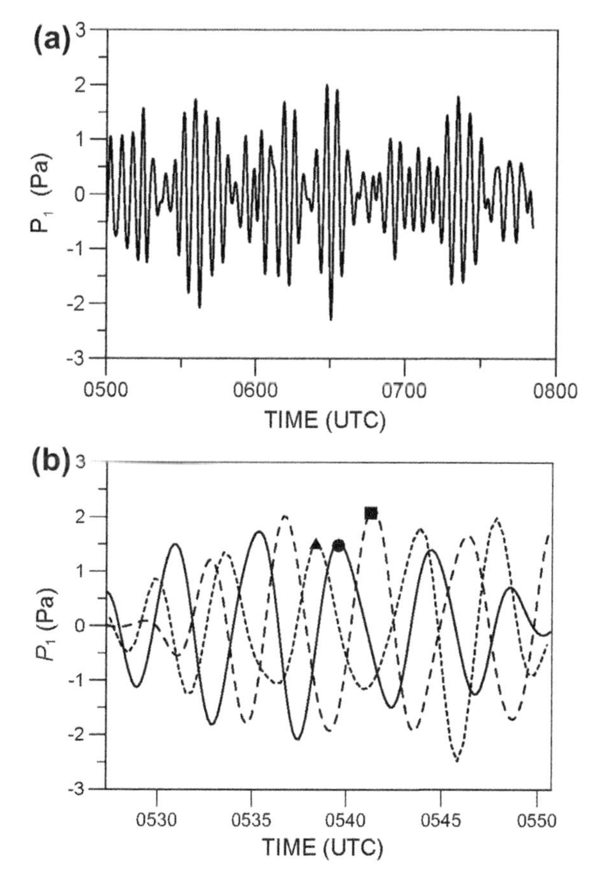

FIGURE 10.7 Band-passed wave signals. (a) Station 1; (b) Stations 7 (filled triangle), 8 (block), and 9 (filled circle) from the period 0525 to 0550.

desired frequency range be $\Delta f = f_1 - f_2 = 1/\tau_1 - 1/\tau_2$ where τ is the wave period. The total range of frequencies is $\Delta F = 1/T$ where T is the time length of the record. The number of waves, n, that pass through the filter is given by

$$n = \frac{\Delta f}{\Delta F} = \frac{\tau_1 - \tau_2}{\tau_1 \tau_2} T. \tag{10.13}$$

As an illustration, if $T = 60$ min, $\tau_1 = 5$ min, and $\tau_2 = 10$ min, then $n = 6$ waves. For the same value of F, if $\tau_1 = 10$ min and $\tau_2 = 15$ min, $n = 2$. It is possible for $n < 1$ which results in peculiar looking plots.

10.2.5 LAG ANALYSIS

If a coherent disturbance moves across a horizontal array of sensors and experiences little change of frequency, wavenumber, and amplitude then the arrival

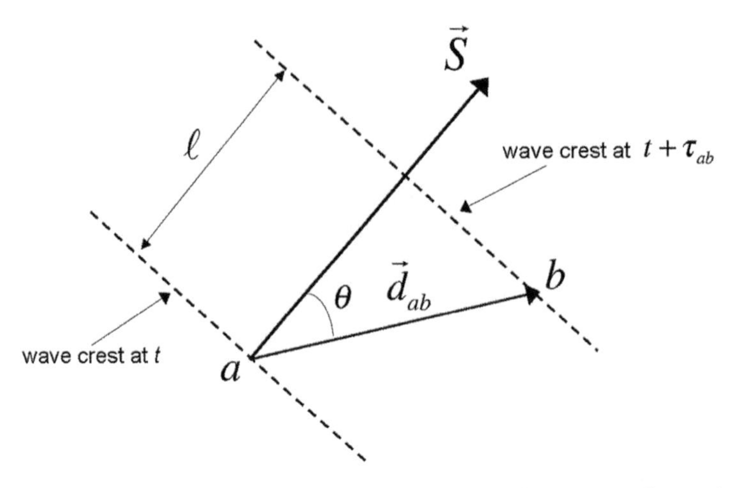

FIGURE 10.8 Wave crest as it passes from station a to station b in time τ_{ab}. The wavelength of the wave is ℓ, and $\vec{S}$ is the slowness vector.

times of the disturbance at each station can be used to calculate the characteristics of the disturbance. Consider such a disturbance, q, observed at the ground surface or at the same altitude so that we do not consider vertical variations of wave phase between sensors. Let us imagine the disturbance is a linear wave[3] with constant amplitude and horizontal wave vector, $\vec{\kappa}_{\mathrm{h}}$, observed at stations a and b whose separation is described by the vector $\vec{d}_{ab}$ as illustrated in Fig. 10.8. Let the horizontal distance between the wave fronts be ℓ, and let the crests of the wave be observed at station a at time t and at station b at time $t + \tau_{ab}$ so that

$$q_a(kx_a + ly_a - \omega t) = q_b\left[kx_b + ly_b - \omega(t + \tau_{ab})\right], \tag{10.14}$$

where k and l are the wavenumbers in the x- and y-directions respectively. Because the wave does not change shape in moving from a to b, we can write

$$kx_a + ly_a - \omega t = kx_b + ly_b - \omega(t + \tau_{ab}), \tag{10.15}$$

which reduces to

$$k(x_b - x_a) + l(y_b - y_a) = \omega\tau_{ba}. \tag{10.16}$$

The left-hand-side of (10.16) is recognized as $\vec{\kappa}_{\mathrm{h}} \bullet \vec{d}_{ab}$, where $\vec{\kappa}_{\mathrm{h}}$ is the horizontal component of the wave vector, and $\vec{d}_{ab}$ is the distance vector between stations b and a. We can write (10.16) as

$$\frac{\vec{\kappa}_{\mathrm{h}}}{\omega} \bullet \vec{d}_{ab} = \tau_{ab}. \tag{10.17}$$

[3] This and the following analysis techniques hold for any coherent disturbance such as a solitary wave, a gust front, a density current, *etc.*

From the definitions of phase velocity we see that

$$\frac{\vec{\kappa}_h}{\omega} = \left(\frac{1}{c_x}\right)\hat{x} + \left(\frac{1}{c_y}\right)\hat{y}. \tag{10.18}$$

We introduce the slowness vector, $\vec{S}$ as

$$\vec{S} = \left(\frac{1}{c_x}\right)\hat{x} + \left(\frac{1}{c_y}\right)\hat{y}, \tag{10.19}$$

with components S_x and S_y. The slowness vector is always parallel to the wave vector, and has magnitude $1/c$. Equation (10.17) now takes the form

$$\vec{S} \bullet \vec{d}_{ba} = \frac{|d_{ba}|}{c}\cos\theta, \tag{10.20}$$

and from Fig. 9.8 we can write:

$$\vec{S} \bullet \vec{d}_{ab} = \frac{\ell}{c} = \tau_{ab}. \tag{10.21}$$

If there is another station, say e, with time lag τ_{ae} between stations e and a, then we have the simultaneous equations

$$S_x(x_b - x_a) + S_y(y_b - y_a) = \tau_{ab}, \tag{10.22}$$

$$S_x(x_e - x_a) + S_y(y_e - y_a) = \tau_{ae}, \tag{10.23}$$

which can be used to solve for S_x and S_y. Once known, the slowness components can be used to estimate the wave speed, c, and the azimuth (degrees from north) of phase propagation ϕ using

$$c = \frac{1}{(S_x^2 + S_y^2)^{1/2}}, \tag{10.24}$$

$$\phi = \tan^{-1}\left(\frac{S_x}{S_y}\right). \tag{10.25}$$

If the wave frequency is known, then the horizontal wavenumber can also be estimated since $k = \omega/c$.

The use of lags between pairs of stations to estimate wave characteristics requires plotting the time series wave perturbations at each station, and noting the times of passage of a particular crest or trough at each station.

10.2.6 CROSS-CORRELATION LAG ANALYSIS

Lag analysis can be a time-consuming process and prone to errors. An alternative approach to estimating time lags was used by Rees and Mobbs (1988). Instead of using graphical data to estimate lag times, they calculated the cross-correlations as functions of lag time using wind speed data between pairs of wind stations in a three-station array. The cross-correlation coefficient is defined as

$$r_{ij}(\tau_{ij}) = \frac{1}{t_2 - t_1}\frac{1}{\sigma_i \sigma_j}\int_{t_1}^{t_2-\tau_{ij}} q_i(t)q_j(t+\tau_{ij})\mathrm{d}t, \tag{10.26}$$

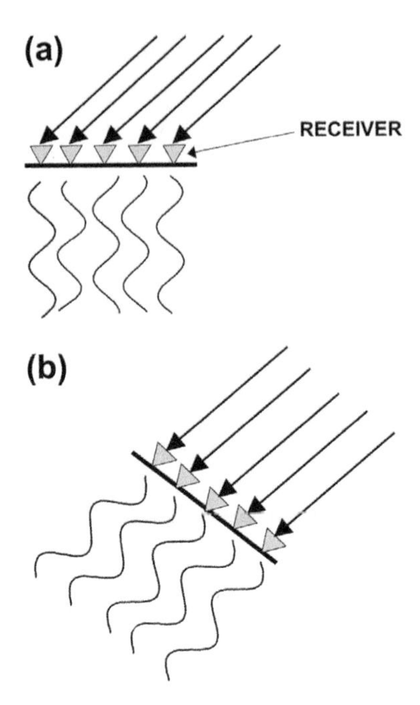

FIGURE 10.9 Sound waves receives by an array of microphones. In (a) each microphone receives a signal out of phase with its neighbor. In (b) all microphones are in phase and the signal amplitude is maximum.

where σ_i and σ_j are the standard deviations of q_i and q_j respectively. Rees and Mobbs (1988) assumed that the time lag between stations i and j is that value of τ_{ij} which gives maximum cross-correlation. With three stations, three maximum lags can be estimated, and these can be used to obtain estimates of S_x and S_y. However, they caution that this method is sensitive to changes in wind speed, wave frequency, and wave amplitudes between stations. To check the accuracy of these lag estimates, Rees and Mobbs (1988) required that the sum of the two shorter lags equal the longest lag, *i.e.*,

$$|\tau_{ij} + \tau_{jk} - \tau_{ik}| < \epsilon, \tag{10.27}$$

where ϵ is a convergence factor which they took to be 15 s. This consistency check requires that the wave speed be constant along the array.

An interesting combination of wavelet analysis and cross-correlation lag analysis was used by Viana, Terradellas, and Yagüe (2010) to calculate the group velocities of gravity waves propagating at the top of a drainage current. They used surface pressures from three microbarometers located at the vertices of a near-isosceles triangle with average side length of about 214 m. These data were collected during the Stable Atmospheric Boundary Layer Experiment in Spain 2006 (SABLES2006) field campaign (Yagüe *et al.*, 2007). The transit time lags of the wave packets between the different pairs of sensors was found from the maximization of the convolution of the wavelet energy density functions. Wavelet cross correlation gives much better estimates than lag analysis because it removes the uncertainty in estimating the arrival times of wave crests at the microbarometers.

10.2.7 BEAMSTEERING

Rees *et al.* (2000) defined *beamsteering*[4] as "any process that uses data from a spatial array of sensors to determine the direction from which a disturbance is propagating." The method has been used by Young and Hoyle (1975), Einaudi, Bedard, and Finnigan (1989), Hauf *et al.* (1996), Denholm-Price and Rees (1999), and Rees *et al.* (2000) to determine horizontal wave speeds and directions from surface pressures observed over a sampling array; Evers (2008) used beamsteering to analyze infrasound. Discussions of beamsteering techniques are found, for example, in Young and Hoyle (1975), Gossard and Hooke (1975), Asten and Henstridge (1984), and Denholm-Price and Rees (1999). To grasp the technique, imagine a downward propagating sound wave striking a horizontal array of microphones as illustrated in Fig. 10.9a. Each microphone receives a wave crest at a different time so that the sound waves recorded by each microphone will be out of phase. These phase differences will cause an interference so that the total amplitude across the array is diminished. If the array is swung about until the sound amplitude is maximum, then the sound waves received at each microphone will be in phase, as illustrated in Fig. 10.9b. The array of microphones is now facing the direction from which the sound is coming.[5] However, the physical movements of the array can be virtually produced by shifting the arrival times of each signal until they are all in phase. In electronics, this is often called a *phased array*. This phase shifting of signals is the principle of beamsteering. The phase differences or time lags of the wave at the array sampling stations could then be used to estimate phase speeds. In applications, the "beam" of the array can be steered either through time, *i.e.*, slowness domain where we seek maximum cross-correlation of signals between all pairs of stations, or through the frequency, *i.e.*, wavenumber domain where we seek maximum cross-power spectral density between all pairs of stations.

10.2.7.1 Beamsteering in the Slowness Domain

Beamsteering in the slowness domain makes use of the cross-correlations of signals between all pairs of stations. The phase lag, τ, between station pairs are varied until maximum cross-correlation over all station pairs is achieved. Consider (10.21) in the form

$$\vec{S} \bullet \vec{d}_{ij} = \tau_{ij}(\vec{S}). \qquad (10.28)$$

[4] In electrical engineering and telecommunications beamsteering is referred to as waveforming.

[5] Before the operational wartime use of radar, such arrays of microphones were used to locate aircraft.

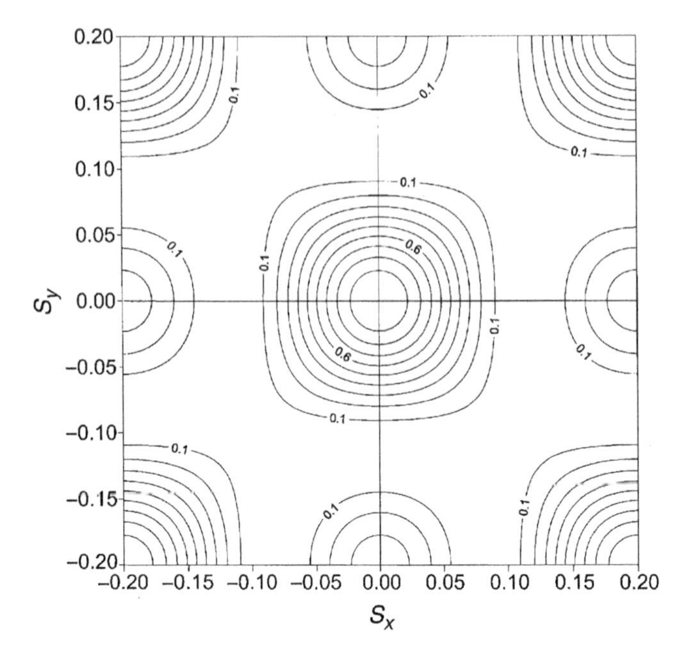

FIGURE 10.10 Array response for the ideal five-sensor array shown in Fig. 8.12.

The phase lag is now a function of the slowness, and using (10.26) we can write

$$r_{ij}(\tau_{ij}) = \frac{1}{t_2 - t_1}\frac{1}{\sigma_i \sigma_j}\int_{t_1}^{t_2-\tau_{ij}} q_i(t)q_j(t + \vec{S} \bullet \vec{d}_{ij})\mathrm{d}t, \tag{10.29}$$

i.e., the cross-correlation is also a function of the slowness vector. Imagine now a *slowness plane* with horizontal axis S_x and vertical axis S_y. Each point in this plane defines a slowness vector, and the scalar product of this vector with $\vec{d}_{ij}$ defines a lag, τ_{ij}, which can be used in (10.29) to calculate the cross-correlations between station pairs. Let these cross-correlations be defined as

$$R_{ij}(\vec{S}) = r_{ij}\left[\tau_{ij}(\vec{S})\right]. \tag{10.30}$$

If there are M stations, then at each point in the slowness plane the average cross-correlation among non-repeated pairs of stations in the array is

$$R(S_x, S_y) = \frac{2}{M(M-1)}\sum_{i=1}^{M-1}\sum_{j=i+1}^{M} R_{ij}(\vec{S}). \tag{10.31}$$

In practice, (10.31) is evaluated at each point in the slowness plane, and the location of the maximum value of R is taken to be the true slowness vector. Estimates of phase speed and direction are obtained using (10.24) and (10.25) respectively. Note that beamsteering in the slowness plane does not require information regarding the angular frequency of the disturbance being tracked. By implication, we know

ω *a priori*. This frequency is determined either directly from observations or analytically from wavelet analysis. Because we know ω and the phase speed, the horizontal wavenumber of the disturbance is $k = c/\omega$.

10.2.7.2 Beamsteering in the Frequency Domain

When beamsteering in the slowness domain we know the wave frequency, and we search for the phase speed that leads to maximum cross-correlation. In the frequency domain, beamsteering searches for the horizontal wave vector and wave frequency that lead to maximum *cross-power spectral density* between station pairs. The cross-power spectral density function between stations i and j, $G_{ij}(\omega)$, is obtained by taking the Fourier transform of the cross-correlation function (10.26). $G_{ij}(\omega)$ consists of a real part, $C(\omega)$, called the *co-spectrum* and an imaginary part $Q(\omega)$ called the *quadrature spectrum, i.e.*, $G_{ij} = C + iQ$. The phase relation between the frequencies being analyzed is

$$\tan(\omega\tau) = \frac{Q(\omega)}{C(\omega)}, \tag{10.32}$$

where τ is the time lag between the signals. The cross-spectral magnitude is

$$|G_{ij}(\omega)| = \sqrt{C^2(\omega) + Q^2(\omega)}. \tag{10.33}$$

Using (10.32) and (10.33), we can write

$$C(\omega) = |G_{ij}|\cos(\omega\tau), \tag{10.34}$$

$$Q(\omega) = |G_{ij}|\sin(\omega\tau). \tag{10.35}$$

We can replace τ in (10.34) and (10.35) using (10.17) to get

$$G_{ij}(\omega, \vec{\kappa}) = C + iQ = |G_{ij}(\omega)|e^{i\vec{\kappa}\bullet\vec{d}_{ij}}. \tag{10.36}$$

Following Gossard and Hooke (1975) we define an *array power spectrum* as the sum of (10.36) over all the combinations of pairs of sensors in an array of M sensors, *i.e.*,

$$\widehat{G}(\omega, \vec{\kappa}) = \sum_{i=1}^{M}\sum_{j=1}^{M} G_{ij}(\omega)e^{i\vec{\kappa}\bullet\vec{d}_{ij}}. \tag{10.37}$$

Equation (10.37) is the wavenumber-domain counterpart to beamsteering in the slowness domain. Now, one beamsteers in the wavenumber plane over various frequencies to search for that particular pair, ω and $\vec{d}_{ij}$, that gives the maximum value of $\widehat{G}$. Because we now know ω and $\vec{\kappa}_H$, we can calculate the slowness components using $S_x = k/\omega$ and $S_y = l/\omega$. Then using (10.24) we can calculate the phase speed.

10.2.7.3 Array Response and Examples

Capon (1969) pointed out that a single monochromatic plane wave of frequency ω_0 and wavenumber $\vec{\kappa}_0$ differs only in phase at each station, and therefore has a cross-power spectral density of the form

$$G_{ij}(\omega) = \delta(\omega - \omega_0)\mathrm{e}^{-i\kappa_0 \bullet \vec{d}_{ij}}. \tag{10.38}$$

Using (10.38) in (10.37) gives

$$\widehat{G}_{ij}(\omega, \vec{\kappa}) = \delta(\omega - \omega_0)\left|\sum_{j=1}^{M} \mathrm{e}^{i(\vec{\kappa}-\vec{\kappa}_0)\bullet \vec{d}_j}\right|^2. \tag{10.39}$$

The function

$$H(\omega, \vec{S}) = \left|\frac{1}{M}\sum_{j=1}^{M} \mathrm{e}^{-i\vec{\kappa}\bullet \vec{d}_j}\right|^2 = \left|\frac{1}{M}\sum_{j=1}^{M} \mathrm{e}^{-\omega \vec{S}\bullet \vec{d}_j}\right|^2, \tag{10.40}$$

is called the *array response* function (also called the *array transfer* function). A result similar to (10.40) is given by Evers (2008). The array response function is an important parameter when considering array design. H has a peak value of unity at the origin in either the wavenumber plane, $(k, \ell) = (0, 0)$, or the slowness plane $(S_x, S_y) = (0, 0)$. The response function is symmetric with respect to reflection through the origin, and it reveals side lobes which are caused by aliasing of wave lengths which are too small to be resolved by the sampling array. Fig. 10.10 shows a plot of the response function for the ideal array shown in Fig. 8.12, and Fig. 10.11 shows a plot of the response for the six-station array shown in Fig. 8.13d. In both cases, $\omega = 0.03\ \mathrm{s}^{-1}$. Side lobes are present, and these indicate the wavenumber or phase speed limits to the beamsteering. The response function plots the sensitivity of the array to a stationary wave. If the wave is moving across the array, then the peak of the response function is moved in the direction from which the wave is coming. This is illustrated in Fig. 10.12 which shows the contours of $R(S_x, S_y)$ resulting from a wave with wavelength of 10 km coming from 45° with a speed of 20 ms^{-1} across the array shown in Fig. 8.13d. Fig. 10.13 shows the $R(S_x, S_y)$ contours for a wave with a 1 km wavelength, coming from 45° with a speed of 5 ms^{-1} across the same array as in Fig. 8.13d. In this case, the side lobes are quite pronounced. Note that the ranges of slowness axes in Fig. 10.13 have been doubled relative to those in Fig. 10.12 in order to "see" the maximum cross-correlation. This illustrates one of the fundamental difficulties in beamsteering. In any real application, we will not know *a priori* the wave characteristics, and so we will not be able to adjust the beamsteering parameters in order to achieve maximum sensitivity. In general, one can expect a trial-and-error approach to the problem. Further complications arise when the waves are nonlinear or changing in frequency or amplitude as they move across the sampling array. In these not unusual circumstances, one must be prepared to accept a degree of uncertainty in the estimated wave characteristics.

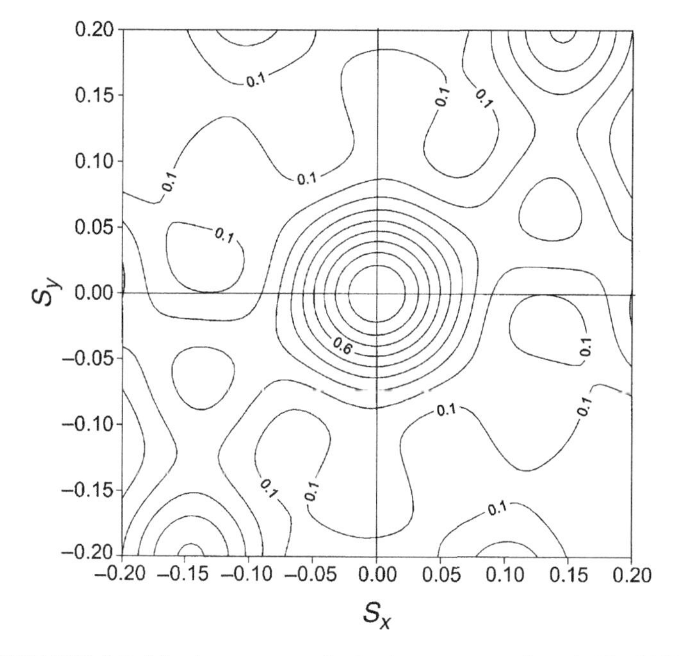

FIGURE 10.11 Array response for the six-sensor array shown in Fig. 8.13d.

10.2.8 IMPEDANCE

The polarization equations (2.22)–(2.25) contain the correlations between wave perturbation quantities. Gossard and Munk (1954) showed how these correlations can be used to determine wave characteristics. In particular, the correlation between pressure and wind speed perturbations can be used since these are usually maximum near the ground surface and can be measured with high precision. This *pressure–wind correlation* is demonstrated in Figs. 10.14 and 10.15. Fig. 10.14 shows band-passed filtered data (2–6 min) recorded on October 14, 1999 during the CASES-99 field campaign. The correlations between pressure, wind speed, and wind direction perturbations are clearly seen. Fig. 10.15 taken from Koch and Golus (1988) shows the absolute surface pressure, the pressure perturbation, and the perturbation velocity associated with a mesoscale gravity wave event that occurred over the north-central United States on July 11–12, 1981. The wave episode lasted about 8 h. The wave period had a mean value of 2.5 h, the wavelength was about 160 km, and the phase speed was about 19 ms^{-1}. The coefficient of correlation between p_1 and u_1 in Fig. 10.15 is about 0.95. This is a good illustration of the utility and applicability of the linear theory. Gossard and Sweezy (1974) showed that the pressure–wind correlation is a direct measure of the wave signal-to-noise ratio. A strong pressure–wind correlation has been used as supporting evidence for gravity waves by, for example, Bosart and Cussen (1973), Bosart and Sanders (1986), and Koch and Golus (1988).

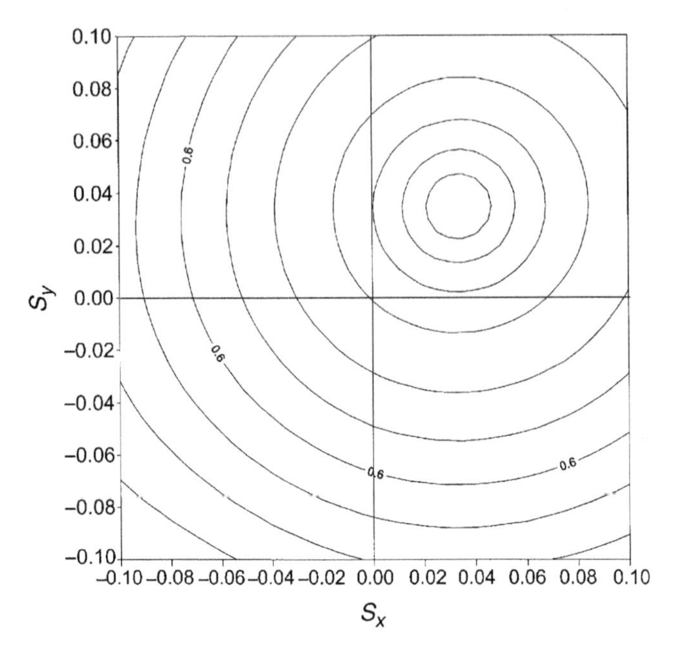

FIGURE 10.12 Contours of cross-correlation in the slowness plane for a wave with wavelength of 10 km moving at 20 ms^{-1} from 45° across the array shown in Fig. 8.13d.

Gossard and Munk (1954) assumed constant wind speed and Brunt–Väisälä frequency and used (2.22) to get

$$u_1 = \frac{p_1}{\rho_0(c - u_0)}. \tag{10.41}$$

They used (10.41) to define an impedance relationship

$$c_I = c - u_0 = \frac{p_1}{\rho_0 u_1}, \tag{10.42}$$

which could be used to estimate wave characteristics using standard wind and pressure observations if the background wind is assumed constant. In (10.42), c_I is the intrinsic phase speed, and we note that because u_0 is assumed constant, c_I could be evaluated at any elevation. Before we can use (10.42) we must know the direction of wave propagation, and this we generally do not know *a priori*. Figure 10.16 illustrates how the wave perturbations to the background horizontal wind can be used to determine the direction of the wave vector. Horizontal wind vectors over a wave cycle are calculated and plotted, and the line connecting the extreme wind vectors lies in the direction of the wave vector. The midpoint of this line marks the magnitude and direction of the mean-wind vector, and the magnitude of u_1 is equal to one-half of this line. The direction of the wave along this line is determined by the location of maximum of p_1 as shown in Fig. 10.16. From the

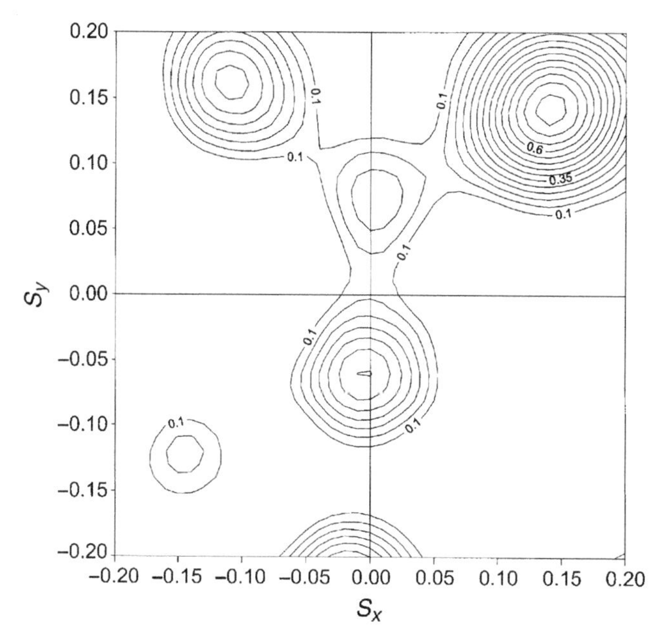

FIGURE 10.13 Contours of cross-correlation in the slowness plane for a wave with wavelength of 1 km moving at 5 ms^{-1} from 45° across the array shown in Fig. 8.13d.

definition of phase velocity (1.19) we see that

$$c_I = c - u_0 = \frac{\omega}{k} = \frac{\lambda_x}{\tau}, \tag{10.43}$$

where λ_x is the wavelength of the wave and τ is the wave period. Equation (10.43) can be used to estimate λ_x.

Although the pressure–wind correlation technique is straightforward to apply it has several shortcomings. Turbulence in the surface layer can obscure wave perturbations, and if the background wind speeds are too low there will be uncertainty in the mean-wind vectors. It has been assumed that linear wave theory is applicable near the ground surface; however, as discussed by Finnigan (1988) gravity waves in the planetary boundary layer are seldom linear. We have also assumed that the background wind and stratification are constant, but this is also not often the case. Indeed, strong wind shears and vertically-changing temperature gradients are often seen near the ground surface. These deviations from ideal flow conditions can result in weak pressure–wind correlations.

10.2.9 ANALYSIS RESULTS

The results of a gravity wave analysis using the methods described above is given as an example. The data were taken on October 14, 1999 (UTC) during the CASES-99 field campaign. Figure 10.17 shows the wavelet energy contours from

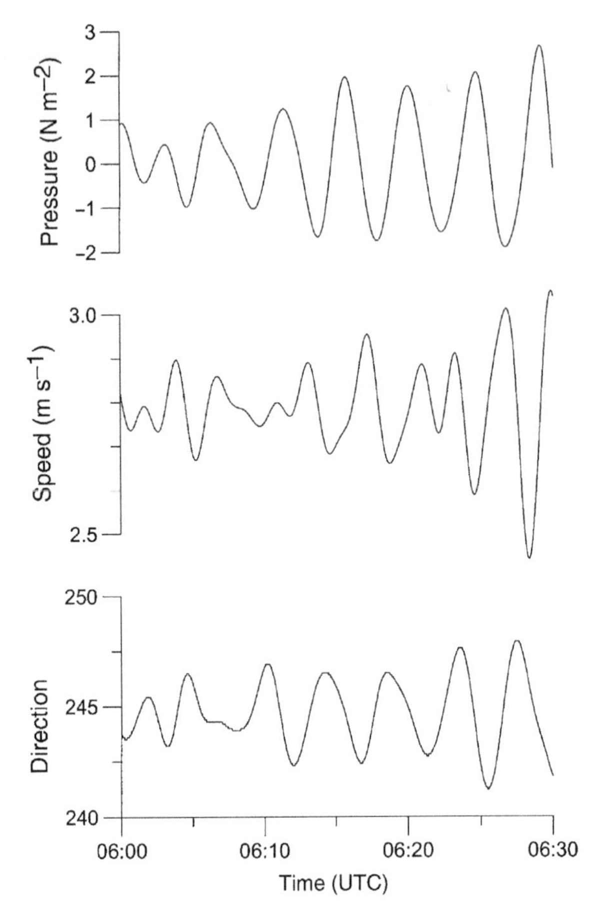

FIGURE 10.14 Band-passed (2–6 min window) pressure, wind speed and wind direction observed on October 14, 1999 during CASES-99 field campaign.

the period 0600 to about 0900. We see three distinct wave signatures along with their estimated speeds and horizontal propagation directions. The average wave periods for the slow, medium, and fast wave speeds are 4, 9, and 22 min respectively. Using the estimated wave speeds, the corresponding horizontal wavelengths are about 2.2, 10.3, and 26.4 km respectively. We see, as expected, that long waves move faster than the short waves. We shall analyze the 4-min period wave.

Lag analysis: Figure 10.7 plots the 2–6 min filtered pressure from 0520 to 0540 for Stations 7–9. The dots mark the crest of the wave as it moved across the pressure array. These times are: 05:30:53 for Station 7, 05:32:47 for Station 8, and 05:30:58 for Station 9. The lag analysis gives $c = 9\ \text{ms}^{-1}$ and $\theta = 324°$, here θ is the azimuth of wave propagation. A similar lag analysis applied to Stations 1, 4, and 5 gives $c = 8.8\ \text{ms}^{-1}$ and $\theta = 325°$.

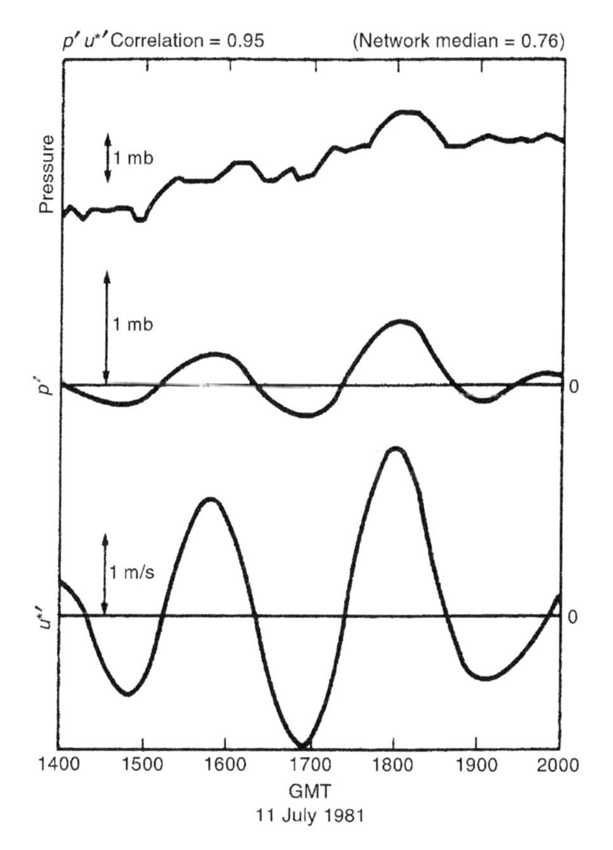

FIGURE 10.15 Time series of pressure (top), perturbation pressure (middle), and perturbation wind speed in the direction of wave propagation. (Taken from Koch and Golus (1988).)

Cross-correlation: The cross-correlation technique was used separately on the inner and outer arrays. The time period used in each case was 0515 to 0530. For the inner array, $c = 6$ ms^{-1} and $\theta = 232°$, and for the outer array $c = 11$ ms^{-1} and $\theta = 247°$. While the phase speeds are relatively consistent with those derived using lag analysis, the propagation directions differ by as much as 90°.

Now consider a different night during CASES-99, *i.e.*, October 18, 1999. Figure 10.18 shows the pressure perturbations observed at the outer stations from 01:00 to 02:00. These data have been band-passed filtered, and represent disturbances with periods between 5 and 25 min. The crosses mark the points which were used to identify the passing of the event at the three stations. The event was observed at Station 7 at 01:27:24, Station 8 at 01:32.36, and Station 9 at 01:34.18. Using lag analysis, it is estimated that $c = 3.6$ ms^{-1} and $\theta = 231°$. Using cross-correlation, we estimate $c = 3.6$ ms^{-1} and $\theta = 236°$. For this night, lag analysis and cross-correlation give essentially the same results. The difference between the results for October 14 and October 18 is attributed to the difference in phase speeds.

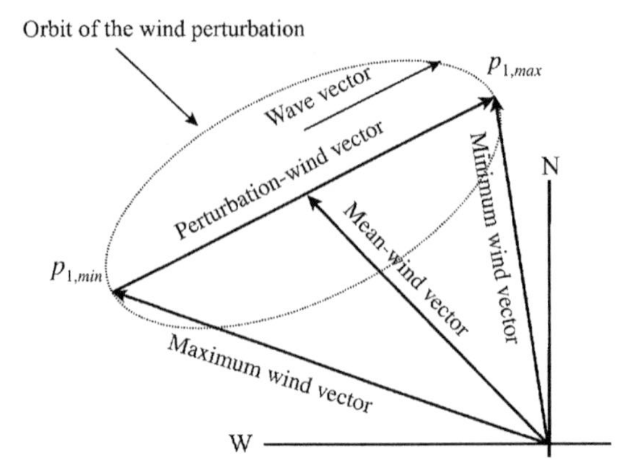

FIGURE 10.16 Mean-wind vector and wave-perturbation-velocity vector. Total velocity vectors at the extreme points are shown by the dashed vectors. The phase velocity is in the direction of maximum pressure perturbation.

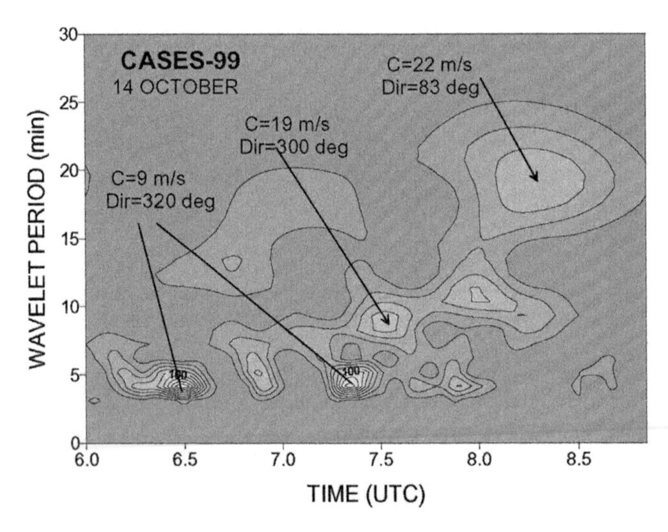

FIGURE 10.17 Wavelet energy density contours for October 14, 1999 at Station 1. Three wave regimes are identified along with their phase speeds and directions.

On October 14 $c \sim 10$ ms^{-1} while on October 18 $c \sim 4$ ms^{-1}. The relatively large sampling period of 10 s, also complicates the cross-correlation analysis. For example, on October 14 the wave crests move about 100 m between each sampling, while on October 18 the crests mover about 36 m between each sampling. Thus, at slow phase speeds the wave horizontal structure is better resolved than for fast wave speeds.

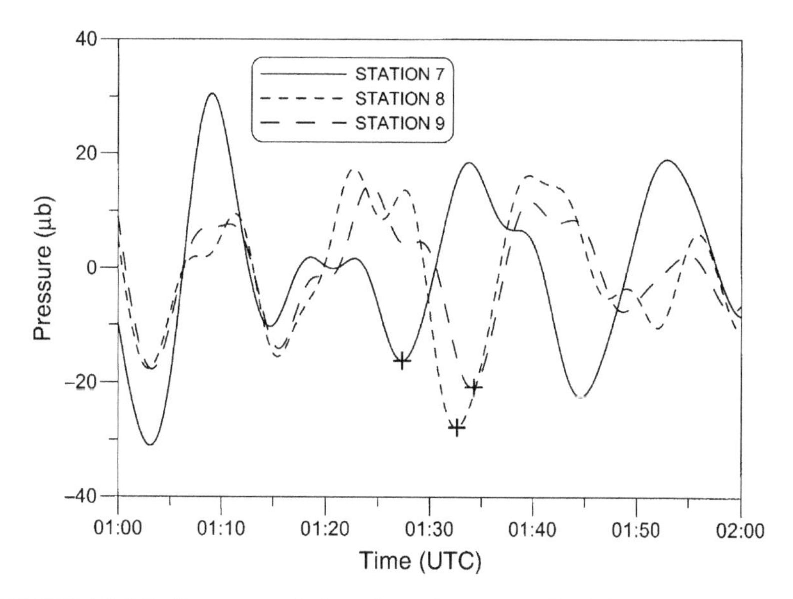

FIGURE 10.18 Pressure perturbations observed on October 18, 1999, during CASES-99 field study. The crosses mark the times used to calculate wave speed and directions using lag analysis.

Beamsteering: Fig. 10.19 shows the results of beamsteering in the slowness plane using all the stations of the surface array. Note that beamsteering gives the direction the disturbance is coming from, which is 144°. The beamsteering results are $c = 9.8$ ms^{-1} and $\theta = 324°$.

Impedance: Fig. 10.20 shows a hodograph of wave-perturbation velocity made using sonic anemometer data at 1.5 m AGL near Station 1. The time period extends from 0615 to 0640, and every fourth point of the time series is plotted. These data were given a linear fit which is represented by the solid line. It is assumed that this line is parallel to the horizontal wave vector. Using the end points of this line, it was determined that the $u_1 = 0.12$ ms^{-1} and $\theta = 318°$. The background wind, u_0, in the direction of θ was 1.2 ms^{-1} and $p_1 = 2$ Pa. Using (10.42), $c = 17$ ms^{-1}. The average of all the estimates are $c = 11.1$ ms^{-1} and $\theta = 308°$.

10.3 GRAVITY WAVE ANALYSES IN THE MUA

Continuous measurements over an extended horizontal domain in the MUA are not yet feasible. This limits the options of techniques for wave analyses. Two techniques dominate the analyses, wavelet analysis, and hodographs.

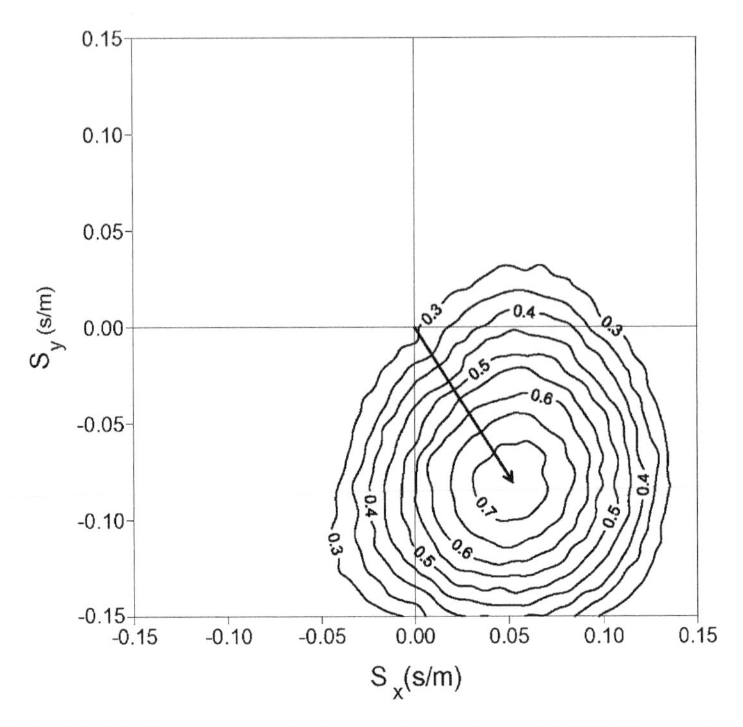

FIGURE 10.19 Beamsteering in the slowness plane for October 14, 1999. The wave is traveling from 144° toward 324° at 9.8 ms^{-1}.

10.3.1 WAVELET ANALYSIS

Wavelets are used for several kinds of analyses. Zhang *et al.* (2001) used wavelet transform to analyze vertical motions in a numerical simulation of a large-amplitude mesoscale gravity wave event. They also give an excellent review of the wavelet technique. Zink and Vincent (2001) used wavelet analysis to detect gravity wave packets in high-resolution radiosonde soundings of horizontal wind and temperature. They applied the wavelet analysis to stratospheric gravity waves over Macquarie Island. They bring up an interesting point. While the Morlet wavelet is very good at recognizing wave fluctuations, it is not orthogonal, and the wavelet coefficients are not independent of time. Thus, the total energy of a superposition of different wavelets is not necessarily equal to the sum of the energies of each wavelet. Boccara *et al.* (2008) used wavelet analysis to identify wave packets from observations on constant-level ballon flights. In their balloon observations, gravity waves often appear as individual wave packets. Wavelet techniques were used because it enables the analyses of a non-stationary wave field.

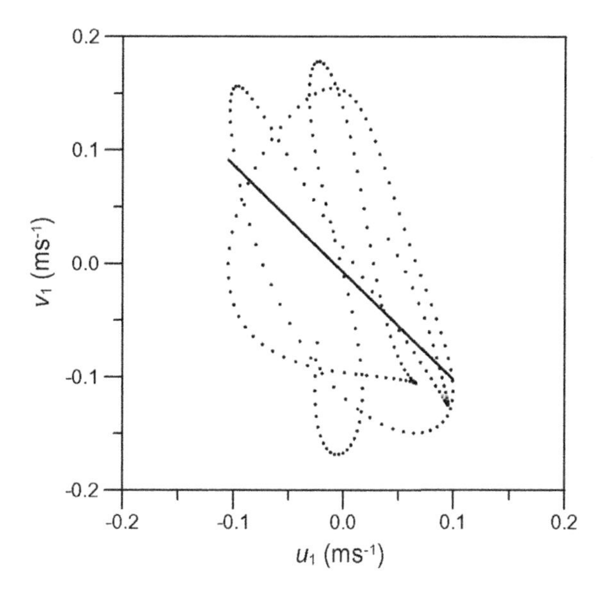

FIGURE 10.20 Hodograph made at 1.5 m using 0.1 Hz data from 0615 to 0640 on October 14, 1999. Every fourth point of the hodograph is plotted. The straight line is a polynomial fit to the hodograph.

10.3.2 HODOGRAPH ANALYSIS

From (9.36), the horizontal wave perturbation of inertio-gravity waves are 90° out of phase, and in the northern hemisphere the horizontal wave vector turns clockwise for an upward moving wave. Gill (1982) gives an explanation of the physics. Using the polarization equations Gill develops

$$u_1 = \frac{\kappa\Omega}{(N^2 - f^2)\cos\beta}\frac{p_1}{\rho_0} = -\tan\beta w_1, \tag{10.44}$$

$$v_1 = \frac{-i\kappa f}{(N^2 - f^2)\cos\beta}\frac{p_1}{\rho_0} = \frac{-if}{\Omega}\tan\beta w_1. \tag{10.45}$$

If $w_1 < 0$, wave energy moves upward ($w_g > 0$) but the phase fronts move downward and so

$$w_1 \propto -e^{-imz}. \tag{10.46}$$

Using (10.46) in (10.44) and (10.45) and taking the real parts gives $u_1 \propto \cos mz$ and $v_1 \propto -\sin mz$. Thus the horizontal wave vector rotates clockwise (anticyclonic) for an upward propagating wave. If the wave is propagating downward then the wave vector would spiral counter clockwise (cyclonic).

Figure 10.21 shows a hodograph taken from Guest *et al.* (2000) above Macquarie Island (54°S, 159°E). The hodograph plots the trajectory of an airsonde, and the numbers indicate the altitude in kilometers. In the southern hemisphere

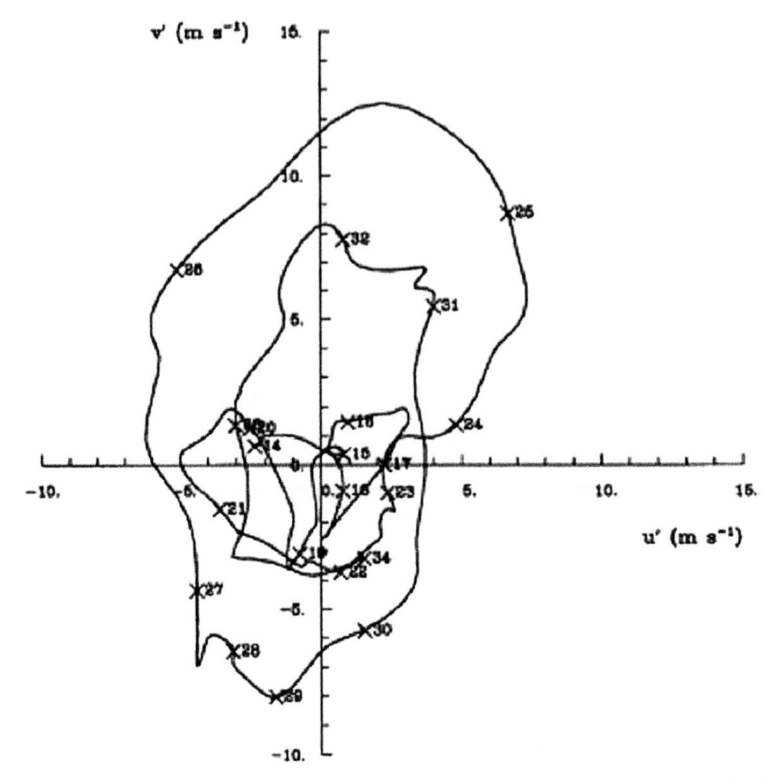

FIGURE 10.21 Typical upper atmospheric hodograph of horizontal perturbation velocities as various heights. The crosses indicate altitude in kilometers. (Taken from Guest *et al.* (2000).)

$f < 0$ which causes the sense of rotation of the wave vector to be opposite to that in the northern hemisphere. We see from Fig. 10.21 that the wave vector is rotating cyclonically with height and so wave energy is propagating upward. Guest *et al.* (2000) suggest that the inertio-gravity wave is generated by a jet-front system southwest of Macquarie Island

PROBLEMS

1. A gravity wave has slowness vector $\vec{S} = 0.1\hat{x} + 0.05\hat{y}$. What is the horizontal phase speed of the wave?
2. The gravity wave in Problem 1 has a period of 10 min. It crosses an equilateral triangle of microbarometers located at the ground surface at stations a, b, and c. The sides of the triangle are 100 m. Calculate the time lags between stations a–b, a–c, and b–c.
3. A gravity wave with amplitude 50 Pa has a wavelength 5 km, a phase speed of 3 ms^{-1}, and a baring of 50° from north. It passes over the surface array

of microbarometers as in Problem 2. Plot on the same graph the time series of pressure at each station when the length of the triangles sides are 100 m, 500 m and 1000 m. Plot these cases on separate graphs.

4. We wish to band-pass filter a time series of pressure data. The filter should pass waves with angular frequencies between 0.021 and 0.007 s^{-1}. We want to have at least three cycles of the wave. How long must be the sampling time?
5. What is the intrinsic phase speed for a gravity wave with pressure perturbation amplitude of 10 Pa and velocity perturbation magnitude of 3 m s^{-1}? The atmospheric density is 1 kgm^{-3}.
6. If the wave in Problem 5 has a horizontal wavelength of 2 km, a bearing of 135° from north and a background zonal wind speed on 5 ms^{-1}, plot the locus of points of the total wind vector.
7. When analyzing a time series of data for perturbations is it best to de-trend before time averaging or time average and then de-trending, and why?
8. Repeat Problem 3 but now the microbarographs are locates 100 m above the ground surface where the background wind in the direction of the wave vector is 6 ms^{-1}.

APPENDIX A

THE HYDROSTATIC ATMOSPHERE

A.1 THE HYDROSTATIC APPROXIMATION

Vertical motions of the atmosphere tend to be small except within relatively small horizontal regions of convection, for example, thermal plumes, cumulus clouds, thunderstorms, hurricanes, *etc.* This suggests there is a balance between the upward-directed pressure force and the weight of the atmosphere. Consider an elemental volume of atmosphere with unit cross section, height δz, and density ρ. The weight of the volume is $g\rho\delta z$, and this force is directed downward. The net pressure force acting on the volume is difference the between the pressures at the top and bottom of the volume. Let this difference δp. The balance between the pressure and gravity forces gives

$$\delta p = -\rho g \delta z. \tag{A.1}$$

In the limit of small δz (A.1) becomes

$$\frac{\partial p}{\partial z} = -\rho g, \tag{A.2}$$

which is generally referred to as the *hydrostatic equation.*

A.2 THE SCALE HEIGHT OF THE ISOTHERMAL ATMOSPHERE

Using the equation of state for an ideal gas in (A.2) leads to

$$\frac{1}{\rho}\frac{\partial \rho}{\partial z} + \frac{1}{T}\frac{\partial T}{\partial z} = -\frac{g}{RT}, \tag{A.3}$$

where R is the universal gas constant for dry air, and T is the temperature. For many applications, we can replace the atmospheric temperature by an average value, T_{A}. Then integrating (A.3) givcs

$$\rho(z) = \rho_{\mathrm{s}}\,\mathrm{e}^{-gz/RT_{\mathrm{A}}}, \tag{A.4}$$

where ρ_{s} is the atmospheric density at the ground surface. We define the isothermal scale height of the atmosphere as $H_{\mathrm{s}} = RT_{\mathrm{A}}/g$. For the standard atmosphere, $H_{\mathrm{s}} = 8$ km.

A.3 ADIABATIC LAPSE RATE

An adiabatic process within a system is one in which there is neither loss nor gain of heat within the system. Then the first law of thermodynamics becomes

$$\mathrm{d}q = c_p\,\mathrm{d}T - \alpha\,\mathrm{d}p = 0, \tag{A.5}$$

where c_p is the specific heat capacity at constant pressure, and $\alpha = 1/\rho = RT/P$ is the specific volume. Using the hydrostatic equation (A.2) in (A.5) leads to

$$c_p\,\mathrm{d}T + g\,\mathrm{d}z = 0. \tag{A.6}$$

We define the adiabatic lapse rate, Γ, as

$$\gamma = -\frac{\mathrm{d}T}{\mathrm{d}z} = \frac{g}{c_p} \approx 0.0098\ \mathrm{K\,m^{-1}}. \tag{A.7}$$

A.4 POTENTIAL TEMPERATURE

The potential temperature is defined as the temperature an air parcel would have if it were expanded or compressed adiabatically from its existing pressure to a pressure of 1000 mbar or 100 hPa. From (A.5) using the definition of specific volume we get

$$\frac{c_p}{R}\frac{\mathrm{d}T}{T} = \frac{\mathrm{d}P}{P}. \tag{A.8}$$

Integration of (A.8) from $P = 1000$ mbar where the temperature is θ to pressure P where the temperature is T gives

$$\frac{c_p}{R}\ln\left(\frac{T}{\theta}\right) = \ln\left(\frac{P}{1000}\right). \tag{A.9}$$

Then the potential temperature is

$$\theta = T\left(\frac{1000}{P}\right)^{R/c_p}, \tag{A.10}$$

where $\frac{R}{c_p} = 0.286$.

A.5 BOUSSINESQ RELATIONS

We can develop (1.65) by first taking the logarithmic differential of the potential temperature (A.10) with respect to z to get

$$\frac{1}{\theta}\frac{\mathrm{d}\theta}{\mathrm{d}z} = \frac{1}{T}\frac{\mathrm{d}T}{\mathrm{d}z} - \frac{R}{pc_p}\frac{\mathrm{d}p}{\mathrm{d}z}. \tag{A.11}$$

The equation of state for the atmosphere is well-approximated by the ideal gas law,

$$p = \rho RT. \tag{A.12}$$

Taking the vertical derivative of (A.12) with respect to z, and dividing this result by p, one gets

$$\frac{1}{T}\frac{\mathrm{d}T}{\mathrm{d}z} = \frac{1}{p}\frac{\mathrm{d}p}{\mathrm{d}z} - \frac{1}{\rho}\frac{\mathrm{d}\rho}{\mathrm{d}z}. \tag{A.13}$$

Substitution of (A.13) into (A.11) leads to

$$\frac{1}{\theta}\frac{\mathrm{d}\theta}{\mathrm{d}z} = \frac{1}{\rho}\left[\left(\frac{\rho}{p}\frac{c_v}{c_p}\right)\frac{\mathrm{d}p}{\mathrm{d}z} - \frac{T\rho}{\mathrm{d}z}\right]. \tag{A.14}$$

The speed of sound is $c_s^2 = \frac{c_p}{c_v}\frac{p}{\rho}$. Thus, we can write (A.14) as

$$\frac{1}{\theta}\frac{\mathrm{d}\theta}{\mathrm{d}z} = \frac{1}{\rho}\left[\frac{1}{c_s^2}\frac{\mathrm{d}p}{\mathrm{d}z} - \frac{\mathrm{d}\rho}{\mathrm{d}z}\right]. \tag{A.15}$$

Because rates of motion of the atmosphere and gravity waves are generally much less than the speed of sound, Eq. (A.15) reduces to

$$\frac{1}{\theta}\frac{\mathrm{d}\theta}{\mathrm{d}z} = -\frac{1}{\rho}\frac{\mathrm{d}\rho}{\mathrm{d}z}. \tag{A.16}$$

Equation (1.66) follows from (A.16) if we identify $\mathrm{d}\theta$ with perturbation θ_1 and $\mathrm{d}\rho$ with perturbation ρ_1.

A.6 THE GEOSTROPHIC WIND

The equations of motion on a rotating earth are (Holton, 2004)

$$\frac{\mathrm{d}u}{\mathrm{d}t} - \frac{uv \tan\Phi}{a} + \frac{uw}{a} = -\frac{1}{\rho}\frac{\partial p}{\partial x} + 2\Omega_{\mathrm{E}} v \sin\Phi - 2\Omega_{\mathrm{E}} \cos\Phi + F_x, \tag{A.17}$$

$$\frac{\mathrm{d}v}{\mathrm{d}t} + \frac{u^2 \tan\Phi}{a} + \frac{vw}{a} = -\frac{1}{\rho}\frac{\partial p}{\partial y} - 2\Omega_{\mathrm{E}} \sin\Phi + F_y, \tag{A.18}$$

$$\frac{\mathrm{d}w}{\mathrm{d}t} - \frac{u^2 + v^2}{a} = -\frac{1}{\rho}\frac{\partial p}{\partial z} - g + 2\Omega_{\mathrm{E}} u \cos\Phi + F_z, \tag{A.19}$$

where

$$+\frac{\mathrm{d}}{\mathrm{d}t} = \frac{\partial}{\partial t} + u\frac{\partial}{\partial x} + v\frac{\partial}{\partial y} + w\frac{\partial}{\partial z}$$

is the total derivative, a is the mean radius of the earth, Ω_{E} is the angular velocity of the earth, and F_x, F_y, and F_z are friction forces in the x-, y-, and z-directions respectively. The terms involving $1/a$ are due to the effects of the earth's curvature. The terms involving Ω_{E} represent the apparent forces due to the earth's rotation. These are the *Coriolis forces*. The Coriolis force acts perpendicular to the velocity vector; thus, it can change the direction of motion, but not the magnitude. In the northern hemisphere, bodies moving horizontally are deflected to the right. In the southern hemisphere (where the angular velocity vector is negative), the motion is directed to the left. Holton (2004) scales Eqs. (A.17)–(A.19) using

$U \sim 10\ \mathrm{m\ s^{-1}}$:	horizontal velocity scale,
$W \sim 1\ \mathrm{cm\ s^{-1}}$:	vertical velocity scale,
$L \sim 10^6\ \mathrm{m}$:	length scale,
$D \sim 10^4\ \mathrm{m}$:	depth scale,
$\Delta P/\rho \sim 10^3\ \mathrm{m^2\ s^{-2}}$:	horizontal pressure fluctuation scale,
$L/U \sim 10^5\ \mathrm{s}$:	time scale.

Table A.1 taken from Holton (2004) shows the scaled magnitudes of the terms in (A.17)–(A.19), where

$$f_0 = 2\Omega_{\mathrm{E}} \sin 45^\circ = 2\Omega_{\mathrm{E}} \cos 45^\circ. \tag{A.20}$$

Retaining only the greatest terms in Table A.1 gives

$$fu = -\frac{1}{\rho}\frac{\partial p}{\partial x}, \tag{A.21}$$

$$fv = \frac{1}{\rho}\frac{\partial p}{\partial y}. \tag{A.22}$$

The horizontal components of the *geostrophic wind* are defined as

$$u_g = -\frac{1}{f\rho}\frac{\partial p}{\partial x}, \tag{A.23}$$

$$v_g = \frac{1}{f\rho}\frac{\partial p}{\partial y}. \tag{A.24}$$

The geostrophic wind represents the balance between the pressure gradient force and the Coriolis force. In the northern hemisphere, this balance results in a wind blowing parallel to lines of constant pressure with low pressure to the left.

A.7 THE CRITICAL LEVEL

In Section 5.4 we developed the means for carrying an analytical solution of the Taylor–Goldstein equation across a critical level. What remains is how to describe $(-\zeta)^{\lambda}$ and $(-\zeta)^{\lambda^*}$ in (5.101) and (5.102). In this section we complete this argument. We pick up the development at (5.97), *i.e.*,

$$w_1(k, z_c + \zeta) = C_0^+ \zeta^{\lambda} \psi_1(\zeta) + C_0^- \zeta^{\lambda^*} \psi_1^*(\zeta) \tag{A.25}$$

and

$$\frac{dw_1}{dz} = C_0^+ \zeta^{\lambda} \psi_2(\zeta) + C_0^- \zeta^{\lambda^*} \psi_2^*(\zeta), \tag{A.26}$$

where

$$\psi_1(\zeta) = 1 + \frac{C_1}{C_0}\zeta + \frac{C_2}{C_0}\zeta^2 \tag{A.27}$$

and

$$\psi_2(\zeta) = \lambda\zeta^{-1} + \frac{C_1}{C_0}(\lambda + 1) + \frac{C_2}{C_0}(\lambda + 2)\zeta. \tag{A.28}$$

In the above equations, C_0^+ refers to values using $\lambda = 1/2 + i\mu$, and C_0^- refers to values using $\lambda = 1/2 - i\mu$.

Using (A.25) and (A.26) to solve for C_0^+ and C_0^- we get

$$C_0^+ = \frac{\zeta_1^* \frac{dw_1}{dz} - \zeta_2^* w_1}{D}, \tag{A.29}$$

$$C_0^- = \frac{\zeta_1 \frac{dw_1}{dz} - \zeta_2 w_1}{D^*}, \tag{A.30}$$

where

$$D = \zeta_1^* \zeta_2 - \zeta_1 \zeta_2^*. \tag{A.31}$$

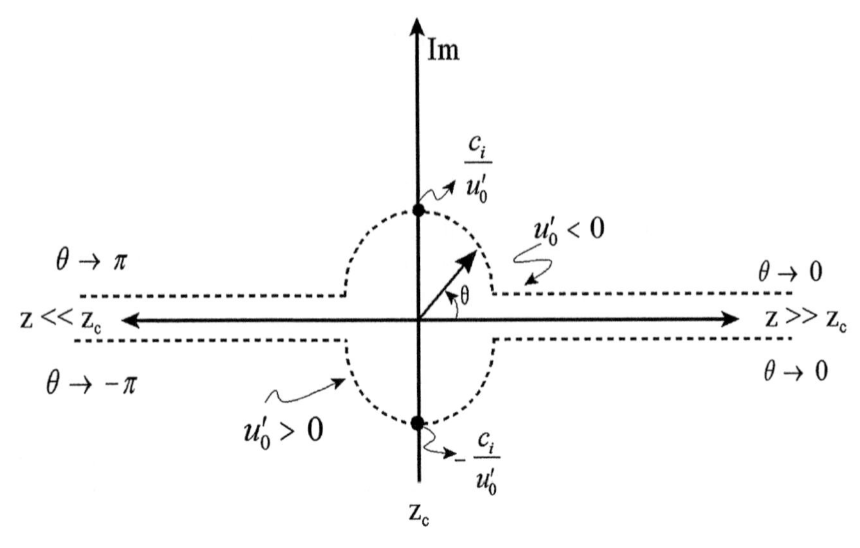

FIGURE A.1 The path over the singularity at $z = z_c$ in the complex plane determines whether -1 equals $\exp(\mathrm{i}\pi)$ or $\exp(-\mathrm{i}\pi)$ when $z = z_c$.

Below the critical level where $\zeta < 0$ we have

$$w_1(k, z_c - \zeta) = C_0^+(-\zeta)^\lambda \psi_1(-\zeta) + C_0^-(-\zeta)^{\lambda^*} \psi_1^*(-\zeta), \tag{A.32}$$

$$\frac{dw_1}{dz}(k, z_c - \zeta) = -C_0^+(-\zeta)^\lambda \psi_2(-\zeta) - C_0^-(-\zeta)^{\lambda^*} \psi_2^*(-\zeta). \tag{A.33}$$

What remains is how to describe $(-\zeta)^\lambda$ and $(-\zeta)^{\lambda^*}$ in (A.32) and (A.33).

To evaluate these functions, we introduce a small imaginary phase speed, *i.e.*,

$$c = c_R + \mathrm{i}c_I. \tag{A.34}$$

Then (5.86) becomes

$$\frac{d^2 w_1}{dz^2} + \left[\frac{N^2}{(c_r + \mathrm{i}c_i - u_0)^2} + \frac{u_0''}{c_r + \mathrm{i}c_i - u_0} - k^2\right] w_1 = 0. \tag{A.35}$$

As the wave approaches the critical level from above, we expand the background wind speed to first order

$$c_r - u_0 = -\left.\frac{du_0}{dz}\right|_{z_c} (z - z_c) = a_1(z - z_c), \tag{A.36}$$

where we have used (5.84). As $z \to z_c$, the buoyancy term in (A.35) dominates so that

$$\frac{d^2 w_1}{dz^2} + \frac{R_c}{(z - z_c - \mathrm{i}c_i/a_1)^2} w_1 = 0. \tag{A.37}$$

Now using (5.90) with $\zeta = z - z_c - \mathrm{i}c_i/a_1$ the solution to (A.37) takes the form

$$w_1(k, z - z_c) = A\left(z - z_c - \mathrm{i}\frac{c_i}{a_1}\right)^{1/2+\mathrm{i}\mu} + B\left(z - z_c - \mathrm{i}\frac{c_i}{a_1}\right)^{1/2-\mathrm{i}\mu}. \quad \text{(A.38)}$$

In polar form,

$$z - z_c - \mathrm{i}\frac{c_i}{a_1} = r\,\mathrm{e}^{\mathrm{i}\theta}, \quad \text{(A.39)}$$

where

$$r = \left[(z - z_c)^2 + \left(\frac{c_i}{a_1}\right)^2\right]^{1/2} \quad \text{(A.40)}$$

and

$$\theta = \tan^{-1}\left[-\frac{c_i}{a_1(z - z_c)}\right]. \quad \text{(A.41)}$$

We now go to the complex plane illustrated in Fig. A.1. For $z \gg z_c$, $\theta \to 0$, but for $z \ll z_c$, θ goes to $\pm\pi$ depending on whether we go over or under the singularity at $z = z_c$. If $a_1 > 0$ (positive wind shear) then from (A.41) $\theta \to -\pi/2$ as $z \to z_c$. Thus, we go under the branch point so that for $z \ll z_c \theta \to -\pi$, *i.e.*,

$$-\zeta = \zeta\,\mathrm{e}^{-\mathrm{i}\pi}. \quad \text{(A.42)}$$

From (A.42) we see that

$$(-\zeta)^\lambda = \zeta^\lambda\,\mathrm{e}^{-\mathrm{i}\pi\lambda}. \quad \text{(A.43)}$$

Using the positive branch of (5.95) for λ in (A.43) gives

$$(-\zeta)^\lambda = \zeta^\lambda\,\mathrm{e}^{-\mathrm{i}\pi(1/2+\mathrm{i}\mu)} = \zeta^\lambda\,\mathrm{e}^{-\mathrm{i}\pi/2}\,\mathrm{e}^{\pi\mu} = -\mathrm{i}\zeta^\lambda\,\mathrm{e}^{\pi\mu}. \quad \text{(A.44)}$$

If $a_1 < 0$ (negative wind shear), then in a similar way we can show that

$$(-\zeta)^\lambda = \mathrm{i}\zeta^\lambda\,\mathrm{e}^{-\pi\mu}. \quad \text{(A.45)}$$

A.8 CONVOLUTION

Integrals of the form

$$g(t) = \int_{-\infty}^{+\infty} h(t) f(t - \tau)\mathrm{d}\tau \quad \text{(A.46)}$$

are basic to digital filtering, and are called convolution integrals. Equation (A.46) can be evaluated numerically, but there is an easier way using Fourier transformation. Take the Fourier transform of (A.45),

$$\int_{-\infty}^{+\infty} g(t)\mathrm{e}^{-\mathrm{i}\omega t}\,\mathrm{d}t = \int_{-\infty}^{+\infty} \mathrm{d}t \int_{-\infty}^{+\infty} f(t - \tau)h(\tau)\mathrm{e}^{-\mathrm{i}\omega t}\,\mathrm{d}\tau. \quad \text{(A.47)}$$

Now let $t' = t - \tau$, and hence $\mathrm{d}t = \mathrm{d}t'$. Using these in (A.47) gives

$$\int g(t)\mathrm{e}^{-\mathrm{i}\omega t}\mathrm{d}t = \int \mathrm{d}t' \int h(t') f(\tau)\mathrm{e}^{-\mathrm{i}\omega(\tau+t')}\,\mathrm{d}\tau, \tag{A.48}$$

where the limits of integration are $\pm\infty$. Separating the exponential on the right-hand side of (A.48) gives

$$\int g(t)\mathrm{e}^{-\mathrm{i}\omega t}\,\mathrm{d}t = \int h(t')\mathrm{e}^{-\mathrm{i}\omega t'}\,\mathrm{d}t' \int f(\tau)\mathrm{e}^{-\mathrm{i}\omega\tau}\,\mathrm{d}\tau. \tag{A.49}$$

But each of these integrals represents the Fourier transform so that

$$\hat{g}(\omega) = \hat{h}(\omega)\hat{f}(\omega), \tag{A.50}$$

where the $\hat{h}$ is the Fourier transform of h, *etc.* Thus, convolution is calculated by multiplying the Fourier transforms of the integrands and then taking the inverse transform of the product.

A.9 THE ECKMAN WIND SPIRAL

In the planetary boundary layer, time scales for variations of the wind range from several minutes to an hour or so. Thus, to first order, and using Table A.1, the mean winds in the boundary layer can be represented by

$$-fv = -\frac{1}{\rho}\frac{\partial p}{\partial x} + \frac{\partial}{\partial z}\left(\frac{\tau_x}{\rho}\right), \tag{A.51}$$

$$fu = -\frac{1}{\rho}\frac{\partial p}{\partial y} + \frac{\partial}{\partial z}\left(\frac{\tau_y}{\rho}\right), \tag{A.52}$$

where τ_x and τ_y are the turbulence stress terms in the x- and y-directions respectively. Representing the turbulence stresses with a constant *eddy diffusivity* K gives, for example, $\tau_x = \rho K\,\partial u/\partial z$. Using this in (A.51) and (A.52) along with the definitions of u_{g} and v_{g} gives

$$K\frac{\partial^2 u}{\partial z^2} + f(v - v_{\mathrm{g}}) = 0, \tag{A.53}$$

$$K\frac{\partial^2 v}{\partial z^2} - f(u - u_{\mathrm{g}}) = 0. \tag{A.54}$$

Equations (A.53) and (A.54) show that the height-varying departure of the local wind from its geostrophic value is due to the turbulence stress. Physically, the stress reduces the wind speed, and this reduces the Coriolis force. However, the pressure gradient force is unchanged, and an imbalance between the forces results. Now, because the pressure gradient force is greater than the Coriolis force, the wind takes on a component directed down the pressure gradient. Now let us assume that at $z =$, $u = 0$ and $v = 0$, and at some upper boundary $u \to u_{\mathrm{g}}$ and $v \to v_{\mathrm{g}}$. For simplicity, we assume that $v_g = 0$. We can then write

$$K \frac{\partial^2}{\partial z^2}(u - u_g) + fv = 0, \tag{A.55}$$

$$K \frac{\partial^2}{\partial z^2} v - f(u - u_g) = 0. \tag{A.56}$$

If we now multiply (A.56) by $\mathrm{i} = \sqrt{-1}$, and add this result to (A.55) we get

$$\frac{\partial^2}{\partial z^2}\left[(u - u_g) + \mathrm{i}v\right] - \frac{\mathrm{i}f}{K}\left[(u - u_g) + \mathrm{i}v\right] = 0. \tag{A.57}$$

The general solution of (A.57) is

$$(u - u_g) + \mathrm{i}v = A\,\mathrm{e}^{\sqrt{\mathrm{i}f/K}z} + B\,\mathrm{e}^{-\sqrt{\mathrm{i}f/K}z}. \tag{A.58}$$

For bounded solutions, we take $A = 0$, and to satisfy the boundary condition at $z = 0$ we set $B = -u_g$. Using the identity $i = \mathrm{e}^{\mathrm{i}\pi/2}$ it follows that $\sqrt{i} = (1 + \mathrm{i})/\sqrt{2}$. Thus, we can write (A.58) as

$$u + \mathrm{i}v = u_g(1 - \mathrm{e}^{-\gamma(1+\mathrm{i})z}), \tag{A.59}$$

where $\gamma = \sqrt{f/2K}$. Separating (A.59) into real and imaginary parts gives the equations of the Ekman spiral:

$$u = u_g(1 - \mathrm{e}^{-\gamma z})\cos(\gamma z), \tag{A.60}$$

$$v = u_g\,\mathrm{e}^{-\gamma z}\sin(\gamma z). \tag{A.61}$$

When $z = \pi/\gamma$, the wind becomes parallel to the isobars. The depth of the Ekman, π/γ is usually associated with the depth of the planetary boundary layer.

A.10 NUMERICAL METHODS

The Taylor–Goldstein equation (TGE) requires a constant and horizontally uniform background flow and stratification. The undetermined constants are evaluated by the vertical boundary conditions. In the cases treated in this book, $u_0(z)$ and $N(z)$ were constant; however, in real applications this will not be the case, and the TGE must be solved numerically. Here we give outlines and suggestions for constructing numerical solution for mountain waves and ducted waves.

A.10.1 MOUNTAIN WAVES

The TGE is

$$\frac{\mathrm{d}^2 w_n}{\mathrm{d}z^2} + \left[\frac{N^2}{u_0^2} - \frac{u_0''}{u_0} - k_n^2\right] w_n = 0, \tag{A.62}$$

where subscript n refers to the nth component of the Fourier transform of w_1; also we have neglected Coriolis forces and compressibility, *i.e.*, $1/2H_s$, but these can be added in a straightforward way. We set the top of the model domain at height $z = z_{\rm top}$, and assume that above this height the background wind speed and stratification are constant. This allows us to use the radiation boundary condition at $z_{\rm top}$. The general solution for $z \geqslant z_{\rm top}$ takes the form

$$w_n(k_n, z) = A_n \, \mathrm{e}^{\mathrm{i}m_{n,\rm top}(z - z_{\rm top})} + B_n \, \mathrm{e}^{-\mathrm{i}m_{n,\rm top}(z - z_{\rm top})}, \tag{A.63}$$

where $m_{n,\rm top}$ is

$$m_{n,\rm top} = \left[\frac{N^2(z_{\rm top})}{u_0^2(z_{\rm top})} - k_n^2 \right]^{1/2}. \tag{A.64}$$

If $m_{n,\rm top}$ is not real, then there is no wave stress associated with that Fourier component. We have seen in Chapter 2, that the A_n solutions in (A.63) corresponds to upward moving wave fronts and downward moving energy, and the B_n solutions corresponds to downward moving wave fronts and upward moving energy. Thus, in order to satisfy the radiation boundary condition at $z = z_{\rm top}$, the solution of (A.63) is

$$w_n(k_n, z_{\rm top}) = B_n(k_n), \tag{A.65}$$

and the vertical derivative at $z_{\rm top}$ is

$$\frac{\mathrm{d}w}{\mathrm{d}z} = -\mathrm{i}m_n(z_{\rm top}) B_n(k_n). \tag{A.66}$$

We introduce the shape factor $\Phi(k_n, z)$ defined by

$$w_n = w_n(k_n, 0) \frac{\Phi(k_n, z)}{\Phi(k_n, 0)}, \tag{A.67}$$

where $w_n(k_n, 0) = \mathrm{i}u_0 k\hat{h}$ is the bottom boundary condition, where $\hat{h}$ is the Fourier transform of the mountain. Then below $z_{\rm top}$ the TGE has the form

$$\frac{\mathrm{d}^2\Phi_n}{\mathrm{d}z^2} + \left[\frac{N^2}{u_0^2} - \frac{u_0''}{u_0} - k_n^2 \right] \Phi_n = 0, \tag{A.68}$$

with upper boundary condition $\Phi_n(k_n, z_{\rm top}) = 1$. We also see that $B_n(k_n) = w_n(k_n, 0)/\Phi_n(k_n, 0)$. Below $z_{\rm top}$, we must account for upward and downward (reflected) waves, and this is done by choosing Φ complex, *i.e.*, $\Phi = \Phi_{\rm R} + \mathrm{i}\Phi_{\rm I}$, where we have dropped the subscript n for convenience. We must solve (A.68) for the real and imaginary parts of Φ.

There are several waves to solve (A.68). Here we describe the extrapolation method based on the IMSL subroutine DREBS which is a modification of the Bulirsch–Stoer ALGOL procedure DESUB (Bulirsch and Stoer, 1966; Gragg, 1965). Now consider the differential equation

$$\Phi'' + Q^2(z)\Phi = 0, \tag{A.69}$$

where primes denote vertical derivatives. DREBS solves the equations

$$y = \Phi', \tag{A.70}$$

$$y' = -Q^2(z)\Phi. \tag{A.71}$$

At some height, z_i, $\Phi(z_i)$ and $\Phi'(z_i)$ have known values, *i.e.*, boundary values. Then (A.70) and (A.71) are solved at some height z_f, which can be above or below z_i, using extrapolation. At any point in the extrapolation, *i.e.*, $z_i \leqslant z \leqslant z_f$, $Q(z)$ will need to be evaluated. This is because the height z is determined as needed by DREBS, and u_0 and N must be analytical functions or piece-wise continuous functions which can be evaluated at z. (In the later case, u_0 is taken to be linear between z_i and z_f and N is held constant.) This requires a user supplied subroutine, FCN, which evaluates $Q^2(z) = \frac{N^2}{u_0^2} - \frac{u_0''}{u_0} - k^2$. Let $\Delta z = z_f - z_i$ be the initial extrapolation distance. DREBS tests for converges at z_f when consecutive values of y and y' differ by a predetermined tolerance. If convergence is not achieved, the extrapolation is repeated but now $z_f = z_i + \Delta z/2$. This processes continues until convergence is reached at z_f. Because of these extrapolation steps, the effects of large changes in Q between z_i and z_f can be resolved. There are no limiting values of z_i and z_f. What is required are continuous and differentiable background values of u_0 and N. Thus, one can calculate wave stress at the ground surface using only the bottom and top boundary conditions. However, if wave perturbation values are wanted as function of height, then Δz must be divided into subintervals or levels which need not be evenly spaced.

We begin the integrations at $z = z_{top}$ where

$$\Phi_R + i\Phi_I = 1, \tag{A.72}$$

$$\Phi_R' + i\Phi_I' = -im, \tag{A.73}$$

and hence

$$\Phi_R = 1, \tag{A.74}$$

$$\Phi_I = 0, \tag{A.75}$$

$$\Phi_R' = 0, \tag{A.76}$$

$$\Phi_I' = -m. \tag{A.77}$$

The integration continues downward to the bottom boundary. If critical levels are encountered or wave breaking occurs, then this will require analytical continuation and wave stress parameterization respectively.

As previously mentioned, two types of background wind and stratification profiles are possible, continuous analytical functions and discrete piece-wise continuous functions. Because subroutine DREBS must evaluate Q^2 at any height, analytical expressions for wind speed and temperature are desirable. However, in many applications we know the background wind speeds and temperatures only at discrete points. An example is the output from an airsonde or a numerical model. It is possible to make these data continuous by either fitting the points to a polynomial expression or using splines. A simple approach is to join these points with straight lines so that within each model layer u_0 varies linearly with z, and N is constant. But this procedure will result in discontinuities in u_0' at the model levels, and this can lead to discontinuities in pressure across the model levels. At each model level, the dynamic and kinematic boundary conditions must hold. Recall from Chapter 4, that the dynamic boundary condition requires continuity in pressure across the boundary, and the kinematic boundary condition requires continuity in vertical velocity across the boundary. Using (2.22), (2.24) and (A.67) and noting that for mountain waves $\Omega = -u_0 k_n$ we write for Φ_{R}

$$u_0 \frac{d\Phi_{\mathrm{R}}}{dz} - \Phi_{\mathrm{R}} \frac{du_0}{dz} = ik \frac{\tilde{p}}{\rho_0}. \tag{A.78}$$

At each model level J, we apply (A.78) above and below J and require that $\tilde{p}_{J+1} = \tilde{p}_{J-1}$ and $\Phi_{\mathrm{R},J+1} = \Phi_{\mathrm{R},J-1}$. Noting that u_0 is continuous at J, we get

$$\Phi'_{\mathrm{R},J} = \Phi'_{\mathrm{R},J+1} - \frac{u'_{0,J+1} - u'_{0,J-1}}{u_{0,J}} \Phi_{\mathrm{R},J+1}, \tag{A.79}$$

where $u'_{0,J+1}$ and $u'_{0,J-1}$ are the background wind shears in the upper and lower layers respectively. $\Phi'_{\mathrm{R},J}$ and $\Phi_{\mathrm{R},J}$ are then calculated by DREBS extrapolation in the upper layer to level J. A similar calculation is made for Φ_{I}.

A.10.2 DUCTED GRAVITY WAVES

We have seen in Chapter 4 that wave ducting is an eigenvalue problem. For a given flow, only those waves associated with a family of wavenumbers and associated frequencies will be ducted. This is because ducting requires wave reflections from vertical boundaries, and wave reflection is a function of stratification, wind speed, phase speed, and wavenumber. Most commonly, one of the boundaries is the ground surface, and the other is an elevated layer above which the real part of the vertical wavenumber $m_{\mathrm{R}} = 0$. The numerical solution of the ducting problem involves the search for waves with eigenvalues (k, c) which are solutions to the TGE and satisfy the boundary conditions. At the ground surface, we require $w = 0$. At the top boundary of the model, the ducted wave must be evanescent because the wave must not transport energy out of the model domain. In this case, the wave energy is confined below the top boundary. Unlike the terrain-generated gravity wave, we do not consider upward and downward propagating waves. This is because if there is wave reflection below the top boundary, then the wave cannot be a trapped. Instead, we seek a horizontally propagating wave with node points at the bottom boundary and some upper reflecting height. Therefore we take Φ real. We cannot determine wave amplitudes and because there is neither stress nor upward propagating energy associated with ducted waves we need not calculate the wave field at various vertical levels. This greatly simplifies the problem, especially if we have analytical expressions for the background wind and temperature profiles. Above $z = z_{\mathrm{top}}$, the background wind and stratification is assumed constant so that

$$\Phi = A\,\mathrm{e}^{q(z-z_{\mathrm{top}})} + B\,\mathrm{e}^{-q(z-z_{\mathrm{top}})}, \tag{A.80}$$

where

$$q^2 = k^2 - \frac{N^2(z_{\mathrm{top}})}{\left[c - u_0(z_{\mathrm{top}})\right]^2}. \tag{A.81}$$

The solution is obtained by finding the eigenvalues (k, c) that result in $A \to 0$ above z_{top}. At the top boundary, (A.80) and its first derivative take the form

$$\Phi = A + B, \tag{A.82}$$

$$\frac{\mathrm{d}\Phi}{\mathrm{d}z} = qA - qB. \tag{A.83}$$

Solving (A.82) and (A.83) for A and setting this result to zero gives

$$A \equiv \mathrm{GROW} = \Phi + \frac{1}{q}\frac{\mathrm{d}\Phi}{\mathrm{d}z} = 0. \tag{A.84}$$

One begins with a large wavenumber and an initial phase speed. Because critical levels are not allowed, the minimum phase speed must be greater than the maximum background wind speed, *i.e.*, $c_{\min} > u_{0,\max}$. From (A.81) we require $k > N(z_{\mathrm{top}})/\left[c - u_0(z_{\mathrm{top}})\right]$ at z_{top} so that the maximum phase speed is $c_{\max} = N(z_{\mathrm{top}})/k + u_0(z_{\mathrm{top}})$. The phase speeds of the ducted waves must lie between $c_{\min}$ and $c_{\max}$. The extrapolation solution of the TGE begins at the ground surface where $\Phi = 0$, $\mathrm{d}\Phi/\mathrm{d}z = 1$, and $c = c_{\min}$. At z_{top}, the value of GROW is calculated using (A.84) and tested to see if it is less than some small value ϵ. If GROW $> \epsilon$ then the calculation is repeated but now with a slightly greater phase speed, $c_{\min} + \Delta c$. The procedure continues until GROW $< \epsilon$.

BIBLIOGRAPHY

Addison, P. S. Wavelet transforms and ECG a review. *Physiol. Meas.* **26**:R155–R199, 2005.

Akyuz, A. A., Henry, L., andHorst, T. Wind evaluation of PAM II pressure ports. *J.Atmos. Oceanic Technol.* **8**:323–330, 1991.

Alexander, M. J., Beres, J. H., and Pfister, L. Tropical stratospheric gravity wave activity and relationship to clouds. *Geophys. Res. Lett.* **105**:22,299–22,309, 2000.

Alexander, M. J., Richter, J. H., and Sutherland, B. R. Generation and trapping of gravity waves from convection with comparison to parameterization. *J. Atmos. Sci.* **63**:2963–2977, 2006.

Alexander, M. J. Chapter 5. Gravity waves in the stratosphere. In *The Stratosphere: Dynamics, Chemistry, Transport. Geophys. Monogr. Ser.*, L. M. Polvani, A. Sobel, and D. W. Waugh, eds., 2010. *Geophys. Monogr. Ser.*.

Alexander, M. J. and Dunkerton, T. J. A spectral parameterization of mean-flow forcing due to breaking gravity waves. *J. Atmos. Sci.* **56**:4167–4182, 1999.

Alexander, M. J., Holton, J. R., and Durran, D. R. The gravity wave response above deep convection in a squall line simulation. *J. Atmos. Sci.* **52**:2212–2226, 1995.

Viana, S., Terradellas, E., and Yagüe, C. Analysis of gravity waves generated at the top of a drainage layer. *J. Am. Meteorol. Soc.* **67**:3949–3966, 2010.

Alexander, M. J. and Vincent, R. A. Gravity waves in the tropical lower stratosphere: A model study of seasonal and interannual variability. *J. Geophys. Res.* **105**:17,983–17,994, 2000.

Alexander, M. J. and Holton, J. R. A model study of zonal forcing in the equatorial stratosphere by convectively induced gravity waves. *J. Atmos. Sci.* **97**:408–419, 1997.

Allen, S. J. and Vincent, R. A. Gravity-wave activity in the lower atmosphere: Seasonal and latitudinal variations. *J. Geophys. Res.* **100**:1327–1350, 1995.

Anderson, P. S., Mobbs, S.D., King, J. C., McConnell, I., and Rees, J. M. A microbarographfor internal gravity wave studies in Antarctica. *Antarct.Sci.* **4(2)**:241–248, 1992.

Andrews, D. G. and McIntyre, K. E. On wave action and its relatives. *J. Fluid Mech.* **89**:647–664, 1978.

Andrews, D. G., Holton, J. R. and Leovy, C. B. *Moddle Atmosphere Dynamics*, 490 pp., 1987. Academic Press.

Andrews, D. G. and McIntyre, M. E. Planetary waves in horizontal and vertical shear: The generalized Eliassen-Palm relations and the mean zonal acceleration. *J. Atmos. Sci.* **33**:2013–2048, 1976.

Angell, J. K. and Pack, D. H.Analysis of some preliminary low-level constant level balloon(tetroon) flights. *Mon. Weather Rev.* **88**:235–248,1960.

Artu, J. Ducic, V. Kanamori, H. andLognonne, P. Ionospheric detection of gravity waves induced by tsunamis. *Geophys. J. Int.* **160**:840–848, 2005.

Asten, M. W. and Henstridge, J. D. Array estimators and the use of microseisms for reconnaissance of sedimentary basins. *Geophysics* **49**:1828–1837, 1984.

Büshler, O., McIntyre, M. E., and Scinocca, J. F. On shear-generated gravity waves that reach the mesosphere. Part I: Wave generation. *J. Atmos. Sci.* **56**:3749–3763, 1999.

Baines, P. G. *Topographic Effects in Stratified Flows*, 1995. New York: Cambridge University Press.

Baines, P. G. and Palmer, T. N. Rational for a new physically based parameterization of sub-grid scale orographic effects. Technical Report, Tech. Memo. 169, 1990. European Center for Medium-Range Weather Forecasts.

Baker, R. D., Schubert, G., and Jones, P. W. Convectively generated internal gravity waves in the lower atmosphere of Venus. Part I: No wind shear. *J. Atmos. Sci.* **57**:184–215, 2000.

Balachandran, N. K. Gravity waves from thunderstorms. *Mon. Wea. Rev.* **108**:804–816, 1980.

Balsley, B.,Jensen, M. L., and Frehlich, R. G. The use of state-of-the-artkites for profiling the lower atmosphere. *Boundary-LayerMeteorol.* **87**:1–24, 1998.

Balsley, B. and Gage, K. TheMST radar technique: Potential for middle atmosphric studues.*Pure Appl. Geophys.* **118**:452–493, 1980.

Balsley, B. B., Frehlich,R. G., Jensen, M. L., Meillier, Y., and Muschinsky, A. Extremegradients in the nocturnal boundary layer: Structure, evolution,and potential causes. *J. Atmos. Sci.* **60**:2496–2508,2003.

Barber, N. S. Design of 'optimum' arrays for direction finding. *Electron. Radio Eng.* **36**:222–232, 1959.

Bartman, F. L., Chaney, L.W., Jones, M., and Liu, V. C. Upper-air density and temperatureby falling-sphere method. J. Appl. Phys. **27**:706–712, 1956.

Bedard, A. J., Canavero, F., and Einaudi, F. Atmospheric gravity waves and aircraft turbulence encounters. *J. Atmos. Sci.* **43**:2838–2844, 1986.

Bedard,A. J. J. Infrasound originating near mountainous regions inColorado. *J. Appl. Meteorol.* **17**:1014–1022, 1978.

Bedard, A. J. and Ramzy, C.Surface meteorological observations in severe thunderstorms.Part I: Design details of TOTO. *J. Climate Appl.Meteorol.* **22**:911–918, 1983.

Bendat, J. S. andPiersol, A. G. *Randon Data: Analysis and MeasurementProcedures*, 407 pp., 1971. New York: Wiley-Interscience.

Bender, C. M. and Orszag, S. A. *Advanced Mathematical Methods for Scientists and Engineers*, 593 pp., 1999. New York: Springer-Verlag.

Beran, D. W., Hooke, W.H., and Clifford, S. F. Acoustic echo-sounding techniques andtheir application to gravity-wave, turbulence, and stabilitystudies. *Boundary-Layer Meteorol.* **4**:133–153, 1973.

Beran, D. W.,Little, C. G., and Willmarth, B. C. Acoustic Dopplermeasurements of vertical velocities in the atmosphere. *Nature* **230**:160–162, 1971.

Bilbro, J., Fichtl, G.,Fitzjarrald, D. F., Krause, M., and Lee, R. Airborne Dopplerlidar wind field measurements. *Bull. Amer. Meteorol. Soc.***65**:348–359, 1984.

Blumen, W. and McGregor, C. D. Wave drag by three-dimensional mountain lee waves in nonplanar shear flow. *Tellus* **28**:287–298, 1976.

Blumen, W., Banta, R., Burns, S. P., Fritts, D. C., Newman, R., Poulos, G. S., and Sun, J. Turbulence statistics of a Kelvin–Helmholtz billow event observed in the night- time boundary layer during the Cooperative Atmospheric-Surface Exchange Study field program. *Dyn. Atmos. Oceans* **34**:189–204, 2001.

Blumen, W. Momentum flux by mountain waves in a stratified rotating atmosphere. *J. Atmos. Sci.* **22**:529–534, 1965.

Blumen, W. and McGregor, C. D. Wave drag by three-dimensional mountain lee-waves in nonplanar shear flow. *Tellus* **28**:287–298, 1976.

Blumen, W. and Hart, J. E.Airborne Doppler lidar wind field measurements of waves in thelee of Mount Shasta. *J. Atmos. Sci.* **45**:1571–1583,1988.

Blumen, W., Banta, R.,Burns, S. P., Fritts, D. C., Newman, R., Poulos, G. S., and Sun,J. Turbulence statistics of a Kelvin–Helmholtz billow eventobserved in the night-time boundary layer during the CooperativeAtmospheric-Surface Exchange Study field program. *Dyn. Atmos. Oceans* **34**:189–204, 2001.

Boccara, G., Hertzog, A., Vincent, R. A., and Vial, F. Estimation of gravity wave momentum flux and phase speeds from quasi-Lagrangian stratospheric balloon flights. Part I: Theory and simulation. *J. Atmos. Sci.* **65**:3042–3055, 2008.

Booker, J. R. and Bretherton, F. P. The critical level for internal gravity waves in a shear flow. *J. Fluid Mech.* **27**:513–539, 1967.

Booker, D. R. and Cooper,L. W. Superpressure balloons for weather research. *J. Appl.Meteorol.* **4**:122–129, 1965.

Bosart, L. F. and Cussen, J. P. Gravity wave phenomena accompanying east coast cyclogenesis. *Mon. Wea. Rev.* **101**:446–454, 1973.

Bosart, L. F. and Sanders, F. Mesoscale structure in the megalopolitan snowstorm of 11–12 January 1983: Part III: A large-amplitude gravity wave. *J. Atmos. Sci.* **43**:924–939, 1983.

Brassure, G. P. and Solomon, S. *Aeronomy of the Middle Atmosphere: Chemistry and Physics of the Stratosphere and Mesosphere*, 658 pp., 2005. Springer.

Bretherton, F. P. Waves and turbulence in stably stratified fluids. *Radio Sci.* **4**:1279–1287, 1969.

Bretherton, F. P. The propagation of groups of internal gravity waves in a shear flow. *Quart. J. Roy. Meteor. Soc.* **92**:466–480, 1966.

Broad, A. S. Momentum flux due to trapped lee waves forced by mountains. *Quart. J. Roy. Meteorol. Soc.* **128**:2167–2173, 2002.

Brown, R. A. Longitudinal instabilities and secondary flow in the planetary boundary layer: A review. *Rev. Geophys. Space Phys.* **18**:683–697, 1980.

Brown, P. R. A. Aircraft measurements of mountain waves and theirassociated momentum flux over the British Isles. *Quart.J. Roy. Meteor. Soc.* **109**:849–865, 1983.

Brown, E. H. and Hall, F. F.Advances in atmospheric acoustics. *Rev. Geophys. Space Phys.***16**:47–110, 1978.

Burridge, R. The acoustics of pipe arrays. *Geophys. J. Roy. Astron.Soc.* **26**:53–69, 1971.

Businger, S. et al. Thepromise of GPS in atmospheric monitoring. *Bull. Amer.Meteorol. Soc.* **77**:5–18, 1996.

Cadet, D. The superpressure balloon sounding technique for the study ofatmospheric meso- and microscale phenomena. *Bull. Amer.Meteorol. Soc.* **59**:1119–1127, 1978.

Campbell, L. J. and Shepherd, T. G. Constraints on wave drag parameterization schemes for simulating the quasi-biennial oscillation, Part I: Gravity wave forcing. *J. Atmos. Sci.* **62**:4178–4195, 2005.

Capon, J. High resolution frequency-wavenumber analysis. *Proc. Inst. Elect. Electron. Eng.* **58**:760–770, 1969.

Carter, D. A., Balsley, B.B., Ecklund, W. L., Gauge, K. S., Riddle, A. C., Garello, R., andCrochet, M. Investigations of internal gravity waves using threevertically directed closely spaced wind profilers. *J.Geophys. Res.* **94**:8633–8642, 1989.

Caughey, S. J. andReadings, C. J. An observation of waves and turbulence in theEarth's boundary layer. *Boundary-Layer Meteorol.* **9**:279–296, 1975.

Chadwick, R. B. andGossard, E. E. Radar probing and measurement of the planetaryboundary layer: Part I. Scattering from refractive indexirregularities. In *Probing the Atmospheric Boundary Layer*,D. H. Lenschow, ed., pp. 163–182, 1986. Boston, MA: AmericanMeteorological Society.

Chadwick, R. B., Moran, K.P., Strauch, R. G., Morrison, G. E., and Campbell, W. C. A newradar for measuring winds. *Bull. Amer. Meteorol. Soc.* **57**:1120–1125, 1976.

Chandrasekhar, S. *Hydrodynamic and Hydromagnetic Stability*, 1961. Oxford: Clarendon Press

Chen, C.-C., Durran, D. R., and Hakim, G. J. Mountain-wave momentum flux in an evolving synoptic-scale flow. *J. Atmos. Sci.* **62**:3213–3231, 2005.

Cheung, T. K. and Little,C. G. Meteorological tower, microbarograph array, and sodarobservations of solitary-like waves in the nocturnal boundarylayer. *J. Atmos. Sci.* **47**:2516–2536, 1990.

Chimonas, G. The stability of a coupled wave-turbulence system in a parallel shear flow. *Boundary-Layer Meteorol.* **2**:444–452, 1972.

Chimonas, G., and Hines, C. O. Doppler ducting of atmospheric gravity waves. *J. Geophys. Res.* **91**:1219–1230, 1986.

Chimonas, G. and Nappo, C. J. Wave drag in the planetary boundary layer over complex terrain. *Bound.-Lay. Meteorol.* **47**:217–232, 1989.

Chimonas, G. Waves and the middle atmosphere wind irregularities. *J. Atmos. Sci.* **54**:2115–2128, 1997.

Chimonas, G., and Nappo, C. J. A thunderstorm bow wave. *J. Atmos. Sci.* **44**:533–541, 1987.

Chimonas, G. and Hines, C. O. Doppler ducting of atmospheric gravity waves. *J. Geophys. Res.* **91**:1219–1230, 1986.

Chimonas, G. and Nappo,C. J. Wave drag in the planetary boundary layer over complexterrain. *Boundary-Layer Meteorol.* **47**:217–232, 1989.

Chimonas,G. The stability of a coupled wave–turbulence system in aparallel shear flow. *Boundary-Layer Meteorol.* **2**:444–452, 1972.

Chimonas, G. Considerations of the stability of certain heterogeneous shear flows including some inflection-free profiles. *J. Fluid Mech.* **65**:65–69, 1974.

Chimonas, G. Surface drag instabilities in the atmospheric boundary layer. *J. Atmos. Sci.* **50**:1914–1924, 1993.

Chimonas, G. Jeffrey's drag instability applied to waves in the lower atmosphere: Linear and non-linear growth rates. *J. Atmos. Sci.* **51**:3758–3775, 1994.

Chimonas, G. Steps, waves and turbulence in the stably stratified planetary boundary layer. *Boundary-Layer Meteorol.* **90**:397–421, 1999.

Chimonas, G. On internal gravity waves associated with the stable boundary layer. *Boundary-Layer Meteorol.* **102**:139–155, 2002.

Chimonas, G. and Grant, J. R. Shear excitation of gravity waves. Part I: Modes of a two-scale atmosphere. *J. Atmos. Sci.* **41**:2269–2277, 1984.

Chimonas,G. Steps, waves and turbulence in the stably stratified planetaryboundary layer. *Boundary-Layer Meteorol.* **90**:397–421,1999.

Chui, C. K. *An Introduction to Wavelets*, 266 pp., 1992. Academic Press.

Chun, H.-Y., Song, I.-S., and Baik, J.-J. Some aspects of internal gravity waves in the multicell-type connvection system. *Meteorol. Atmos. Phys.* **69**:205–222, 1999.

Chun, H.-Y. and Baik, J.-J. An updated parameterization of convectively forced gravity wave drag for use in large-scale models. *J. Atmos. Sci.* **59**:1006–1017, 2002.

Chun, H.-Y. and Baik, J.-J. Momentum flux by thermally induced internal gravity waves and its approximation for large-scale models. *J. Atmos. Sci.* **55**:3299–3310, 1998.

Chun, H.-Y., Song, M.-D., Kim, J.-W., and Baik, J.-J. Effects of gravity wave drag induced by cumulus convection on the atmospheric general circulation. *J. Atmos. Sci.* **58**:302–319, 2001.

Chun, H.-Y., Song, I.-S., Baik, J.-J., and Kim, Y.-J. Impact of a convectively forced gravity wave drag parameterization in NCAR CCM3. *J. Climate* **18**:3530–3547, 2004.

Chun, H. Y., Kim, Y.-H., Choi, H.-J., and Kim, J.-Y. Influence of gravity waves in the tropical upwelling: WACCM simulations. *J. Atmos. Sci.* **68**:2599–2612, 2011.

Chun, H.-Y., Song, I.-S., and Baik, J.-J. Some aspects of internal gravity waves in the multicell-type convection system. *Meteorol. Atmos. Phys.* **69**:205–222, 1999.

Clark, T. L. andPeltier, W. R. On the evolution and stability of finite amplitudemountain waves. *J. Atmos. Sci.* **34**:1715–1730, 1977.

Clark, R. D. M. Atmospheric micro-oscillation. *J. Meteorol.* **7**:70–75, 1950.

Cohen, A. and Doron, E.Mountain lee waves in the Middle East: Theoreticalcalculations compared with satellite pictures. *J. Appl.Meteor.* **6**:669–673, 1967.

Coleman, T. A., Knupp, K. R., and Herzmann, D. E. An undular bore and gravity waves illustrated by dramatic time-lapse photography. *J. Atmos. Oceanic Technol.* **27**:1355–1361, 2010.

Coleman, T. A., Knup, K.R., and Herzmann, D. The spectacular undular bore in Iowa on 2October 2007. *Mon. Weather Rev.* **137**:495–503, 2009.

Collis, R. T. H.,Fernald, F. G., and Alder, J. E. Lidar observations ofSierra-wave conditions. *J. Appl. Meteorol.* **7**:227–233, 1968.

Combes, J. M., Grossman, A., and Tchamitchian, P. *Wavelets*, 315 pp., 1989. Springer-Verlag.

Conover, J. H. Lee-wave clouds photographed from an aircraft and asatellite. *Weather* **19**:79–92, 1964.

Corby, G. A. A preliminary study of atmospheric waves using radiosondedata. *Quart. J. Roy. Meteor. Soc.* **83**:49–60, 1957.

Cucurull, L. Global positioning system (GPS) radio occultation (RO)data assimilation. In *JCSDA DA Colloquium*, 2009, Stevenson,WA, July 2009.

Cunning, J. B. J. The analysis of surface pressure perturbations within themesoscale range. *J. Appl. Meteorol.* **13**:325–330,1974.

Dörnbrack, A. Turbulent mixing by breaking gravity waves. *J. Fluid Mech.* **375**:113–142, 1998.

Dörnbrack, A. and Nappo, C. J. A note on the application of linear wave theory at a critical level. *Boundary-Layer Meteorol.* **82**:399–416, 1997.

Dörnbrack, A. et al. Evidence for inertia-gravity waves forming polar stratospheric clouds over Scandinavia. *J. Geophys. Res.*, 2001

Döös, B. R. Theoretical analysis of lee wave clouds observed by TYROSI. *Tellus* **14**:301–309, 1962.

A. Dörnbrack et al., Evidence for inertia-gravity waves formingpolar stratospheric clouds over Scandinavia. *J. Geophys.Res.*, 2001.

Daniels, F. B. Noise-reducing line microphone for frequencies below 1 cps.*J. Acoustical Soc. Amer.* **31**:529–531, 1959.

Davies, H. C. and Phillips, P. D. Mountain drag along the Gotthard Section during ALPEX. *J. Atmos. Sci.* **42**:2093–2109, 1985.

De Baas, A. F.and Driedonks, G. M. Internal gravity waves in the stablystratified boundary layer. *Boundary-Layer Meteorol.* **31**:303–323, 1985.

dela Torre, A.and Alexander, P. The interpretation of wavelengths and periodsas measured from atmospheric balloons. *J. Appl. Meteor.* **34**:2747–2754, 1995.

De Silva, I. P. D.,Fernando, H. J. S., Eaton, F., and Hebert, D. Evolution ofKelvin–Helmholtz billows in nature and laboratory. *EarthPlanet. Sci. Lett.* **143**:217–231, 1996.

Denholm-Price, J. C. W. and Rees, J. M. Detecting waves using an array of sensors. *Mon. Weather Rev.* **127**:57–69, 1999.

Devarajan, M. Reddy,C. A., and Reddi, C. R. Rocket observations of Kelvin waves in theupper stratosphere over India. *J. Atmos. Sci.* **42**:1873–1871, 1985.

Dewan, E. and Good, R. E. Saturation and the "universal" spectrum for vertical profiles of horizontal scalar winds in the atmosphere. *J. Geophys. Res.* **91**:2741–2748, 1986.

DiMarzio, C., Harris, C.,Bilbro, J. W., Weaver, E. A., Burnham, D. C., and Hallock, J. N.Pulsed laser Doppler measurements of wind shear. *Bull.Amer. Meteorol. Soc.* **60**:1061–1066, 1979.

Donelan, M. A., Madsen, N.,Kahma, K. K., Tsanis, I. K., and Drennan, W. M. Apparatus foratmospheric surface layer measurements over waves. *J. Atmos.Oceanic Technol.* **16**:1172–1182, 1999.

Doran, J. C., Fast, J. D., and Horel, J. The VTMX 2000 campaign. *Bull. Amer. Meteor. Soc.* 537–551, 2002.

Doviak, R. J., and Ge, R. An atmospheric solitary gust observed with a doppler radar, a tall tower and a surface network. *J. Atmos. Sci.* **41**:2559–2573, 1984.

Drazin, P. G. *Introduction to Hydrodynamic Stability*, pp. 258, 2002. Cambridge University Press.

Duck, T. J. and Whiteway, J. A. The spectrum of waves and turbulence at the tropopause. *Geophys. Res. Lett.* **32**:1–4, 2005.

Dunkerton, T. J. Inertia-gravity waves in the stratosphere. *J. Atmos. Sci.* **41**:3396–3404, 1984.

Dunkerton, T. J. Wave transience in a compressible atmosphere. Part I: Transient internal wave, mean-flow interactions. *J. Atmos. Sci.* **38**:281–297, 1981.

Dunkerton, T. J. Observations of 3-6 daymeridional wind oscillations over the tropical Pacific, 1973-1992. *J. Atmos. Sci.* **50**:3292–3307, 1993.

Dunkerton, T. J. and Butchart, N. Propagation and selective transmission of internal gravity waves in a sudden warming. *J. Atmos. Sci.* **84**:1443–1460, 1984.

Durran, D. R. Pseudomomentum diagnostics for two-dimensional stratified compressible flow. *J. Atmos. Sci.* **52**:4010–4029, 1995.

Durran, D. R. Another look at downslope windstorms. Part I: The development of analogs to supercritical flow in an infinitely deep, continuously stratified fluid. *J. Atmos. Sci.* **43**:2527–2543, 1986.

Eaton, F.D., McLaughlin, S. A., and Hines, J. R. A new frequency-modulatedcontinuous wave radar for studying planetary boundary layermorphology. *Radio Sci.* **30**:75–88, 1995.

Eichinger, W. E., Cooper,D. I., Forman, P. R., Griegos, J., Osborn, M. A., Richter, D.,Tellier, L. L., and Thornton, R. The development of a scanningRaman water vapor lidar for boundary layer and tropospheric observations. *J. Atmos. Oceanic Technol.* **11**:1753–1766, 1999.

Eichinger, W. E.,Eichinger, H. L., Cooper, D. I., Krieger, J., and Carlson, E. Highaltitude activity associated with intermittent turbulence in astable atmosphere.

Einaudi, F. and Finnigan, J. J. The interaction between an internal gravity wave and the planetary boundary layer. *Quart. J. Roy. Meteor. Soc.* **107**:793–806, 1981.

Einaudi, F., Bedard, A. J., and Finnigan, J. J. A climatology of gravity waves and other coherent disturbances at the boulder atmospheric observatory during March–April 1984. *J. Atmos. Sci.* **46**:303–329, 1989.

Einaudi, F. and Hines, C. O. WKB approximation in application to acoustic-gravity waves. *Can. J. Phys.* **48**:1458–1471, 1970.

Einaudi, F. and Finnigan, J. J. The interaction between an internal gravity wave and the planetary boundary layer. *Quart. J. Roy. Meterol. Soc.* **107**:793–806, 1981.

Einaudi, F. and Finnigan, J. J. Wave–turbulence dynamics in the stably stratifiedboundary layer. *J. Atmos. Sci.* **50**:1841–1864, 1993.

Einaudi, F. and Lalas, D. Some new properties of Kelvin–Helmholtz waves in an atmosphere with and without condensation effects. *J. Atmos. Sci.* **31**:1995–2007, 1974.

Einaudi, F., Bedard, A. J.,and Finnigan, J. J. A climatology of gravity waves and othercoherent disturbances at the boulder atmospheric observatoryduring March-April 1984. *J. Atmos. Sci.* **46**:303–329, 1989.

Eliassen, A. and Palm, E. On the transfer of energy in stationary waves. *Geofys. Publ.* **22**:1–23, 1960.

Elliott, J. A. Instrumentation for measuring static pressure fluctuationswithin the atmospheric boundary layer. *Boundary-LayerMeteorol.* **2**:476–495, 1972.

Eloranta, E. W.,King, J. M., and Weinman, J. A. The determination of wind speed inthe boundary layer by monostatic lidar. *J. Appl. Meteorol.***14**:1485–1489, 1975.

Emmanuel, C. B. Richardson number profiles through shear instability waveregions observed in the lower planetary boundary layer. *Boundary-Layer Meteorol.* **5**:19–27, 1973.

Emmanuel, C. B., Bean, B.R., McAllister, L. G., and Pollard, J. R. Observations ofHelmholtz waves in the lower atmosphere with an acousticsounder. *J. Atmos. Sci.* **29**:886–892, 1972.

Ernst, J. A. SMS-1 nighttime infrared imagery of low level mountainwaves. *Mon. Weather Rev.* **104**:207–209, 1976.

Etling, D. and Raasch, S. Numerical simulation of vortex roll development during a cold air outbreak. *Dyn. Atmos. Oceans* **10**:277–290, 1987.

Evers, L. Infrasound monitoring in the Netherlands. *NAG J.* **176**:1–11, 2005.

Eymard, L. and Weill, A. Astudy of gravity waves in the planetary boundary layer by acousticsounding. *Boundary-Layer Meteorol.* **17**:231–245, 1979.

Farge, M. Wavelet transforms and their applications to turbulence. *Ann. Rev. Fluid Mech.* **24**:359–457, 1992.

Farley, D. T., Balsley, B.B., Swartz, W. E., and Hoz, C. L. Tropical winds measured by theArecibo radar. *J. Appl. Meteorol.* **18**:227–230,1979.

Ference, M. J., Stroud, W.G., Walsh, J. R., and Weisner, A. G. Measurement of temperaturesat elevations of 30 to 80 kilometers by the rocket-grenadeexperiment. *J. Meteorol.* **13**:5–12, 1959.

Finnigan, J. J. Kinetic energy transfer between internal gravity waves and turbulence. *J. Atmos. Sci.* **45**:486–505, 1988.

Finnigan, J. J. and Einaudi, F. The interaction of an internal gravity wave and theplanetary boundary layer. II: Effects of the wave on theturbulence structure. *Quart. J. Roy. Meteor. Soc.* **107**:807–832, 1981.

Finnigan, J. J., Einaudi, F., and Fua, D. The interaction between an internal gravity wave and turbulence in a stably-stratified nocturnal boundary layer. *J. Atmos. Sci.* **41**:2409–2436, 1984.

Fiorino, S. T. andCorreia Jr. J. Analysis of a mesoscale gravity wave event usingempirical orthogonal functions. *Earth Interact.* **6**:1–19, 2002.

Foken, T. *Micrometeorology*, 306 pp., 2008. Springer-Verlag.

Foufoula-Georgiou, E. and Kumar, P. *Wavelets in Geophysics*, 372 pp., 1994. Academic Press

Fovell, R., Durran, D., and Holton, J. R. Numerical simulations of convectively generated stratospheric gravity waves. *J. Atmos. Sci.* **49**:1427–1442, 1992.

Frehlich, R., Meillier,Y., Jensen, M. L., and Balsley, B. Turbulence measurements withCIRES tethered lifting system during CASES-99: Calibrationand spectral analysis of temperature and velocity. *J. Atmos.Sci.* **60**:2487–2495, 2003.

Frehlich, R. G.,Meillier, Y. P., Jensen, M. L., and Balsley, B. B. A statisticaldescription of small-scale turbulence in the low-level nocturnaljet. *J. Atmos. Sci.* **61**:1079–1085, 2004.

Fritts, D. C., Nappo, C., Riggin, D. M., Balsley, B. B., Eichinger, W. E., and Newsom, R. Analysis of ducted motions in the stable nocturnal boundary layer during CASES-99. *J. Atmos. Sci.* **60**:2450–2472, 2003.

Fritts, D. C. and Alexander, M. J. Gravity wave dynamics and effects in the middle atmosphere. In *Reviews of Geophysics, Vol. 41*, pp. 1–59, 2003. American Geophysical Union.

Fritts, D. C. A review of gravity wave saturation processes, effects, andvariability in the middle atmosphere. *Pure Appl. Geophys.***130**:343–371, 1989.

Fritts, D. C. and Dunkerton, T. J. A quasi-linear study of gravity wave saturation and self acceleration. *J. Atmos. Sci.* **41**:3272–3289, 1984.

Fritts, D. C. Gravity wave saturation in the middle atmosphere. a review of theory and observations. *Rev. Geophys. Space Phys.* **22**:275–308, 1984.

Fritts, D. C., Wang, L., Werne, J., Lund, T., and Wan, K. Gravity wave instability dynamics at high reynolds numbers. PartI: Wave field evolution at large amplitudes and high frequencies. *J. Atmos. Sci.* **66**:1126–1148, 2009.

Fritts, D. C. and Nastrom, G. D. Sources of mesoscale variability of gravity waves. Part II: Frontal, convective, and jet stream excitation. *J. Atmos. Sci.* **49**:111–127, 1992.

Fritts, D. C. and Alexander, M. J. Gravity wave dynamics and effects in the middle atmosphere. In *Reviews of Geophysics, Vol. 41*, pp. 1–59, 2003. American Geophysical Union.

Fritts, D. C. and Lund, T. Gravity waveinfluences in the thermosphere and ionosphere: Observations and recent modeling. In *Aeronomy of the Earth's Atmosphere and Ionosphere,* M. Abdu and D. Pancheva, eds., pp. 109–130, 2011. Springer.

Fritts, D. C. and VanZandt, T. E. Effects of Doppler shifting on the frequency spectra of atmospheric gravity waves. *J. Geophys. Res.* **92**:9723–9732, 1987.

Fritts, D. C., Tsuda, T., Sato, T., Fukao, S., andKato, S. Observational evidence of a saturated gravity wave spectrum in the troposphere and lower stratosphere. *J. Atmos. Sci.* **45**:1741–1759, 1988.

Fritts D. C. and Geller, M. A. Viscous stabilization of gravity wave critical level flows. *J. Atmos. Sci.* 33:2276–2284, 1976.

Fritts, D. C.,Blanchard, R. C., and Coy, L. Gravity wave structure between 60and 90 km inferred from space shuttle reentry. *J. Atmos.Sci.* **46**:423–434, 1989.

Fritts, D. C., Wang, D.Y., and Blanchard, R. C. Gravity wave and tidal structures between60 and 140 km inferred from space shuttle reentry data. *J.Atmos. Sci.* **50**:837–849, 1993.

Fritts, D. C. and Isler, J.R. Mean motions and tidal and two-day structure and variability inthe mesosphere and lower thermosphere over Hawaii. *J.Atmos. Sci.* **51**:2145–2164, 1994.

Fritts, D. C., Isler, J. R.,Hecht, J. H., Waletscheid, R. L., and Andreassen, O. Wave breakingsignatures in sodium densities and OH nightglow. 2. Simulationof wave and instability structures. *J. Geophys. Res.* **102**:6669–6684, 1997.

Fritz, S. The significance of mountain lee waves as seen from satellitepictures. *J. Appl. Meteorol.* **4**:31–37, 1965.

Fua, D. and Einaudi, F. On the effects of dissipation on shear instabilities in the stable atmospheric boundary layer. *J. Atmos. Sci.* **41**:888–900, 1984.

Fua, D., Einaudi, F., and Lalas, D. P. The stability analysis of an inflection-free velocity profile and its applications to the night-time boundary layer in the atmosphere. *Boundary-Layer Meteorol.* **10**:35–54, 1976.

Fua, D., Chimonas, G., Einaudi,F., and Zeman, O. An analysis of wave–turbulence interaction.*J. Atmos. Sci.* **39**:2450–2463, 1982.

Gardner, C. S. and Gardner, N. F. Measurement distortion in aircraft, space shuttle,and balloon observations of atmospheric density and temperatureperturbation spectra. *J. Geophys. Res.* **98**:1023–1033,1993.

Gardner, C. S. and Shelton, J. D. Density response of neutral atmospheric layers togravity wave perturbations. *J. Geophys. Res.* **90**:1745–1754, 1985.

Gardner, C. S. and Taylor, M. J. Observational limits for lidar, radar, and airglowimager measurements of gravity wave parameters. *J. Geophys.Res.* **103**:6427–6437, 1998.

Gardner, C. S. and Voelz,D. G. Lidar studies of the nighttime sodium layer over Urbana,Illinois. *J. Geophys. Res.* **92**:4673–4694, 1987.

Garratt, J. R. *The Atmospheric Boundary Layer*, 316 pp., 1992. Cambridge University Press.

Gauge, K. S. and Balsley, B. B. Doppler radar probing of the clear atmosphere. *Bull. Amer. Meteorol. Soc.* **59**:1074–1093, 1978.

Georges, T. M. and Greene, G. E. Infrasound from convective storms. Part IV. Isit useful for storm warning? *J. Appl. Meteorol.* **14**:1303–1316, 1975.

Gerding, M.,Rauthe, M., and Höffner, J. Temperature soundings from 1 to105 km altitude by combination of co-located lidars, and itsapplication for gravity wave examination. In *22nd InternationLaser Radar Conference (ILRC 2004), Proceedings of the Conferenceheld 12–16 July, 2004 in Matera, Italy*, (ESA SP- 561) Gelsomina PappalardoAldo Amodeo, eds., 2004. Paris European SpaceAgency.

Gill, A. E. *Atmosphere-Ocean Dynamics*, 662 pp., 1982. Academic Press.

Goldstein, S. On the stability of superposed streams of fluids of different densities. *Proc. R. Soc. Lond. A* **132**:524–548, 1931.

Gossard, E. E. Vertical flux of energy into the lower ionosophere from internal gravity waves generated in the troposphere. *J. Geophys. Res.* **67**:745–758, 1962.

Gossard, E. E. and Hooke, W. H. *Waves in the Atmosphere*, 456 pp., 1975. Elsevier.

Gossard, E. E. and Munk, W. On gravity waves in the atmosphere. *J. Meteorol.* **11**:259–269, 1954.

Gossard, E. E. and Sweezy, W. B. Dispersion and spectra of gravity waves in the atmosphere. *J. Atmos. Sci.* **31**:1540–1548, 1974.

Gossard, E. E. Vertical flux of energy intothe lower ionosphere from internal gravity waves generated in the troposphere. *J. Geophys. Res.* **67**:745–758, 1962.

Gossard, E. E. and Hooke, W. H. *Waves in the Atmosphere* 456 pp., 1975. Elsevier.

Grasso, L. D. The differentiation between grid spacing and resolution and their applications to numerical modeling. *Bull. Amer. Meteor. Soc.* **81**:579–580, 2000.

Greenhow, J. S. and Neufeld, E. L. Measurements of turbulence in the 80- to 100-km region from the radio echo observations of meteors. *J. Geophys. Res.* **64**:2129–2133, 1959.

Grisogono, B., Pryor, S. C., and Keislar, R. E. Mountain wave drag over double bell-shaped orography. *Quart. J. Roy. Meterol. Soc.* **119**:199–206, 1993.

Grisogono, B. Dissipation of waves in the atmospheric boundary layer. *J. Atmos. Sci.* **51**:1237–1243, 1994.

Grisogono, B. Wave drag effects in a mesoscale model with a higher-order closure turbulence scheme. *J. Appl. Meteor.* **34**:942–954, 1995.

Grivet-Talocia, S.,Einaudi, F., Clark, W. L., Diennett, R. D., Nastrom, G. D., andVanZandt, T. E. A 4-yr climatology of pressure disturbances usinga barometer network in Central Illinois. *Mon. WeatherRev.* **127**:1613–1629, 1999.

Grund, C. L., Banta, R. M.,George, J. L., Howell, J. N., Post, M. J., Richter, R. A., andWeickmann, A. M. High-resolution Doppler lidar for boundarylayer and cloud research. *J. Atmos. Oceanic Technol.* **18**:376–393, 2001.

Guest, F. M., Reeder, M. J., Marks, C. J., and Karoly, D. J. Inertia-gravity waves observed in the lower stratosphere over Macquarie Island. *J. Atmos. Sci.* **57**:737–752, 2000.

Haak, T., Wang, C., Garrett, S., Glazer, A., Mailhot, J., and Marshall, R. Mesoscale modeling of boundary layer refractivity and atmospheric ducting. *J. Appl. Meteorol. Climatol.* **49**:2437–2457, 2010.

Hamilton, K., Vincent, R. A., and May, P. T. Darwin area wave experiment(DAWEX) field campaign to study gravity wave generation andpropagation. *J. Geophys. Res.* **109**, 2004.

Hanna, S. R. andHoecker, W. H. The response of constant-density balloons tosinusoidal variations of vertical wind speeds. *J. Appl.Meteorol.* **10**:601–604, 1971.

Hardiman, S. C., Butchart, N., Osprey, S. M., Gray, L. J., and Bushell, A. C. The climatology of the middle atmosphere in a vertically extended version of the Met Office's Climate Model. Part I: Mean state. *J. Atmos. Sci.* **67**:1509–1525, 2010.

Hargreaves, J. K. Random fluctuations invery low frequency signals reflected obliquely from the ionosphere. *J. Atmospheric and Terrest. Phys.* **20**:155–166, 1961.

Haubrich, R. A. Array design. *Bull. Seismol. Soc. Amer.* **58**:977–991, 1968.

Hauf, T., Finke, U., Neisser, J., Bull, G., and Stangenberg, J.-G. A ground-based network for atmospheric pressure fluctuations. *J. Atmos. Oceanic Technol.* **13**:1001–1023, 1996.

Hazel, P. The effect of viscosity and heat conduction on internal gravity waves at a critical level. *J. Fluid Mech.* **30**:775–784, 1967.

Hazel, P. Numerical studies of inviscid stratified shear flows. *J. Fluid Mech.* **51** :39–61, 1972.

Hecht, J. H., Waltersheid, R.L., Fritts, D. C., Isler, J. R., Senft, D. C., Gardner, C. S., andFranke, S. J. Wave breaking signatures in OH airglow and sodiumdensities and temperatures. 1. Airglow imaging, Na lidar, andMF radar observations. *J. Geophys. Res.* **102**:6655–6668, 1997.

Held, I. M. Pseudomomentum and orthogonality of modes in shear flow. *J. Atmos. Sci.* **42**:2280–2288, 1985.

Herron, T. J. andTolstoy, I. Tracking jet stream winds from ground level pressuresensors. *J. Atmos. Sci.* **26**:266–269, 1969.

Hertzog, A., Boccara, G., Vincent, R. A., Vial, F., and Cocquerez, P. Estimation of gravity wave momentum flux and phase speeds from quasi-Lagrangian stratospheric balloon flights. Part II: Results from the Vorcore Campaign in Antartica. *J. Atmos. Sci.* **65**:3056–3070, 2008.

Hicks, J. J. and Angell,J. K. Radar observations of breaking gravity waves in the visuallyclear atmosphere. *J. Appl. Meteorol.* **7**:114–121,1968.

Hines, C. O. The Upper Atmosphere in Motion, American Geophysical Union, Washington, D.C., 1974.

Hines, C. O. The upper atmosphere in motion. *Quart. J. Roy.Meteorol. Soc.* **89**:1–42, 1963.

Hines, C. O. Atmospheric gravity waves: A new toy for the wave theorist. *Radio Sci.* **69D**:375–380, 1965.

Hines, C. O. *The Upper Atmosphere in Motion*, 1974. Washington, DC: American Geophysical Union.

Hines, C. O. A modeling of atmospheric gravity waves and wave drag generated buy isotropic and anisotropic terrain. *J. Atmos. Sci.* **45**:309–322, 1988.

Hines, C. O. Hydromagnetic resonance in ionospheric waves. *J. Atmos. Terr. Phys.* **7**:14–30, 1955.

Hines, C. O. On ray paths in mountain waves. *J. Atmos. Sci.* **45**:323–326, 1988.

Hines, C. O. Internal atmospheric gravity waves at ionospheric heights. *Can. J. Phys.* **38**:1441–1481, 1960.

Hines, C. O. Tropopasual mountain waves over Arecibo: A case study. *J. Atmos. Sci.* **46**:476–488, 1989.

Hines, C. O. Tidal oscillations, shorter period gravity waves and shear waves. *Meteorol. Monogr.* **9**:114–121, 1968.

Hines, C. O. On ray paths in mountain waves. *J. Atmos. Sci.* **45**:323–326, 1988.

Hines, C. O. Earlier days of gravity waves revisited. *Pure Appl.Geophys.* **130**:151–170, 1989.

Hines, C. O. Earlier days of gravity waves revisited. *Pure Appl. Geophys.* **130**:151 – 170, 1989a.

Hines, C. O. A modeling of atmospheric gravitywaves and wave drag generated buy isotropic and anisotropic terrain. *J. Atmos. Sci.* **45**:309–322, 1988.

Hines, C. O. The upper atmosphere in motion. *Quart. J. Roy. Meterol. Soc.* **89**:1–42, 1963.

Hines, C. O. Generalization of the Richardson criterion for the onset of atmospheric turbulence. *Quart. J. Roy. Meterol. Soc.* **97**:429–439, 1971.

Hirsh, J. H. and Booker,D. R. Response of supperpressure balloons to vertical air motions.*J. Appl. Meteorol.* **5**:226–229, 1966.

Hiscox, A. L., Nappo, C. J., and Miller, D. R. On the use of lidar images of smoke plumes to measure dispersion parameters in the stable boundary layer. *J. Atmos. Oceanic Technol.* **23**:1150–1154, 2006.

Hodges, R. R. Generation of turbulence in the upper atmosphere by internal gravity waves. *J. Geophys. Res.* **72**:3455–3458, 1967.

Hodges, R. R. Eddy diffusion coefficients due to instabilities in internal gravity waves. *J. Geophys. Res.* **74**:4087–4090, 1969.

Holton, J. R. The role of gravity wave induced drag and diffusion in the momentum budget of the mesosphere. *J. Atmos. Sci.* **39**:791–799, 1982.

Holton, J. R. *An Introduction to Dynamic Meteorology*, 535 pp., 2004. New York: Academic Press.

Holton, J. R. and Lindzen, R. S. An updated theory for the quasi-biennial cycle of the tropical stratosphere. *J. Atmos. Sci.* **29**:1076–1080, 1972.

Holton, J. R., Haynes, P.H., McIntyre, E., Douglass, A. R., Rood, R. B., and Pfister, L.Stratosphere-troposphere exchange. In *Reviews of Geophysics,Vol. 33*, pp. 403–439, 1995. American Geophysical Union.

Hooke, W. H. and Jones, R. M. Dissipative waves excited by gravity-wave encounters with the stably stratified planetary boundary layer. *J. Atmos. Sci.* **42**:2048–2060, 1986.

Hooke, W. F., Hall, F. F.,and Gossard, E. E. Observed generation of an atmospheric gravitywave by shear instability in the mean flow of the planetaryboundary layer. *Boundary-Layer Meteorol.* **5**:29–42,1973.

Horinouchi, T. Tropical cumulus convection and upward-propagating waves in middle-atmospheric GCMs. *J. Atmos. Sci.* **60**:2765–2782, 2003.

Howard, L. N. Note on a paper of John W. Miles, *J. Fluid Mech.* **10**:509–5126, 1961.

Hunt, J. C. R., Kaimal, J. C., and Gaynor, J. E. Some observations of turbulence structure in stable layers. *Quart. J. Roy. Meteor. Soc.* **111**:793–815, 1985.

Hussain, A. K. M. F. and Reynolds, W. C. The mechanics of an organized wave in a turbulent shear flow. *J. Fluid Mech.* **54**:241–258, 1972.

Jacobs,W. C. Atmospheric waves on isentropic surfaces as evidenced byinter-frontal ceiling oscillations. *Mon. Weather Rev.* 9–12,1937.

Jefferys, H. The flow of water in an inclined channel of rectangular cross section. *Philos. Mag.* **49**:793–807, 1925.

Jiang, Q., Doyle, J. D., and Smith, R. B. Interaction between trapped waves and boundary layers. *J. Atmos. Sci.* **63**:617–633, 2006.

Jiang, Q. and Doyal, J. D. Gravity wave breaking over the central Alps: Role of complex terrain. *J. Atmos. Sci.* **61**:2249–2266, 2008.

Jiang, Q., Doyle, J. D.,and Smith, R. B. Blocking, descent and gravity waves:Observations and modelling of a MAP northerly föhn event.*Quart. J. Roy. Meteor. Soc.* **131**:675–701, 2005.

Johny, C. J.,Sarkar, S. K., and Punyasesdu, D. Atmospheric phenomena deducedfrom radiosonde and gps occultation measurements for variousapplication related studies. *J. Earth Syst. Sci.* **118**:49–59, 2009.

Jones, W. L. and Houghton, D. D. The coupling of momentum between internal gravity waves and mean flow: A numerical study. *J. Atmos. Sci.* **28**:604–608, 1971.

Jones, W. Reflection and stability of waves in stably stratified flows shear flow: A numerical study. *J. Fluid Mech.* **34**:609–624, 1968.

Jordan,A. R. Atmospheric gravity waves from winds and storms. *J.Atmos. Sci.* **29**:445–456, 1972.

Julian, V., Kellogg, W., andSuomi, V. The TWERL experiment. *Bull. Amer. Meteorol. Soc.***58**:936–948, 1977.

Justus, C. G. and Woodrum, A. Upper atmospheric planetary-wave and gravity wave observations. *J. Atmos. Sci.* **30**:1267–1275, 1973.

Köpp,F., Schwiesow, R. L., and Werner, C. Remote measurements ofboundary-layer wind profiles using a CW Doppler lidar. *J.Clim. Appl. Meterol.* **23**:148–154, 1984.

Karacostas, T. S.and Marwitz, J. D. Turbulent kinetic energy budgets overmountainous terrain. *J. Appl. Meteorol.* **19**:163–174,1980.

Karoly, D. J., Roff, G. L., and Reed, M. J. Gravity wave activity associated with tropical convection detected in TOGA COARE sounding data. *Geophys. Res. Lett.* **23**:261–264,1996.

Keller, T. L. Implications of the hydrostatic assumption on atmospheric gravity waves. *J. Atmos. Sci.* **51**:1915–1929, 1994.

Kim, J. and Mahrt, L. Momentum transport by gravity waves. *J. Atmos. Sci.* **49**:735–748, 1992.

Kjelaas, A. G., Beran, D.W., Hooke, W. H., and Bean, B. R. Waves observed in the planetaryboundary layer using an array of acoustic sounders. *J. Atmos.Sci.* **31**:2040–2045, 1974.

Koch, S. E. and Golus, R. E. A mesoscale gravity wave event observed during CCOPE. Part I: Multiscale statistical analysis of wave characteristics. *Mon. Wea. Rev.* **116**:2527–2544, 1988.

Koch, S. E. *et al.* Turbulence and gravity waves within an upper-level front. *J. Atmos. Sci.* **62**:3885–3908, 2005.

Koppel, L. L., Bosart, L. F., and Keyser, D. A 25-yr climatology of large amplitude hourly surface pressure changes over the conterminous Unites States. *Mon. Wea. Rev.* **128**:51–68, 2000.

Kropfli,R. A. Radar probing and measurement of the planetary boundarylayer: Part II. Scattering from particluates. In *Probing the atmospheric boundary layer*, pp. 183–199, 1986.Boston, MA: American Meteorological Society.

Kwon, K. H. and Gardner,C. S. Airborne sodium lidar measurements of gravity wave intrinsicparameters. *J. Geophys. Res.* **95**:20,457–20,467, 1990.

Lac, C., Lafore, J.-P., and Redelsperger, J.-L. Role of gravity waves in triggering deep convection during TOGA COARE. *J. Atmos. Sci.* **59**:1293–1316, 2002.

Lalas, D. P. and Einaudi, F. On the characteristics of gravity waves generated by atmospheric shear layers. *J. Atmos. Sci.* 33:1248–1259, 1976.

Lalas, D. P. and Einaudi, F. Tropospheric gravity waves: Their detection by andinfluence on rawinsonde balloon data. *Quart. J. Roy. Meteor.Soc.* **106**:855–864, 1980.

Lane, T. P., Reeder, M. J.,Morton, B. R., and Clark, T. L. Observations and numericalmodelling of mountain waves over the Southern Alps. *Quart. J. Roy. Meteor. Soc.* **126**:2765–2788, 2000.

Laprise, J. P. R. An assessment of the WKBJ approximation to the vertical structure of linear mountain waves: Implications for gravity-wave drag parameterizations. *J. Atmos. Sci.* **50**:1469–1487, 1993.

LePichon, A.,Blanc, E., and Hauchecorne, A. *Infrasound Monitoring forAtmospheric Studies*, pp. 3–140, 2010. Dordrecht, TheNetherlands: Springer Science and Business Media.

Lee, X. and Barr, A. G. Climatology of gravity waves in a forest. *Quart. J. Roy. Meteorol. Soc.* **124**:1403–1419, 1998.

Lee, X., Neumann, H. H., Hartog, G. D., Fuentes, J. D., Black, T. A., Yang, P. C., and Blanken, P. D. Observation of gravity waves in a boreal forest. *Boundary-Layer Meteorol.* **84**:383–398, 1997.

Leutbecher, M.and Volkert, H. The propagation of mountain waves into thestratosphere: Quantitative evaluation of three-dimensionalsimulations. *J. Atmos. Sci.* **57**:3090–3108, 2000.

Levanon, N., Oehlkers, R.A., Ellington, S. D., Massman, W. M., and Suomi, V. E. On thebehavior of superpressure balloons at 150 mb. *J. Appl.Meteorol.* **13**:494–504, 1974.

Lighthill, M. J. *Waves in Fluids*, 504 pp., 1978. CambridgeUniversity Press.

Liller, W. and Whipple, F. L. High-altitude winds by meteor-train photography. In *Rocket Exploration of the Upper Atmosphere*, Vol. 1 Spec. Supp., J. Atmos. Terr. Phys., pp. 112–118, 1954.

Lilly, D. K. A severe downslope windstorm and aircraft turbulence event induced by a mountain wave. *J. Atmos. Sci.* **35**:59–77, 1978.

Lilly, D. K. and Zipser, E. J. The front range windstorm of 11 January 1972—a meteorological narrative. *Weatherwise* **25**:56–63, 1972.

Lilly, D. K. Wave momentum flux-a GARP problem. *Bull. Amer. Meteorol. Soc.* **53**:17–23, 1972.

Lilly, D. K. A severe downslope windstorm and aircraft turbulence event induced by a mountain wave. *J. Atmos. Sci.* **35**:59–77, 1978.

Lilly, D. K. and Gal-Chen, T. *Mesoscale Meteorology - Theories, Observations and Models*, 796 pp., 1983. Kluwer.

Lilly, D. K. and Kennedy, P. J. Observations of a stationary mountain wave and its associated momentum flux and energy dissipation. *J. Atmos. Sci.* **30**:1135–1152, 1973.

Lilly,D. K. Observations of mountain-induced turbulence. *J.Geophys. Res.* **76**:6585–6588, 1971.

Lilly,D. K. Wave momentum flux-a GARP problem. *Bull. Amer.Meteorol. Soc.* **53**:17–23, 1972.

Lilly, D. K., Pann, Y.,Kennedy, P., and Tootenhoofd, W. Data catalog for the 1970Colorado Lee Wave Observational Program Technicalreport, Tech. Note NT/STR-72, p. 190, 1971. Boulder, CO:National Center for Atmospheric Research.

Lin, Y.-L. and Chun, H.-Y. Effects of diabatic cooling in a shear flow with a critical level. *J. Atmos. Sci.* **48**:2476–2491, 1991

Lindzen,R. S. Thermally driven diurnal tide in the atmosphere. *Quart.J. Roy. Meteorol. Soc.* **93**:18–42, 1967.

Lindzen, R. S. Turbulence and stress owing to gravity wave and tidal breakdown. *J. Geophys. Res.* **86**:9707–9714, 1981.

Lindzen, R. S. Supersaturation of vertically propagating internal gravity waves. *J. Atmos. Sci.* **45**:705–711, 1987.

Lindzen, R. S. and Holton, J. R. A theory of the quasi-biennial oscillation. *J. Atmos. Sci.* **25**:1095–1107, 1968.

Little,C. G. Acoustic methods for the remote probing of the loweratmosphere. *Proc. IEEE* **57**:571–578, 1969.

Long, R. R. Some aspects of the flow of stratified fluids. I. A theoretical investigation, *Tellus* **5**:42–57, 1953.

Lott, F. and Miller,M. J. A new subgrid-scale orographic drag parameterization: Itsformulation and testing. *Quart. J. Roy. Meteorol. Soc.* **123**:101–127, 1997.

Lott, F. The reflection of a stationary gravity wave by a viscus boundary layer. *J. Atmos. Sci.* **64**:3363–3371, 1998.

Lott, F. and Teitelbaum, H. Topographic waves generated by a transient wind. *J. Atmos. Sci.* **50**:2607–2624, 1993.

Ludlum, F. H. Characteristics of billow clouds and their relation to clear-air turbulence. *Quart. J. Roy. Meterol. Soc.* **93**:419–435, 1967.

Mahrt, L. Vertical structure and turbulence in the very stable boundary layer. *J. Atmos. Sci.* **42**:2333–2349, 1985.

Malkus, J. S. and Stern, M. E. The flow of a stable atmosphere over a heated island. Part I. *J. Appl. Meteor.* **10**:30–41, 1953.

Marks, C. J. and Eckermann, S. D. A three-dimensional nonhydrostatic ray-tracing model for gravity waves: formulation and preliminary results for the middle atmosphere. *J. Atmos. Sci.* **52**:1959–1984, 1995.

Martyn, D. F. Cellular atmospheric waves in the ionosphere andtroposphere. *Proc. Roy. Soc.* **A 201**:216–234, 1950.

Martyn, D. F. Cellular atmospheric waves in the ionosphere andtroposphere. *Proc. Roy. Soc. A* **201**:216–234, 1950.

Mason, P. J. and Sykes, R. I. A two-dimensional numerical study of horizontal roll vortices in an inversion capped planetary boundary layer. *Quart. J. Roy. Meterol. Soc.* **108**:801–823, 1982.

Massman, W. J. On the nature of vertical oscillations of constant volumeballoons. *J. Appl. Meteorol.* **17**:1351–1356, 1978.

Mastrantonio, G., Einaudi, F., and Fua, D. Generation of gravity waves by jet streams in the atmosphere. *J. Atmos. Sci.* **33**:1730–1738, 1976.

Matsuno, T. Quasi-geostrophic motions in themiddle atmosphere. *J. Meteorol. Soc. Jpn.* **44**:25–32, 1966.

McAllister, L. G.,Pollard, J. R., Mahoney, A. R., and Shaw, P. J. R. Acousticsounding: A new approach to the study of atmospheric structure.*Proc. IEEE* **57**:597–587, 1969.

McFarlane, N. A. The effect of orographically excited gravity wave drag on the general circulation of the lower stratosphere and troposphere. *J. Atmos. Sci.* **44**:1775–1800, 1987.

McIntyre, M. E. On the 'wave momentum' myth. *J. Fluid Mech.* **106**:331–347, 1981.

McLandres, C. On the importance of gravity waves in the middle atmosphere and their parameterization in general circulation models. *J. Atmos. Sol. Terr. Phys.* **60**:1357 1383, 1998.

McLandress, C. and Scinocca, J. F. The GCM response to current parameterizations of nonorographic gravity wave drag. *J. Atmos. Sci.* **62**:2394–2413, 2005.

Meek, C. E. and Manson, A. H. Vertical motions in the upper middle atmosphere from the Saskatoon (52°N, 107°W) M.F.Radar. *J. Atmos. Sci.* **46**:849–858, 1989.

Meillier, Y. P., Frehlich, R. G., Jones, R. M., and Balsley, B. B. Modulation of small-scale turbulence by ducted gravity waves in the nocturnal boundary layer. *J. Atmos. Sci.* **65**:1414–1427, 2008.

Melbourne, W. G. et al.The application od spaceborne GPS to atmospheric limb soundingand global change monitoring Technical report, 1994. Pasadena, CA:National Aeronautics and Space Administration, Jet PropulsionLaboratory, Calif. Instit. Tech. (JPL Publication 94-18).

Meneveau, C. Analysis of turbulence in the orthonormal wavelet representation. *J. Fluid Mech.* **232**:469–520, 1991.

Mentes, S. S., and Kaymaz, Z. Investigation of surface duct conditions over Istanbul, Turkey. *J. Appl. Meteorol. Climatol.* **46**:318–337, 2007.

Merrill,J. T. Observational and theoretical study of shear instability inthe airflow near the ground. *J. Atmos. Sci.* **34**:911–921, 1977.

Mikesh, R. C. *Japan's World War II Balloon Bomb Attach on NorthAmerica*, p. 85, 1990. Smithsonian Institution Press.

Miksad, R. W. An omni-directional static pressure probe. *J. Appl.Meteorol.* **15**:1215–1225, 1976.

Miles, J. W. On the stability of heterogeneous shear flows. *J. Fluid Mech.* **10**:496–508, 1961.

Miller, M. J., Palmer, T. N., and Swinbank, R. Parameterization and influence of subgrid-scale orography in general circulation and numerical weather prediction models. *Meteorol. Atmos. Phys.* **40**:84–109, 1989.

Miyhara, S. Suppression of stationaryplanetary waves by internal gravity waves in the mesosphere. *J. Atmos. Sci.* **42**:100–107, 1985.

Mohankumar, K. *Stratosphere Troposphere Interactions. An introduction*, 416 pp., 2008. Springer.

Molina, A. Sodium nightglow and gravity waves. *J. Atmos. Sci.* **40**:2444–2450, 1983.

Monserrat, S., and Thorpe, A. J. Use of ducting theory in an observed case of gravity waves. *J. Atmos. Sci.* **53**:1724–1736, 1996.

Moustaoui, M.,Teitelbaum, H., van Velthoven, P. F. J., and Kelder, H. Analysisof gravity waves during the POLINAT experiment and someconsequences of stratospheric-tropospheric exchange. *J.Atmos. Sci.* **56**:1019–1030, 1999.

Munasinghe, G., Hur,H., Huang, T. Y., Bhattachryya, A., and Tuan, T. F. Application ofthe dispersion formula to long-and short-period gravity waves:Comparisons with ALOHA-93 data and an analytical model. *J. Geophys. Res.* **103**:6467–6481, 1998.

Murakami, H., Y. Wang, H. Yoshimura, R. Mizuta, M. Sugi, E. Shindo, Y. Adachi, S. Yukimoto, M. Hosaka, S. Kusunoki, T. Ose, A. Kitoh, 2012: Future changes in tropical cyclone activity projected by the new high-resolution MRIAGCM. *J. Climate*, in press.

Murayama, Y., Igarashi,K., Yamazak, R., Oyama, K.-I., Tsuda, T., Nakamura, T., Fukao,S., Widdel, H.-U., and Schlegel, K. Cooperative wind observationin the upper mesosphere and lower thermosphere with foil chafftechnique, the MU radar, and Yamagawa MF radar. *EarthPlanet. Space* **51**:719–729, 1999.

Muschinski, A. et al. Fine-scale measurements of turbulence in the lowertroposphere: An intercomparison between a kite-andballoon-borne, and a helicopter-borne measurement system. *Boundary-Layer Meteorol.* **98**:219–250, 2001.

Nakamura, T., Aono, T., Tsuda, T., Admiranto, A. G., Achmad, E., and Suranto, Mesospheric gravity waves over a tropical region observed by OH airglow imagining in Indonesia. *Geophys. Res. Lett.* **30**, 2003

Nappo, C. J., Miller, D. R., and Hiscox, A. L. Wave-modified flux and plume dispersion in the stable boundary layer. *Boundary-Layer Meteorol.* **129**:211–220, 2008.

Nappo, C. J., Chun, H.-Y., and Lee, H.-J. A parameterization of wave stress in the planetary boundary layer for use in mesoscale models. *Atmos. Environ.* **38**:2665–2675, 2004.

Nappo, C. J., Miller, D. R., and Hiscox, A. L. Wave-modified flux and plume dispersion in the stable boundary layer. *Boundary-Layer Meteorol.* **129**:211–220, 2008.

Nappo, C. J. and Chimonas, G. Wave exchange between the ground surface and a boundary-layer critical level. *J. Atmos. Sci.* **49**:1075–1091, 1992.

Nappo, C. J. and Physick, W. Gravity wave stress parameterization in a mesoscale sea breeze model. In *11th Conference on Air Pollution Meteorology with the A and WMA, 9–14 January, Long Beach CA*, pp. 263–268, 2000. Boston, MA: Am. Meteorol. Soc.

Nappo, C. J. and Svensson, G. A parameterization of wave stress over three-dimensional topography. In *18th Symposium on Boundary Layer and Turbulence. Stockholm Sweden,* 2008. Boston, MA: Am. Meteorol. Soc.

Nappo, C. J. and Chimonas, G. Wave exchange between the ground surface and a boundary-layer critical level. *J. Atmos. Sci.* **49**:1075–1091, 1992.

Nappo, C. J., Miller, D. R., and Hiscox, A. L. Wave-modified flux and plume dispersion in the stable boundary layer. *Boundary-Layer Meteorol.* **129**:211–220, 2008.

Nappo, C. J. and Rao, K. S. A model study of pure katabatic flows. *Tellus* **39A**:61–71, 1987.

Nappo, C. J. and Johansson, P.-E. Summary of the Lövånger internal workshop on turbulence and diffusion in the stable planetary boundary layer. *Boundary-Layer Meteorol.* **90**:345–374, 1999.

Nappo, C. J., Eckman,R. M., and Coulter, C. L. An episode of wave-generated turbulencein the stable boundary layer. In *Tenth Symposium onTurbulence and Diffusion*, pp. 118–122, 1992. Boston, MA:American Meteorological Society.

Nappo, C. J., Crawford, T.L., Eckman, R. M., and Auble, D. L. A high-precision sensitiveelectronic microbarograph network. In *7th Symposium onMeteorological Observations and Instrumentation*, pp. j179–j181,1991. Boston, MA: American Meteorological Society.

Nappo, C. J., Auble, D. L.,Dumas, E., Cuxart, J., Morales, G., Yague, C., and Terradellas, E.Coherent pressure disturbances during cases-99. In *14thSymposium on Boundary Layer and Turbulence, 7–11 August,Aspen CO*, 2000. Boston, MA: American Meteorological Society.

Nastrom, G. D., Balsley, B. B., and Carter, D. A. Mean meriodinal winds in the mid- and high latitude summer mesosphere. *Geophys. Res. Lett.* **9**:139, 1982.

Nastrom,G. D. The response of supperpressure balloons to gravity waves.*J. Appl. Meteorol.* **19**:1013–1019, 1980.

Neff, W. D. and Coulter,R. L. Acoustic remote sensing. In *Probing the AtmosphericBoundary Layer*, D.H. Lenschow, ed., pp. 201–239, 1986. Boston,MA: American Meteorological Society.

Newsom, R. K. and Banta, R. M. Shear-flow instability in the stable nocturnal boundary layer as observed by Doppler lidar during CASES-99. *J. Atmos. Sci.* **60**:16–33, 2003.

Newsom, R. K., Banta, R.M., Otten, J., Eberhard, W. L., and Lundquist, J. K. Doppler lidarobservations on internal gravity waves, shear instability andturbulence during CASES-99. In *14th Symposium on BoundaryLayer and Turbulence, 7–11 August 2000, Aspen, CO*, pp. 362–365,2000. Boston, MA: American Meteorological Society.

Ottersten, H.,Hardy, K. R., and Little, C. G. Radar and sodar probing of wavesand turbulence in statically stable clear-air layers. *Boundary-Layer Meteorol.* **4**:47–89, 1973.

Palm, E. Multiple-layer mountain wave models with constant stability and shear. Technical Report, Sci. Rep., 3, Contract No. AF 19(604)-728, 1955. Air Force Cambridge Research Center.

Palmer, T. N.,Shutts, G. J., and Swinbank, R. Alleviation of a systematicwesterly bias in general circulation and numerical weatherprediction through an orographic gravity wave dragparameterization. *Quart. J. Roy. Meteorol. Soc.* **112**:1001–1040, 1986.

Peltier, W. R. and Calrk, T. L. The evolution and stability of finite amplitude mountain waves. Part II. Surface wave drag and severe downslope windstorms. *J. Atmos. Sci.* **36**:1498–1529, 1979.

Peterson, A. W.and Keiffaber, L. M. Infrared photography of OH airglowstructures. *Nature* **242**:321–322, 1973.

Pham, H. T. and Sarkar, S. Internal waves and turbulence in a stable stratified jet. *J. Fluid Mech.* **648**:297–324, 2010.

Phillips, O. O. *The Dynamics of the Upper Ocean*, 336 pp., 1977. Cambridge University Press.

Piani, C., Durran, D., Alexander, M. J., and Holton, J. R. The propagation of a gravity-inertia wave in a positively sheared flow. *J. Atmos. Sci.* **57**:3689–3702, 2000.

Pielke, R. A. *Mesoscale Meteorological Modeling*, 612 pp., 1984. Academic Press.

Pierce, A. D. and Coroniti, S. C. A mechanism for the generation of acoustic-gravity waves during thunderstorm formation. *Nature* **210**:1209–1210, 1960.

Pitteway, M. L. and Hines, C. O. The reflection and ducting of atmospheric acoustic-gravity waves. *Can. J. Phys.* **43**:2222–2243, 1965.

Poulos, G. S., Blumen, W., Fritts, D. C., Lundquist, J. K., Sun, J., Burns, S. P., Nappo, C., Banta, R., Newsom, R., Cuxart, J., Terradellas, E., Balsley, B., and Jensen, M. CASES-99: A comprehensive investigation of the stable nocturnal boundary layer. *Bull. Amer. Meteor. Soc.* 2001.

Prandtl, L. andTeitjens, O. G. *Applied Hydro- and Aeromechanics*, p. 311,1937. New York: Dover Publications.

Preusse, P. S. D. E. and Ern, M. Transparency of the atmosphere to short horizontal wavelength gravity waves. *J. Geophys. Res.* **113**:D24104, 2008.

Queney, P. The problem of air flow over mountains: A summary of theoretical studies. *Bull. Am. Meteorol. Soc.* **29**:16–26, 1948.

Radok,U. A procedure for studying mountain effects at low levels. *Bull. Amer. Meteorol. Soc.* **35**:412–416, 1954.

Ralph, F. M., Neiman, P. J.,Keller, T. L., Levinson, D., and Fedor, L. Observations,simulations, and analysis of nonstationary trapped lee waves. *J. Atmos. Sci.* **54**:1308–1333, 1997.

Randall, D. A. *General Circulation Model Development*, 2000. San Diego: Academic Press.

Rayleigh, J. W. S. *The Theory of Sound, Vol.* II, 504 pp., 1945. New York: Dover (reprint of second edition of 1894).

Rees, J. M., Denholm-Price, J. C. W., King, J. C., and Anderson, P. S. A clomatological study of internal gravity waves in the atmospheric boundary layer overlying the Brunt Ice Shelf, Antarctica. *J. Atmos. Sci.* **57**:511–526, 2000.

Rees, J. M. and Mobbs, D. D. Studies of internal gravity waves at Hally Station, Antarctica, using wind observations. *Quart. J. Roy. Meteor. Soc.* **114**:939–966, 1988.

Rees, J. M.,Staszewski, W. J., and Winkler, J. R. Case study of a wave eventin the stable atmospheric boundary layer overlying an Antarticice shelf using orthogonal wavelet transform. *Dyn. Atmos.Oceans* **34**:245–261, 2001.

Reid, S. J. An observational study of lee waves using radiosonde data.*Tellus* **24**:593–596, 1972.

Reynolds, W. C. and Hussain, A. K. M. F. The mechanics of an organized wave in turbulent shear flow. Part III: Theoretical models and comparisons with experiments. *J. Fluid Mech.* **54**:263–288, 1972.

Richter, J. H. High resolution tropospheric radar sounding. *RadioSci.* **4**:1261–1268, 1969.

Rontu, L. *Studies on orographic effects in a numerical weather prediction model*, p. 206, 2007. Helsinki: Finnish Meteorological Institute.

Rosenhead, L. The formation of vortices from a surface discontinuity. *Proc. Roy. Soc. A* **143**:170–192, 1931.

Rottman, J. W. and Smith, R. B. A laboratory model of severe downslope winds. *Tellus* **41A**:401–415, 1989.

Rottman, J. W. andEinaudi, F. Solitary waves in the atmosphere. *J. Atmos. Sci.***50**:2116–2136, 1993.

Schoeberl, M. R. A model of stationary gravity wave breakdown with convective adjustment. *J. Atmos. Sci.* **45**:980–992, 1988.

Schoeberl, M. R. A ray tracing model ofgravity wave propagation and breakdown middle atmosphere. *J. Geophys. Res.* **90**:7999–8010, 1985.

Schwiesow,R. L. Lidar measurements of boundary-layer variables. In *Probing the Atmospheric Boundary Layer*, D. H. Lenschow, pp.139–162, 1986. Boston, MA: American Meteorological Society.

Scorer, R. S. Theory of waves in the lee of mountains. *Quart. J. Roy. Meteor. Soc.* **75**:41–56, 1949.

Shutts, G. Gravity-wave drag parameterization over complex terrain: The effect of critical-level absorption in directional wind-shear. *Quart. J. Roy. Meteorol. Soc.* **121**:1005–1021, 1995.

Shutts, G. J., Healey,P., Mobbs, S. D. A multiple sounding technique for the study ofgravity waves. *Quart. J. Roy. Meteor. Soc.* **120**:59–79,1994.

Shutts, G. J., Kitchen,M., Hoare, P. H. A large amplitude gravity wave in the lowerstratosphere detected by radiosonde. *Quart. J. Roy. Meteor.Soc.* **114**:579–594, 1988.

Simkhada, D. B., Snively, J. B., Taylor, M. J., and Franke, S. J. Analysis and modeling of ducted and evanescent gravity waves observed in the Hawaiian airglow. *Ann. Geophys.* **27**:3213–3224, 2009.

Sinclair, P. C. Some preliminary dust devil measurements. *Mon. WeatherRev.* **22**:363–367, 1937.

Skamarock, W. Evaluating mesoscale NWP models using kinetic energy spectra. *Mon. Weather Rev.* **132**:3019–3032, 2004.

Smith, R. B., Jiang, Q., and Doyle, J. D. A theory of gravity wave absorption by a boundary layer. *J. Atmos. Sci.* **63**:774–781, 2006.

Smith, R. B. A measurement of mountain drag. *J. Atmos. Sci.* **35**:1644–1654, 1978.

Smith, R. B. The generation of lee waves by the Blue Ridge. *J. Atmos. Sci.* **33**:507–519, 1976.

Smith, R. B. The steepening of hydrostatic mountain waves. *J. Atmos. Sci.* **34**:1634–1654, 1977.

Smith, R. B. The influence of mountains on the atmosphere. In *Advances in Geophysics, Vol.* 21, pp. 87–230, 1979.

Smith, R. B. Linear theory of stratified hydrostatic flow past an isolated mountain. *Tellus* **32**:348–364, 1980.

Smith, R. B. and Lin, Y.-L. The addition of heat to a stratified airstream with application to the dynamics of orographic rain. *Quart. J. Roy. Meterol. Soc.* **108**:353–378, 1982.

Smith, R. B., Skubis, S., Doyal, J. D., Broad, A. S., Kiemle, C., and Volkert, H. Mountain waves over Mont Blanc: Influence of a stagnant boundary layer. *J. Atmos. Sci.* **59**:2073–2092, 2002.

Smith, R. B. The generation of lee waves by the Blue Ridge. *J. Atmos. Sci.* **33**:507–519, 1976.

Smith, R. B. On severe downslope winds. *J. Atmos. Sci.* **42**:2597–2603, 1985.

Smith, S. A., Fritts,D. C., and VanZandt, T. E. Evidence of a saturation spectrum ofatmospheric gravity waves. *J. Atmos. Sci.* **44**:1404–1410, 1989.

Smith,R. B. The generation of lee waves by the Blue Ridge. *J.Atmos. Sci.* **33**:507–519, 1976.

Smith, R. B., Doyle, J.D., Jiang, Q., and Smith, S. A. Alpine gravity waves: Lessonsfrom MAP regarding mountain wave generation and breaking. *Quart. J. Roy. Meteor. Soc.* **133**:917–936, 2007.

Snyder, W. H., Thompson, R. S., Eskridge, R. E., and Lawson, R. E. The structure of strongly stratified flows over hills: Dividing-streamline concept. *J. Fluid Mech.* **152**:249–288, 1985.

Spiegel, E. A. and Veronis, G. On the Boussinesq approximation for a compressible fluid. *Astrophys. J.* **131**:442–447, 1960.

Sroga, J. T., Eloranta,E. W., and Barber, T. Lidar measurements of wind velocity profilesin the boundary layer. *J. Appl. Meteorol.* **19**:598–605,1980.

Stern, W. F. and Pierrehumbert, R. T. The impact of an orographic gravity wave drag parameterization on extended range predictions with a GCM. In *Eighth Conference on Numerical Weather Prediction*, pp. 745–750, 1988. Am. Meteorol. soc.

Stewart, R. W. Turbulence and waves in a stratified atmosphere. *Radio Sci.* **4**:1269–1278, 1969.

Stobie, J. G., Einaudi, F., and Uccellini, L. W. A case study of gravity waves - convective storms interactions: 9 may 1997, *J. Atmos. Sci.* **40**:2804–2830, 1983.

Strauch, R. G., Campbell,W. C., Chadwick, R. B., and Moran, K. P. Microwave FM-CW radarfor boundary layer probing. *Geophys. Res. Lett.* **3**:193–196, 1976.

Stull, R. B. *An Introduction to Boundary Layer Meteorology*, 666 pp., 1988. Dordrecht, The Netherlands: Kluwer Academic Publishers.

Sun, J., Lenschow, D.H., Burns, S.B., Banta, R.M., Coulter, R., Frasier, S., Ince, T., Nappo, C., Balsley, B.B., Jenson, M., Mahrt, L., Miller, D. and Skelly, B. Atmospheric disturbances that generate intermittent turbulence in nocturnal boundary layers. *Boundary-Layer Meteorol.* **110**:255–279, 2004 .

Sutherland, B. R., Caulfield, C. P., and Peltier, W.R. Internal gravity wave generation and hydrodynamic instability. *J. Atmos. Sci.* **51**:3261–3280, 1994.

Sutton, O. G. *Micrometeorology*, 1953. New York: McGraw-Hill Book Company, Inc.

Swenson, G. R., Alexander, M. J., and Haque, R. Dispersion imposed limits on atmospheric gravity waves in the mesosphere: Observations from OH airglow. *Geophys. Res. Lett.* **27**:875–878, 2000.

Swenson, G. R. andGardner, C. S. Analytical models for the response of themesospheric OH* and Na layers to atmosphericgravity waves. *J. Geophys. Res.* **103**:6271–6294, 1998.

Symons,G. J. On barometric oscillations during thunderstorms, and on thebarometer, an instrument designed to facilitate their study. *Proc. Roy. Soc. London* **48**:59–68, 1890.

Tanaka, H. A slowly varying model of the lowerstratospheric zonal wind minimum induced by mesoscale mountain wave breaking. *J. Atmos. Sci.* **43**:1881–1892, 1996.

Taylor, G. I. Effect of variation in density on the stability of superposed streams of fluid. *Proc. R. Soc. Lond. A* **201**:499–523, 1931.

Taylor, M. J., Turnbull,D. N., and Lowe, R. P. Spectrometric and imaging measurements of aspectacular gravity wave event observed during the ALOHA-93campaign. *Geophys. Res. Lett.* **22**:2848–2852, 1995.

Teixeira, M. A. C., Miranda, P. M. A., and Valente, M. A. An analytical model of mountain wave drag for wind profiles with shear and curvature. *J. Atmos. Sci.* **61**:1040–1054, 2004.

Teixeira, M. A. C. and Miranda, P. M. A. On the momentum flux associated with mountain waves in directionally sheared flows. *J. Atmos. Sci.* **66**:3419–3433, 2009.

Thomas, L., Worthington, R. M., and McDonald, A. J. Inertia-gravity waves in the troposphere and lower stratosphere associated with a jet stream exit region. *Ann. Geophys.* **17**:115–121, 1999.

Thorpe, S. A. On the reflection of a train of finite-amplitude internal waves from a uniform slope. *J. Fluid Mech.* **178**:279–302, 1987.

Thorpe, S. A. An experimental study of critical levels. *J. Fluid Mech.* **32**:693–704, 1981.

Tjernström, M. and Mauritsen, T. Mesoscale variability in the summer Arctic boundary layer. *Boundary-Layer Meteorol.* **230**:383–406, 2009.

Torrence, C. T. and Compo, G. P. A practical guide to wavelet analysis. *Bull. Amer. Meteor. Soc.* **79**:61–78, 1998.

Treviño, G. and Andreas, E. L. On wavelet analysis of nonstationary turbulence.*Boundary-Layer Meteorol.* **81**:271–288, 1996.

Tsuda, T., Murayama, Y., Wiryosumarto, H., Harijono, S.-W. B., and Kato, S. Radiosonde observations of equatorial atmosphere dynamics over indonesia, Part I: Equatorial waves anddiurnal tides. *J. Geophys. Res.* **99**:10491–10505, 1994.

Tsuda, T. and Hocke, K. Application of GPS radio occultation data for studies ofatmospheric waves in the middle atmosphere and ionosphere. *J.Meteorol. Soc. Jpn.* **82**:419–426, 2004.

Tsuda, T. Inoue, T., Fritts,D. C., VanZandt, T. E., Kato, S., Sato, T., and Fukao, S. MSTradar observations of a saturated gravity wave spectrum. *J.Atmos. Sci.* **46**:2440–2447, 1989.

Tsuda, T., Ratnam, M. V.,May, P. T., Alexander, M. J., Vincent, R. A., and MacKinnon, A.Characteristics of gravity waves with short vertical wavelengthsobserved with radiosonde and GPS occultation during DAWEX (DarwinArea Wave Experiment). *J. Geophys. Res.* **109**:1–13,2004.

Turner, J. S. *Buoyancy Effects in Fluids*, 1973. Cambridge University Press.

Uccellini, L. W. A case study of apparent gravity wave initiation of severe convective storms. *Mon. Wea. Rev.* **103**:497–513, 1975.

Uccellini, L. W. and Kock, S. E. The synoptic setting and possible energy sources for mesoscale wave disturbances. *Mon. Weather Rev.* **115**:721–729, 1987.

Vargas, F., Gobbi, D., Takahashi, H., and Lima, L. M. Gravity wave amplitudes andmomentum fluxes inferred from OH airglow intensities and meteorradar winds during SpreadFEx. *Ann. Geophys.* **27**:2361–2369, 2009.

Vergeiner, I. and Lilly, D. K. The dynamic structure of lee wave flow as obtained from balloon and air plane observations. *Mon. Weather Rev.* **98**:220–238, 1970.

Vincent, R. A., Allen, S. J., and Eckermann, S. D. Gravity wave parameters in the lower stratosphere. In *Gravity Wave Processes: Their Parameterization in Global Climate Models*. pp. 7–25, 1997. Springer-Verlag.

Vincent, R. A. and Alexander, M. J. Gravity waves in the tropical lower troposphere: An observational study of seasonal and interannual variability. *J. Geophys. Res.* **105**:17,971–17,982, 2000.

Vincent, R. A. Gravity wave coupling from below: A review. In *Climate and Weather of the Sun-Earth System (CAWSES): Selected Papers from the 2007 Kyoto Symposium*, T. Tsuda, R. Fujii, K. Shibata, and M. A. Geller, eds., pp. 279–293, 2009. Tokyo: TERRAPUB.

Vincent, R. A. and Reid, I.M. HF Dopplar measurements of mesocale gravity wave momentumfluxes. *J. Atmos. Sci.* **40**:1321–1333, 1983.

Walker, K. T. andHedlin, A. H. Chapter 5: A review of wind-noise reductionmethodologies. In *Infrasound Monitoring for AtmosphericStudies*, A. Le Pichon, E. Blanc, and A. Hauchecorne, eds., pp.141–182, 2010. Springer Science, Geosciences.

Wallace, J. M. and Kousky, V. E. Observational evidence of Kelvin waves in the tropical stratosphere. *J. Atmos. Sci.* **25**:900–907, 1968.

Walterscheid, R. L., Schubert, G., and Brinkman, D. G. Small-scale gravity waves in the upper mesosphere and lower thermosphere generated by deep tropical convection. *J. Geophys. Res.* **106**:31825–31832, 2001.

Wang, S. and Zhang, F.Sensitivity of mesoscale gravity waves to the baroclinicity ofjet-front systems. *Mon. Weather Rev.* **135**:670–688,2007.

Ware, R. et al. GPSsounding of the atmosphere from low Earth orbit: Primaryresults. *Bull. Amer. Meteorol. Soc.* **77**:19–40, 1996.

Warner, T. T.*Numerical Weather and ClimatePrediction*, 548 pp., 2011. Cambridge University Press.

Waters, J. W. Microwave limb-sounding of Earth's upper atmosphere. *Atmos. Res.* **23**:391–410, 1989.

Waters, J. W., Kunzi, K.F., Pettyjohn, R. L., Poon, R. K. L., and Staelin, D. H. Remotesensing of atmospheric temperature profiles with the Nimbus 5microwave spectrometer. *J. Atmos. Sci.* **32**:1953–1969,1975.

Waters, J. W. et al. TheURS and EOS microwave limb sounder MLS experiments. *J.Atmos. Sci.* **56**:194–218, 1999.

Wehrbein, W. M. and Levoy, C. B. An accurate radiative heating and cooling algorithm for use in a dynamic model of he middle atmosphere. *J. Atmos. Sci.* **39**:1532–1544, 1982.

Wilczak, J. M. and Bedard, A. J. A new turbulence microbarometer and its evaluationusing the budget of horizontal heat flux. *J. Atmos. Sci.***21**:1170–1181, 2004.

Willmarth, W.W. and Wooldridge, C. E. Measurements of fluctuating pressure atthe wall beneath a thick turbulent boundary layer. *J. FluidMech.* **14**:187–210, 1962.

Wu, D. L. and Eckermann, S. Global gravity wave variances from aura:Characteristics and interpretation. *J. Atmos. Sci.* **65**:3695–3718, 2008.

Wu, Y.-F. and Widdle, H.-U.Saturated gravity wave spectrum in the pole summer lowertroposphere observed by foil chaff during campaign "Sodium 88".*J. Appl. Meteorol.* **49**:1781–1789, 1992.

Wu, Y.-F., Xu, J., Widdel,H.-U., and Lübeken, F.-J. Mean characteristics of thespectrum of horizontal velocity in the polar summer mesosphere andlower thermosphere observed by foil chaff. *J. Atmos. SolarTerr. Phys.* **63**:1831–1839, 2001.

Wunsch, C. On the propagation of internal waves on a slope, *Deep Sea Res.* **18**:588–591, 1968.

Wyngaard, J. C.,Seigel, A., and Wilczak, J. M. On the response of aturbulent-pressure probe and the measurement of pressuretransport. *Boundary-Layer Meteorol.* **69**:379–396,1994.

Xing-sheng, L.,Nai-ping, L., Gaynor, J. E., and Kaimal, J. C. A method formeasuring the phase speed and azimuth of gravity waves in theboundary layer using an optical triangle. In *Studies ofNocturnal Stable Layers at BAO*, pp. 93–107, 1983. Boulder, CO:National Atmospheric and Oceanic Administration, NOAA/ERL.

Xu, Q. Modal and nonmodal symmetric perturbations. Part I: Completeness of normal modes and constructions of nonmodal solutions. *J. Atmos. Sci.* **64**:1745–1763, 2007.

Xue, M., Droegemeier, K. K., Wong, V., Shapiro, A., and Brewster K. ARPS version 4.0 User's Guide. Technical report, 380 pp., 1995. Center for Analysis and Prediction of Storms, University of Oklahoma.

Yague, C., Viana, S.,Maqueda, G., Lazcano, M. F., Morales, G., and Rees, J. M. A studyof the nocturnal atmospheric boundary layer: SABLES2006.Fís. Tierra **19**:37–53, 2007.

Yanai, M. and Maruyama, T. Stratospheric wave disturbances propagating over the equatorial pacific. *J. Meteorol. Soc. Jpn.* **44**:291–294, 1966.

Young, J. M. and Hoyle, W. A. Computer programs for multidimensional spectra array processing. Technical Report, NOAA Technical Report ERL 345-WPL 43, 1975. US Dept. of Commerce, National Oceanic and Atmospheric Administration.

Yunck, T., Liu, C.-H., andWare, R. A history of GPS sounding. *TAO* **11**:1–20,2000.

Zamora, R. L. Richardson number computations in the planetary boundary layer. In *Studies of Nocturnal Stable Layers at BAO*, pp. 109–129, 1983. Boulder, CO: National Atmospheric and Oceanic Administration, NOAA/ERL.

Zhang, F., Koch, S. E., Davis, C.A., and Kaplan, M. L. Wavelet analysis and the governing dynamics of a large-amplitude mesoscale gravity-wave event along the Eeat Coast of the Unites States. *Quart. J. Roy. Meteor. Soc.* **127**:2209–2245, 2001.

Zink, F. and Vincent, R. A. Wavelet analysis of stratospheric gravity wave packets over Macquarie Island, I Wave parameters. *J. Geophys. Res.* **106**:10,275–10,298, 2001.

Wu, [illegible] and [illegible]: [illegible] gravity wave [illegible] [illegible], 2008.
W[illegible] and Wilde, [illegible]: [illegible] gravity wave spectra [illegible] during [illegible] [illegible].
W[illegible], [illegible], Wilde, [illegible] and [illegible]: [illegible] characteristics [illegible] of the polar [illegible] [illegible].
W[illegible]: On the propagation of internal waves [illegible] slope, Deep-Sea [illegible], [illegible].
Wyngaard, J. C., Siegel, A., and Wilczak, J. M.: On the response of a turbulent-pressure probe and the measurement of pressure transport, Boundary-Layer Meteorol. 69, [illegible], 1994.
[illegible]: A method for estimating the phase [illegible] of gravity waves [illegible] [illegible].
[illegible] and [illegible]: [illegible] [illegible].

INDEX

M

Z

International Geophysics Series

EDITED BY

Volume 1	BENO GUTENBERG. Physics of the Earth's Interior. 1959*
Volume 2	JOSEPH W. CHAMBERLAIN. Physics of the Aurora and Airglow. 1961*
Volume 3	S. K. RUNCORN (ed.). Continental Drift. 1962*
Volume 4	C. E. JUNGE. Air Chemistry and Radioactivity. 1963*
Volume 5	ROBERT G. FLEAGLE AND JOOST A. BUSINGER. An Introduction to Atmospheric Physics. 1963*
Volume 6	L. DEFOUR AND R. DEFAY. Thermodynamics of Clouds. 1963*
Volume 7	H. U. ROLL. Physics of the Marine Atmosphere. 1965*
Volume 8	RICHARD A. CRAIG. The Upper Atmosphere: Meteorology and Physics. 1965*
Volume 9	WILLIS L. WEBB. Structure of the Stratosphere and Mesosphere. 1966*
Volume 10	MICHELE CAPUTO. The Gravity Field of the Earth from Classical and Modern Methods. 1967*
Volume 11	S. MATSUSHITA ANDWALLACE H. CAMPBELL (eds.). Physics of Geomagnetic Phenomena (In two volumes). 1967*
Volume 12	K. YA KONDRATYEV. Radiation in the Atmosphere. 1969*
Volume 13	E. PALMN AND C. W. NEWTON. Atmospheric Circulation Systems: Their Structure and Physical Interpretation. 1969*
Volume 14	HENRY RISHBETH AND OWEN K. GARRIOTT. Introduction to Ionospheric Physics. 1969*
Volume 15	C. S. RAMAGE. Monsoon Meteorology. 1971*
Volume 16	JAMES R. HOLTON. An Introduction to Dynamic Meteorology. 1972*
Volume 17	K. C.YEH AND C. H. LIU. Theory of Ionospheric Waves. 1972*
Volume 18	M. I. BUDYKO. Climate and Life. 1974*
Volume 19	MELVIN E. STERN. Ocean Circulation Physics. 1975
Volume 20	J. A. JACOBS. The Earth's Core. 1975*
Volume 21	DAVID H. MILLER. Water at the Surface of the Earth: An Introduction to Ecosystem Hydrodynamics. 1977
Volume 22	JOSEPH W. CHAMBERLAIN. Theory of Planetary Atmospheres: An Introduction to Their Physics and Chemistry. 1978*
Volume 23	JAMES R. HOLTON. An Introduction to Dynamic Meteorology, Second Edition. 1979*
Volume 24	ARNETT S. DENNIS. Weather Modification by Cloud Seeding. 1980
Volume 25	ROBERT G. FLEAGLE AND JOOST A. BUSINGER. An Introduction to Atmospheric Physics, Second Edition. 1980

Volume 26 Kuo-Nan Liou. An Introduction to Atmospheric Radiation. 1980
Volume 27 David H. Miller. Energy at the Surface of the Earth:An Introduction to the Energetics of Ecosystems. 1981
Volume 28 Helmut G. Landsberg. The Urban Climate. 1991
Volume 29 M. I. Budkyo. The Earth's Climate: Past and Future. 1982*
Volume 30 Adrian E. Gill. Atmosphere-Ocean Dynamics. 1982
Volume 31 Paolo Lanzano. Deformations of an Elastic Earth. 1982*
Volume 32 Ronald T. Merrill and Michael W. McElhinny. The Earth's Magnetic Field: Its History, Origin, and Planetary Perspective. 1983
Volume 33 John S. Lewis and Ronald G. Prinn. Planets and Their Atmospheres: Origin and Evolution. 1983
Volume 34 Rolf Meissner. The Continental Crust: A Geophysical Approach. 1986
Volume 35 M. U. Sagitov, B. Bodki, V. S. Nazarenko, and Kh. G. Tadzhidinov. Lunar Gravimetry. 1986
Volume 36 Joseph W. Chamberlain and Donald M. Hunten. Theory of Planetary Atmospheres, 2nd Edition. 1987
Volume 37 J. A. Jacobs. The Earth's Core, 2nd Edition. 1987
Volume 38 J. R. Apel. Principles of Ocean Physics. 1987
Volume 39 Martin A. Uman. The Lightning Discharge. 1987
Volume 40 David G. Andrews, James R. Holton, and Conway B. Leovy. Middle Atmosphere Dynamics. 1987
Volume 41 Peter Warneck. Chemistry of the Natural Atmosphere. 1988
Volume 42 S. Pal Arya. Introduction to Micrometeorology. 1988
Volume 43 Michael C. Kelley. The Earth's Ionosphere. 1989
Volume 44 William R. Cotton and Richard A. Anthes. Storm and Cloud Dynamics. 1989
Volume 45 William Menke. Geophysical Data Analysis: Discrete Inverse Theory, Revised Edition. 1989
Volume 46 S. George Philander. El Nio, La Nia, and the Southern Oscillation. 1990
Volume 47 Robert A. Brown. Fluid Mechanics of the Atmosphere. 1991
Volume 48 James R. Holton. An Introduction to Dynamic Meteorology, Third Edition. 1992
Volume 49 Alexander A. Kaufman. Geophysical Field Theory and Method.
Part A: Gravitational, Electric, and Magnetic Fields. 1992
Part B: Electromagnetic Fields I. 1994
Part C: Electromagnetic Fields II. 1994
Volume 50 Samuel S. Butcher, Gordon H. Orians, Robert J. Charlson, and Gordon V. Wolfe. Global Biogeochemical Cycles. 1992
Volume 51 Brian Evans and Teng-Fong Wong. Fault Mechanics and Transport Properties of Rocks. 1992
Volume 52 Robert E. Huffman. Atmospheric Ultraviolet Remote Sensing. 1992
Volume 53 Robert A. Houze, Jr. Cloud Dynamics. 1993
Volume 54 Peter V. Hobbs. Aerosol-Cloud-Climate Interactions. 1993
Volume 55 S. J. Gibowicz and A. Kijko. An Introduction to Mining Seismology. 1993
Volume 56 Dennis L. Hartmann. Global Physical Climatology. 1994
Volume 57 Michael P. Ryan. Magmatic Systems. 1994
Volume 58 Thorne Lay and Terry C. Wallace. Modern Global Seismology. 1995
Volume 59 Daniel S. Wilks. Statistical Methods in the Atmospheric Sciences. 1995
Volume 60 Frederik Nebeker. Calculating the Weather. 1995
Volume 61 Murry L. Salby. Fundamentals of Atmospheric Physics. 1996
Volume 62 James P. McCalpin. Paleoseismology. 1996
Volume 63 Ronald T. Merrill, Michael W. McElhinny, and Phillip L. McFadden. The Magnetic Field of the Earth: Paleomagnetism, the Core, and the Deep Mantle. 1996
Volume 64 Neil D. Opdyke and James E. T. Channell. Magnetic Stratigraphy. 1996
Volume 65 Judith A. Curry and Peter J.Webster. Thermodynamics of Atmospheres and Oceans. 1998

Volume 66	Lakshmi H. Kantha and Carol Anne Clayson. Numerical Models of Oceans and Oceanic Processes. 2000
Volume 67	Lakshmi H. Kantha and Carol Anne Clayson. Small Scale Processesin Geophysical Fluid Flows. 2000
Volume 68	Raymond S. Bradley. Paleoclimatology, Second Edition. 1999
Volume 69	Lee-Lueng Fu and Anny Cazanave. Satellite Altimetry and Earth Sciences: A Handbook of Techniques and Applications. 2000
Volume 70	David A. Randall. General Circulation Model Development: Past, Present, and Future. 2000
Volume 71	Peter Warneck. Chemistry of the Natural Atmosphere, Second Edition. 2000
Volume 72	Michael C. Jacobson, Robert J. Charleson, Henning Rodhe, and Gordon H. Orians. Earth System Science: From Biogeochemical Cycles to Global Change. 2000
Volume 73	MichaelW. McElhinny and Phillip L. McFadden. Paleomagnetism: Continents and Oceans. 2000
Volume 74	Andrew E. Dessler. The Chemistry and Physics of Stratospheric Ozone. 2000
Volume 75	Bruce Douglas, Michael Kearney, and Stephen Leatherman. Sea Level Rise: History and Consequences. 2000
Volume 76	Roman Teisseyre and Eugeniusz Majewski. Earthquake Thermodynamics and Phase Transformations in the Interior. 2001
Volume 77	Gerold Siedler, John Church, and John Gould. Ocean Circulation and Climate: Observing and Modelling the Global Ocean. 2001
Volume 78	Roger A. Pielke Sr. Mesoscale Meteorological Modeling, 2nd Edition. 2001
Volume 79	S. Pal Arya. Introduction to Micrometeorology. 2001
Volume 80	Barry Saltzman. Dynamical Paleoclimatology: Generalized Theory of Global Climate Change. 2002
Volume 81A	William H. K. Lee, Hiroo Kanamori, Paul Jennings, and Carl Kisslinger. International Handbook of Earthquake and Engineering Seismology, PartA. 2002[NYP]
Volume 81B	William H. K. Lee, Hiroo Kanamori, Paul Jennings, and Carl Kisslinger. International Handbook of Earthquake and Engineering Seismology, Part B. 2002[NYP]
Volume 82	Gordon G. Shepherd. Spectral Imaging of the Atmosphere. 2002[NYP]
Volume 83	Robert P. Pearce. Meteorology at the Millennium. 2001
Volume 84	Kuo-Nan Liou. An Introduction to Atmospheric Radiation, 2nd Ed. 2002[NYP]
Volume 85	Carmen J. Nappo. An Introduction to Atmospheric Gravity Waves. 2002[NYP]

[*]Out of print.
[NYP]Not yet published.

Printed and bound by CPI Group (UK) Ltd, Croydon, CR0 4YY

08/05/2025

01864827-0002